Introduction
to
Statistical Methods

Volume I

Introduction

to

Statistical Methods

Volume I

Jagdish S. Rustagi

Rowman & Allanheld

PUBLISHERS

ROWMAN & ALLANHELD

Published in the United States of America in 1984
by Rowman & Allanheld, Publishers
(A division of Littlefield, Adams & Company)
81 Adams Drive, Totowa, New Jersey 07512

Library of Congress Cataloging in Publication Data

Rustagi, Jagdish S.
    Introduction to statistical methods.

    Includes index.
    1. Statistics. I. Title.
QA276.12.R87   1984   519.5   84–6911
ISBN 0–86598–127–2 (v. 1)

84 85 86 /10 9 8 7 6 5 4 3 2 1

Printed in the United States of America

Photoset at Thomson Press (India) Limited, New Delhi

To the memory of my parents

Contents of Volume I

Inference for Multivariate Normal Distribution, **313**; Test of Hypothesis of the Mean of Normal Population, **314**; Linear Discriminant Functions, **319**; Appendix: Determinants and Matrices, **322**; Summary, **328**; References, **328**

Contents of Volume II

List of Figures

List of Tables

Appendix Tables

Preface

Statistical methods are being used more and more by experimental scientists, medical clinicians, government administrators, business managers, industrial engineers, and market researchers. Statistics now plays a major role in policy decisions, quality assurance, public health, and clinical trials. Students in the behavioral sciences, biomedical sciences, agricultural sciences, and business administration are required to take statistics courses. With easy access to pocket calculators and microcomputers, statistical computations have become painless.

This book provides an introduction to statistical methods. The first volume covers the basic topics in probability, statistical inference, and regression analysis. It also includes a brief introduction to the design and analysis of experiments. The second volume covers topics of survey sampling, quality assurance, survival analysis, statistical bioassay, and sequential experimentation. Topics on clinical trials, life tables, and compartmental analysis are also included.

The mathematical background necessary is high-school algebra. We do not give derivations of results. Statistical concepts are explained with as little mathematics as possible. Examples from real-life situations are given wherever possible. Simple examples with artificial data are used to illustrate many statistical concepts. We do not stress numerical computations. Statistical packages such as SAS, SPSS, and BMDP contain almost all the commonly used statistical techniques and are now used in classrooms. Instructors are encouraged to use such packages to further illustrate a statistical notion.

The purpose of the book is to introduce the reader to the statistical philosophy and commonly used methods in statistics. For the use of statistical techniques in complicated situations, a statistician should be consulted. The first volume of the book can be covered in a two-quarter course meeting three times a week. Some topics need to be omitted if the course is to be covered in a semester. Topics from the second volume can be covered in a one-semester course.

Acknowledgments

I am grateful to several generations of students at Carnegie Institute of Technology, Michigan State University, Aligarh Muslim University, The University of Cincinnati College of Medicine, and The Ohio State University, where I have taught. I am highly obliged to the authors of textbooks and journal articles who are quoted throughout the book. Professor Paul Switzer was a source of constant encouragement during the preparation of the manuscript. Professor Abhaya Indrayan provided extensive suggestions and corrections on the first draft. The book was improved considerably as a result of a critical review of Professor Eswar G. Phadia, who made many suggestions. I am also obliged to Professor Jean Powers for reading the manuscript. The manuscript was completed during a sabbatical from The Ohio State University spent at Stanford University and The India Statistical Institute, Delhi Center, both of which I would like to thank.

For typing the manuscript I thank Bach-Hong Tran, Dolores Wills, and Mona Swanger. Professor Herman Chernoff has always been a source of inspiration and encouragement, and I am grateful to him. I am obliged to my wife, Kamla, and my family for providing a peaceful environment during the writing of the book. I am grateful to the staff of Rowman & Allanheld, Publishers, especially to senior editor Paul Lee, who has been very helpful in various aspects of the project.

I shall be greatly obliged to readers for their suggestions and comments.

Introduction
to
Statistical Methods
Volume I

chapter one

Introduction

Every real phenomenon has uncertain elements in it. We do not know how many patients will be admitted to a hospital on a given day, nor do we know how long a cancer patient will survive. The response of a new drug is not known and the success of a new business enterprise cannot be determined in advance. We do not know if the price of a stock will go up tomorrow or if the yield of a given variety of wheat will increase under a new fertilizer. The percentage of defectives produced by a machine and the number of auto accidents in a city cannot be determined with surety. Whether the U.S. unemployment rate will go down or whether the mortgage interests rates will go up depends on many uncertain factors.

Modern statistics provides various methods to answer these problems. Although statistics is usually regarded as being concerned with the collection, analysis, and interpretation of data, it can be described more broadly as the science of making decisions under uncertainty. The mathematical model of uncertainty is the theory of probability, and probability is an essential component in the study of statistics.

Descriptive statistics provides condensation of numerical information in the form of graphical displays, tables, and charts and is an essential aspect of data analysis. Modern computers have made these analyses easily accessible, and the use of statistics has grown as a result of this convenience.

The theory of probability provides a model of uncertainty. By a model we mean a mathematical description of a phenomenon. Just as the model of a building provides essentials of its dimensions and form but not all the details, a mathematical model describes the essentials of a phenomenon. Mathematical models involving concepts of probability are called *stochastic*; otherwise, we call them *deterministic*.

Statistical methods are being used in manufacturing and production to improve quality; in sales and marketing to develop strategies of product distribution, and in field experimentation and clinical trials to determine the efficacy of new varieties of treatments and drugs. They are used in surveys of populations, agricultural censuses, and in estimating unemployment rates and the gross national product in order to formulate national economic policies. There are hardly any areas of modern life where statistics is not applied. A collection of important examples of the application of statistics is in *Statistics: A Guide to the Unknown* edited by Tanur et al. (1972). Mathematical models in genetics are given by Elandt–Johnson (1971), in social sciences by Kemeny and Snell (1962), and in medicine by Rustagi (1971).

Example 1.1: The Instant Game of the Ohio Lottery

More than a dozen states in the United States and several countries around the world have legalized lotteries. Several kinds of games are in use. In an instant game the player is declared winner or loser immediately on the purchase of the ticket. In other games the player has to wait for a day, as in a "numbers" game, or a week to find out if he is a winner.

In one game in the Ohio Lottery the cost of a ticket is one dollar. A player can win instantly prizes of $2, $5, $200, $1,000, and $10,000 according to the following chances.

Prizes (in dollars)	2	5	200	1,000	10,000
Chance of winning	1/10	3/200	1/20,000	1/100,000	1/100,000

The winning combination on a ticket is usually covered with a security-proof film and when the film is uncovered, the player knows if he is a winner and also the amount of prize.

If 100,000 tickets are sold on a given day, what is the "expected" number of winners? Intuitively, we see that since a $10,000 prize has a chance of one in 100,000, we "expect" one prize in 100,000 tickets. When we study probability theory later, this important concept of "expected value" will become more precise. The expected number of prizes in various categories is given by the following.

Prizes (in dollars)	2	5	200	1,000	10,000
Expected number of prizes	10,000	1,500	5	1	1

In practice, the actual number of prizes will be different, since tickets with prizes are *randomly* distributed among all the tickets sold.

The total "expected" prize value is:

$$(10,000 \times 2) + (1,500 \times 5) + (5 \times 200) + (1 \times 1,000) + (1 \times 10,000) = \$39,500$$

Since the total cost of 100,000 tickets is \$100,000, the expected return on one dollar is $39,500/100,000 = .395$. That is, the expected return on one dollar invested is 39.5 cents. In the actual game the expected return is increased a little by introducing additional prizes. However, the maximum expected return in any state lottery game is 50 cents, since the states run the lotteries for profit.

Several questions arise when a player takes part in this lottery game. What is the chance that a player will not win a prize until 100 tickets are bought? Are the prizes being given according to the claimed prize structure? That is, are the lotteries honest? What is the chance that by investing, say, \$20 dollars a player may win a prize of \$10,000? Questions such as these will be answered by probability theory and statistical methods discussed later in the book.

Example 1.2: Drug Screening

Thousands of drugs are screened to see if they are effective in treating certain diseases. Consider a drug that has been developed by a pharmaceutical company to cure cancer tumors. The data for a drug tested on animals are given by Dunnett (1972) for the weight of a tumor for three treated and five untreated (controls) animals.

Treated: 0.96   1.59   1.14
Controls: 1.29, 1.60, 2.27, 1.33, 1.88

We note the high *variability* in the data, so to decide if the drug is effective in reducing tumors is not easy. Essentially this is a problem of "decision making." Should we accept the drug (perhaps for further testing) or reject it? There is also the possibility of discarding an effective drug or accepting an ineffective drug. These problems will be addressed in "testing hypotheses" later in the book.

Once the drug is accepted, we would like to determine the amount of dose; Thus we shall be *experimenting* with several possible doses. A problem of this kind is discussed in "statistical bioassay" in the second volume of the book.

Example 1.3: Olympic Races

Athletes have always been concerned with finding the best attainable time in a race. A study of four Olympic races was done by Chatterji and Chatterji (1982). The data for all the men's races for 1900–1980 are given in Table 1.1.

Table 1.1   Winning Times in Men's Running Events in Olympic Games (1900–1980) and Altitude

Year	100 meters	200 meters	400 meters	800 meters	Altitude (feet)
1900	10.80	20.20	49.40	121.40	25
1904	11.00	21.60	49.20	116.00	455
1908	10.80	22.60	50.00	112.80	8
1912	10.80	21.70	48.20	111.90	46
1920	10.80	22.00	49.60	113.40	3
1924	10.60	21.60	47.60	112.40	25
1928	10.80	21.80	47.80	111.80	8
1932	10.30	21.20	46.20	109.80	340
1936	10.30	20.70	46.50	112.90	115
1948	10.30	21.10	46.20	109.20	8
1952	10.40	20.70	45.90	109.20	25
1956	10.50	20.60	46.70	107.70	3
1960	10.20	20.50	44.90	106.30	66
1964	10.00	20.30	45.10	105.10	45
1968	9.90	19.80	43.80	104.30	7349
1972	10.14	20.00	44.66	105.90	1699
1976	10.06	20.23	44.26	103.50	104
1980	10.25	20.19	44.60	105.40	—

A mathematical model for winning times involving the races and the years was developed to answer the following questions. Does altitude affect the winning time of the race? Is the time for the 200-meter race double of the time for the 100-meter race? What are the winning times for the 1984 race? Using the theory of "regression analysis," several questions were answered. The following predictions are made for 1984.

	100 meters	200 meters	400 meters	800 meters
Predicted winning times	10.1	19.98	44.16	104.38

The reader is encouraged to compare them with the actual times!

Example 1.4: Water Fluoridation and Cancer

Claims were made that the fluoridation of city water supplies was associated with an increased incidence of cancer. Data were collected for ten U.S. cities with fluoridation and ten cities without. The fluoridated cities were San Francisco, Milwaukee, Pittsburgh, Philadelphia, Cleveland, Buffalo, St. Louis, Baltimore, Chicago, and Washington D.C. The unfluoridated cities were Los Angeles, Atlanta, New Orleans, Boston, Newark, Cincinnati,

Columbus, Portland, Seattle, and Kansas City, Missouri. The following table is given by Oldham (1982):

	Cancer deaths		Population		Rate per 100,000	
	1950	1970	1950	1970	1950	1970
Fluoridated	21,485	23,405	11,855,844	10,766,632	181.21	217.38
Unfluoridated	11,257	14,487	6,290,077	7,347,712	178.96	197.16

The rates from 1950 to 1970 show an apparent increase in the death rates from cancer per 100,000 of population, and therefore fluoridation was given as the cause. However, since cancer death rates differ for different age groups and races, the aggregate deaths hide the different composition of the cities. For example, the death rates for malignant neoplasms for the United States by age, sex, and race are available. For ages 45 to 50, they are:

	Rates per 100,000 of population
White males	150.8
White females	185.8
Nonwhite males	207.4
Nonwhite females	273.3

Since the two groups of cities are not homogeneous and have different age, sex, and race distributions, the risks are not the same. Oldham's analysis shows that there is no difference among the rates if adjustments to the heterogeneity of the population are made.

Population and Sample

Statistics essentially studies *populations.* By population we mean not only populations of humans, animals, and trees but also of all objects about which certain decisions are to be made. We have the population of scores of all college students, the population of heights of all individuals, the population of all possible responses to a drug, and so forth. The population may consist of only a finite number of elements or it may be infinite. Since it is almost impossible to study the whole population because of cost and time constraints, a smaller group of items in the population is selected for study. This is a *sample.* To get valid inferences about the population from a sample, it is imperative that the sample be representative—in other words, a *random sample.* Based on the sample, inferences about the population can be made. This kind of inference, going from the sample to the population, is *inductive*. Statistics is concerned with inductive inference and it deviates from mathematics in this important aspect, since mathematics is *deductive*. An important area of statistics concerned with the methods of obtaining a random sample is *sample surveys*. Similarly, the area of *design of experiments* is concerned with obtaining observations so that the results of an experiment can be used to make inferences about the wider population represented by the experiment.

Plan of the Book

Techniques of data summarization and other graphical methods of data representation are given in Chapter 2. Chapter 3 introduces the concept of probability. The concept of random variables is defined in Chapter 4 and several important discrete probability distributions are given. Continuous probability density functions are given in Chapter 5. Problems of statistical estimation and confidence intervals are introduced in Chapter 6. Chapter 7, discusses the testing of hypotheses for *parametric* models. Inference in nonparametric models is given in Chapter 8 and several techniques are discussed. One of the most commonly applied statistical techniques is that of regression, which is given in Chapter 9. Analysis of variance techniques are discussed in Chapter 10 and are applied to the design of experiments. Some elementary designs are discussed in Chapter 11. In Chapter 12 graphical methods of multivariate data display and techniques of inference for data in several dimensions are given.

Summary

The study of statistics requires a good understanding of probability theory, which provides the *stochastic* model of a real phenomenon. Several examples of the applications of statistics are discussed. We need to make decisions about *populations* with the help of a *sample*. The study of statistics is the study of generalization from the sample to the population, or *inductive inference*.

References

Chatterji, Samprit, and Chatterji, Sangit. New lamps for old: An exploratory analysis of running times in Olympic Games, *App. Statist., 1982, 31,* 14–22.
Dunnett, Charles, W., "Drug Screening: The never-ending search for new and better drugs," in *Statistics: A Guide to the Unknown,* Edited by Tanur, J. M., et al. San Francisco: Holden-Day, 1972.

Elandt-Johnson, Regina G. *Probability Models and Statistical Methods in Genetics.* New York: John Wiley & Sons, 1971.

Kemeny, John G., and Snell, J. Laurie. *Mathematical Models in Social Sciences.* Boston: Ginn & Co., 1962.

Oldham, P. D. Fluoridation of water supplies and cancer: A possible association? *App. Statist., 1977, 26,* 125–135.

Rustagi, J. S. Mathematical models in medicine, *J. Math. Ed. Sc. Tech.,* 1971, 193–203.

Tanur, Judith M., et al. *Statistics: A Guide to the Unknown,* San Francisco: Holden-Day, 1972.

chapter two

Data Analysis

Digesting numerical information is an important function of data analysis. Before discussing the theoretical models that help in the interpretation of available data, let's describe the various methods of organizing and summarizing data—for example, the frequency distribution and several graphical methods. In addition to the description of a histogram, frequency polygon, and so on, we give some recently introduced methods of data display, such as box plots and stem-and-leaf plots. These methods are useful in a preliminary look at the data. Experienced statisticians generally look at the data in some of these forms before making attempts at further analysis.

Frequency Distribution

The task of data analysis has been made much simpler by computers. The most fundamental notion in data analysis is the *frequency distribution*, which describes the population that gave rise to a sample of data available for analysis. Individual observations are replaced by the number of counts in an interval. Consider the following example, where a sample of 75 cholesterol levels are given for residents of Columbus, Ohio. As such, the data do not provide much insight into the blood cholesterol levels of the population. However, once we formulate a frequency distribution, we are able to say much more about the cholesterol levels of Columbus residents.

Example 2.1: The cholesterol levels of 75 residents are given in Table 2.1. The object here is to provide a frequency distribution of the data using 10 class intervals.

Table 2.1   Blood Cholesterol Levels (mg/L)

253	281	230	215	193
171	275	290	368	294
217	246	289	215	226
296	299	295	201	290
279	258	257	266	246
247	175	212	235	287
291	254	180	294	301
245	293	247	241	290
177	285	241	265	214
255	184	241	291	286
240	242	286	341	237
255	260	301	252	341
244	248	288	219	299
203	279	293	248	231
223	234	222	193	277

The frequency distribution is obtained in various steps:

1. We find the smallest and largest observations, which are 171 and 368, respectively. The range of values is $368 - 171 = 197$, and since we want 10 classes, the class length or interval length is chosen to be 21. The endpoints of the intervals are chosen in such a way that an observation falls in only one interval; therefore, the endpoints have one more decimal point. The intervals should have a midpoint, which does not have a decimal. This is one reason why 21 was chosen as the interval length. So if the interval is 170.5–191.5, the midpoint is 181. However, if we had chosen 20 as the class length, the midpoint of 170.5–190.5 would have been 180.5, which is not

Table 2.2   Classes and Midpoints

Classes	Midpoints
170.5–191.5	181
191.5–212.5	202
212.5–233.5	223
233.5–254.5	244
254.5–275.5	265
275.5–296.5	286
296.5–317.5	307
317.5–338.5	328
338.5–359.5	349
359.5–380.5	370

Table 2.3 Frequency Distribution of Blood Cholesterol Levels

Interval	Tally	Frequency	Relative Frequency	Cumulative Relative Frequency
170.5–191.5	1꓿꓿1	5	0.067	0.067
191.5–212.5	1꓿꓿꓿	5	0.067	0.134
212.5–233.5	1꓿꓿꓿ 11꓿꓿	10	0.133	0.267
233.5–254.5	11꓿꓿ 1꓿꓿꓿ 1꓿꓿1 1111	19	0.253	0.520
254.5–275.5	1꓿꓿1 1111	8	0.107	0.627
275.5–296.5	1꓿꓿꓿ 1꓿꓿1 1꓿꓿1 1꓿꓿1 1	21	0.280	0.907
296.5–317.5	11꓿꓿	4	0.053	0.960
317.5–338.5		0	0.000	0.960
338.5–359.5	11	2	0.027	0.987
359.5–380.5	1	1	0.013	1.000
Total		75	1.000	

in the same units as the data. Table 2.2 gives the class intervals with endpoints and midpoints.

2. After the class intervals are decided, we use a tally sheet to find the number of observations that fall in a given interval. For example, 253 falls in the interval 233.5–254.5. In this way we get the tallies in Table 2.3.

3. The total number of tallies are counted and noted in the column under frequency. The relative frequencies are obtained by dividing each frequency by the total number of observations. The cumulative relative frequency is obtained by adding relative frequencies successively. Table 2.3 provides the calculations for Example 2.1.

Keep in mind the following when obtaining frequency distributions from data.

1. The number of classes should be between six and ten in order to obtain five or six observations in a given class. If there are many classes that contain no observations, you should reassign the class intervals.
2. The length of the classes should be the same. However, many times the two end classes are taken to be open-ended.
3. The frequency distribution reclassifies data in such a way that the individual observations in an interval are all replaced by midpoint of the interval. This fact is used in obtaining averages and other sample statistics from a frequency distribution. The length and position of the intervals should be such as not to distort the original data.

Table 2.4 Frequency Distribution of Weights

Class Interval	Midpoint	Tally	Frequency	Relative Frequency	Cumulative Relative Frequency
16.25–26.35	21.3	TTHL 1111	9	0.18	0.18
26.35–36.45	31.4	TTHL	5	0.10	0.28
36.45–46.55	41.5	TTHL 11	7	0.14	0.42
46.55–56.65	51.6	TTHL 1	6	0.12	0.54
56.65–66.75	61.7	1111	4	0.08	0.62
66.75–76.85	71.8	1111	4	0.08	0.70
76.85–86.95	81.9	TTHL 111	8	0.16	0.86
86.95–97.05	92.0	TTHL 11	7	0.14	1.00

Note: The total of the tally and frequency columns is 50.

4. The number of observations should be sufficient to give about five or six observations per class on the average. The frequency distribution obtained from a smaller number of observations is not useful.

Example 2.2: The weights in pounds are given for a sample of 50 children below. The class intervals, midpoints, tally sheet, frequencies, and corresponding relative frequencies and cumulative relative frequencies are given in Table 2.4. We have used eight classes for this data.

79.8	24.4	21.3	97.0	52.2
96.5	22.3	16.4	47.0	47.0
31.2	27.4	75.2	38.1	77.4
16.7	24.9	26.0	83.2	43.8
36.8	61.8	46.7	24.2	23.6
88.9	47.0	87.5	97.0	72.9
92.8	88.9	55.7	59.2	82.1
46.5	80.3	39.2	75.8	43.0
39.5	78.8	82.7	64.4	61.8
30.7	29.3	83.7	36.1	71.2

Notice that the smallest observation is 16.4 and the largest is 97.0. The range is $97 - 16.4 = 80.6$. We intend to make eight classes and hence choose the length of each class to be 10.1. The lowest endpoint is chosen to be 16.25, having one more decimal point, so that the classification of any observation in a class interval does not create a problem. The intervals are made in such a way that the midpoint still has only the accuracy of the data—that is, with one decimal point.

Exercises

1. The waiting times of forty patients at a medical clinic are given in minutes. Using six classes, find the frequency distribution of waiting times.

34	19	17	10
24	23	27	5
7	25	14	20
15	18	45	24
32	11	16	12
25	13	52	35
18	28	32	25
37	8	15	5
23	25	12	60
15	7	33	37

2. The maximum daily ozone concentrations in a given city are recorded in parts per billion. Using eight classes, find the frequency distribution.

52	66	52	49	64	68	26	86
43	75	87	188	118	71	14	47
31	51	47	40	31	140	103	71
82	103	118	188	87	75	169	38
71	23	32	122	182	143	125	72
111	64	119	59	73	27	87	68
102	141	174	169	24	80	155	82
170	64	28	38	61	33	52	71
99	80	38	66	146	113	89	99
37	52	94	41	36	12	19	14

3. Daily sales (in thousands of dollars) are given for a department store for a period of 50 days. Construct two frequency tables, one using six classes and another using eight classes.

75	38	53	22	10
17	81	92	32	42
73	82	15	23	27
83	85	17	56	65
78	84	32	35	48
27	56	68	27	18
57	83	91	101	27
27	28	15	28	53
17	83	85	93	82
18	17	91	37	81

4. The number of auto accidents in a city is given for a period of 50 days.

Construct a frequency distribution. Use the digits as the points and do not use class intervals.

```
0  5  3  0  3  4  0  2  0  1
1  3  0  3  3  0  2  1  2  0
0  2  7  0  1  6  2  4  3  2
1  0  3  8  3  4  0  0  2  1
2  6  9  2  0  1  3  2  0  0
```

Stem-and-Leaf Plots

Several methods of organization of moderate-size data sets have been introduced recently by John Tukey in *Exploratory Data Analysis* (Addison–Wesley, 1976). One such method is the stem-and-leaf plot. In these plots data are left in the original form, and the organization provides a frequency distribution as well as a diagram.

Consider the data given in Exercise 1. Table 2.5 provides the stem-and-leaf plot for the data in Exercise 1. The *stem* is chosen to be digits 0 to 9 and forms the first column of a vertical line. The unit digits in each observation are recorded across from the digit in the first column, which is regarded as a ten's digit. The ten's digit is common to all observations and is not repeated. The first entry is 34, and hence 4 is entered across 3, forming a *leaf.* Similarly we record 9 across 1 for 19, 7 in line for 1 for 17, and so on.

The advantages of stem-and-leaf plots are many. The data are retained in their original form, unlike the frequency distribution, where individual data points are lost. The writing of data in this form gives frequencies as well as a histogramlike plot, which depicts the form of the distribution. The plot also allows us to find the smallest, largest, and middle observations, for example.

Example 2.3: Consider the data in Exercise 2. Since the data here have

Table 2.5   Stem-and-Leaf Plot

Stem	Leaf
0	7 8 7 5 5
1	7 5 8 5 9 8 1 3 4 6 5 2 0 2
2	4 5 3 3 5 8 5 7 0 4 5
3	4 2 7 2 3 5 7
4	5
5	2
6	0

Table 2.6   Stem-and-Leaf Plot

Stem	Leaf
1	4 2 9 4
2	6 3 8 7 4
3	1 1 8 2 8 3 8 7 6
4	9 3 7 7 0 1
5	2 2 1 9 2 2
6	6 4 8 4 8 4 1 6
7	5 1 1 5 1 2 3 1
8	6 7 2 7 7 0 2 0 9
9	9 9 4
10	3 3 2
11	8 8 1 9 3
12	2 5
13	
14	0 3 1 6
15	5
16	9 9
17	4 0
18	8 8 2

three digits, the stems are chosen from 1 to 18. The resulting stem-and-leaf plot is given in Table 2.6.

Essentially a stem-and-leaf plot gives a standard frequency distribution where classes are from 0–9, 10–19, 20–29, and so on. In Example 2.3, one can combine two classes into one, so that in place of 18 classes we can have 9 classes. Table 2.7 provides the frequencies for the stem-and-leaf plot of Table 2.6 with combined classes.

Table 2.7   Frequency Distribution

Class Interval	Frequency
10– 29	9
30– 49	15
50– 69	14
70– 89	17
90–109	6
110–129	7
130–149	4
150–169	3
170–189	5

Another use of stem-and-leaf analysis can be made in nonnumerical data. Suppose that the literacy measure (standardized in some way) for the following ten states is given, based on Wainer (1983).

State	Literacy Measure
Iowa	1.4
Oregon	.7
South Dakota	1.4
California	− 1.2
Vermont	.7
New Hampshire	.2
Ohio	− .2
Pennsylvania	− .9
Alabama	− 2.6
Mississippi	− 2.8

We order the literacy rate and write it as a stem, then write the state adjacent to the median. Such an analysis shows the relative position of the states based on this measure.

Stem-and-Leaf Plot

Stem	Leaf
− 2.8	Mississippi
− 2.6	Alabama
− 1.2	California
− .9	Pennsylvania
− .2	Ohio
.2	New Hampshire
.7	Oregon, Vermont
1.4	Iowa, South Dakota

Exercises

5. Give a stem-and-leaf plot of the data in Exercise 3.
6. Compare the stem-and-leaf plot of the data in Exercise 4 with its frequency distribution.
7. The pulse rate in beats per minute of a sample of 60 heart patients is

given below. Give a stem-and-leaf plot of the data and also a frequency distribution using six intervals.

88	64	80	90	88	80
72	64	72	80	76	78
62	96	92	108	60	72
80	78	60	80	66	84
60	94	100	104	64	70
72	80	80	84	64	80
72	64	84	60	52	60
75	72	60	80	78	72
80	62	68	88	68	108
84	64	76	96	80	64

Graphical Representation of Data

In data analysis, graphs and charts play an important role in describing the phenomenon that generated the data. There is a considerable volume of work on graphical data analysis in statistics, and with the availability of modern computers capable of graphical displays, this method is increasingly being used in statistical inference. A recent collection of several methods in use is given by Chambers, Cleveland, Kleiner, and Tukey in their book *Graphical Methods of Data Analysis* Wadsworth International Group, 1983.

The graphical methods in the book will appear in various chapters according to the topics under discussion. This chapter provides several graphical methods used in descriptive statistics, such as scatter plots, histograms, frequency polygons, cumulative relative frequency polygons, and box plots.

One-Dimensional Scatter Plots

This method of plotting points along a line recognizes the ordering and clustering of a given set of observations. The data are ordered and plotted along a line, by dots. We stack points one above the other if there are repeated observations. However, scatter plots can be made in various other ways. Scatter plots allow us to see the concentration of points as well as provide the smallest, largest, and other points that divide the data set. Figure 2.1 gives the scatter plot of data in Exercise 1.

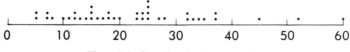

Figure 2.1. One-dimesional scatter plot.

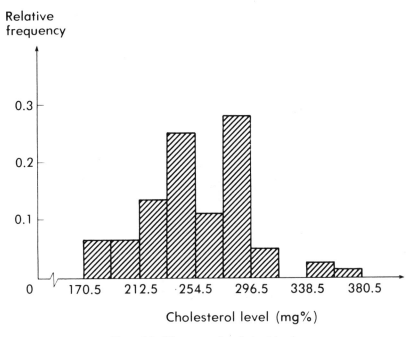

Figure 2.2. Histogram of cholesterol levels.

Histogram

The graph of the frequency distribution obtained by erecting rectangles of a height equivalent to the relative frequencies over the class intervals is called a histogram. The histogram is one of the oldest forms of graphical representation. Figure 2.2 shows a histogram for the frequency distribution discussed in Example 2.1 for the cholesterol levels of 75 patients. Similarly, the histogram for the frequency distribution in Table 2.4 is given in Figure 2.3.

Frequency Polygon

A frequency polygon is used to describe the frequency distribution, but with piecewise lines. The points whose *x*-coordinate is the midpoint of the class interval in the frequency distribution and whose *y*-coordinate is the frequency or relative frequency are plotted. These points are joined by straight lines. The resulting curve, made up of broken lines, is called a *frequency* or *relative frequency polygon*. Figure 2.4 illustrates the relative frequency polygon for the frequency distribution of Table 2.3.

Another relative frequency polygon is sketched for the frequency distribution of weights in Table 2.4. This is given in Figure 2.5. The relative

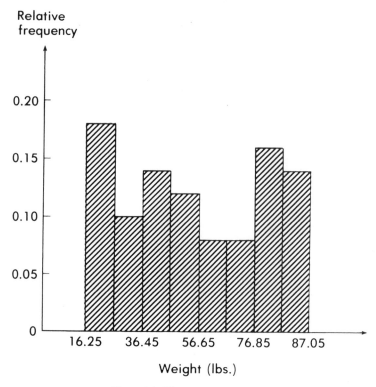

Figure 2.3. Histrogram of weights.

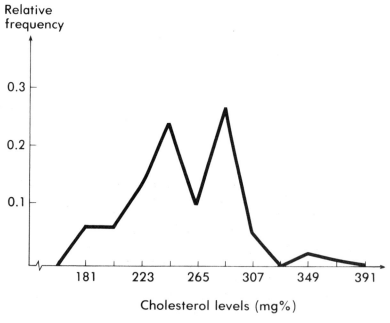

Figure 2.4. Relative frequency polygon of cholesterol levels.

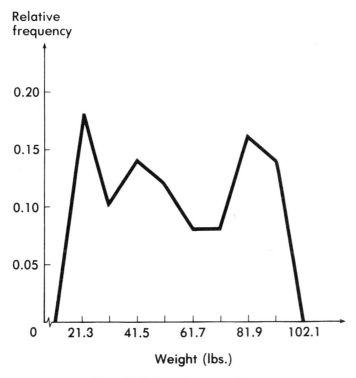

Figure 2.5. Relative frequency polygon.

frequencies are plotted against the midpoints of the interval in the frequency distribution and then joined by straight lines. The polygon reflects the same behavior as that given by the histogram.

Cumulative Frequency Polygon

When we plot the cumulative frequencies against the midpoints of class intervals and join them with straight lines, we get a *cumulative frequency polygon*. Similarly, the plotting of cumulative relative frequencies gives the *cumulative relative frequency polygon*. These polygons can be utilized in finding graphically values of certain statistics such as percentiles and quantiles, discussed later. Figures 2.6 and 2.7 give the cumulative relative frequency polygons for the frequency distributions in Tables 2.3 and 2.4.

Exercises

8. The daily levels of chromium in the air (micrograms per cubic meter) obtained at Kettering Laboratory, Cincinnati, over a certain period have the

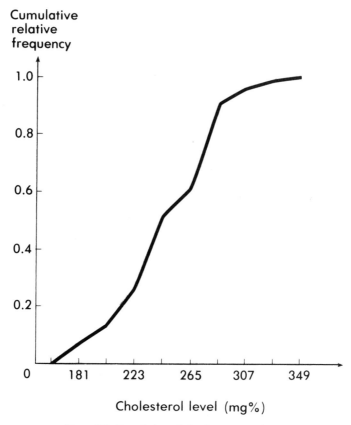

Figure 2.6. Cumulative relative frequency polygon.

following frequency distributions:

Class Interval	Frequency
0.000–0.019	143
0.020–0.039	64
0.040–0.059	16
0.060–0.079	6
0.080–0.099	2
0.100–0.119	1
0.120–0.139	1
0.140–0.159	1
0.160–0.179	1

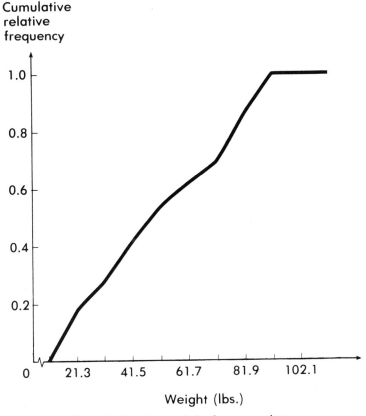

Figure 2.7. Cumulative relative frequency polygon.

(a) Obtain the relative frequency distribution and plot the histogram
(b) Plot the relative frequency polygon
(c) Plot the cumulative relative frequency polygon

9. For the frequency distribution in Table 2.4, plot:

(a) Histogram
(b) Relative frequency polygon
(c) Cumulative relative frequency polygon
(d) One-dimensional scatter plot from the data

10. For the frequency distribution of daily sales in Exercise 3, plot:

(a) Histogram
(b) Relative frequency polygon
(c) Cumulative relative frequency polygon
(d) One-dimensional scatter plots from the data

11. For the frequency distribution of ozone concentrations in Exercise 2, plot:

    (a) Histogram
    (b) Relative frequency polygon
    (c) Cumulative relative frequency polygon
    (d) One-dimensional scatter plots from the data

12. The frequency distribution of the number of cars per hour passing at an intersection in Palo Alto, California, during January 1983 observed for 200 hours is given below. Sketch for the data:

    (a) Histogram
    (b) Relative frequency polygon
    (c) Cumulative relative frequency polygon

Class Interval	Frequency
101–125	10
126–150	25
151–175	33
176–200	50
201–225	47
226–250	26
251–275	5
276–300	4
Total	200

13. In an examination at Ohio State University, the grades A, B, C, D, and E are given according to the rule:

Grade	Score
A	91–100
B	76–90
C	56–75
D	46–55
E	0–45

In a given examination taken by 40 students, the following scores were obtained. Give the frequency distribution and histogram of the grades.

30	72	58	79	85
70	59	70	37	65
71	53	93	53	69
55	48	78	52	78
71	5	98	48	85
63	95	69	67	49
57	89	72	83	17
63	76	97	95	28

Sample Statistics

Although the frequency distribution gives an overall view of the sample, many other quantities can be obtained from it. We call these quantities *sample statistics*. Some of the most commonly used statistics are defined below.

The sample average of $X_1, X_2, \ldots, X_n$ is defined as:

$$\bar{X} = \frac{1}{n} \sum_{i=1}^{n} X_i$$

$\sum$ is the capital Greek letter *sigma* and is used for denoting summation. $\sum_{i=1}^{n} X_i$ means that all values of X starting with subscript 1 to n—that is, $X_1, X_2, \ldots, X_n$—are added, so that:

$$\sum_{i=1}^{n} X_i = X_1 + X_2 + \cdots + X_n$$

Similarly,

$$\sum_{i=1}^{n} X_i^2 = X_1^2 + X_2^2 + \cdots + X_n^2$$

and

$$\sum_{i=1}^{n} (X_i - a)^2 = (X_1 - a)^2 + (X_2 - a)^2 + \cdots + (X_n - a)^2$$

Let the data be given in the form of a frequency table, with k classes having midpoints and frequencies as given below.

Midpoints	Frequency
Y_1	f_1
Y_2	f_2
$\vdots$	$\vdots$
Y_k	f_k

The *sample average* for a frequency distribution is:

$$\bar{Y} = \frac{\sum_{i=1}^{k} f_i Y_i}{\sum_{i=1}^{k} f_i} = \frac{\sum_{i=1}^{k} f_i Y_i}{n} \tag{2.1}$$

where $n = \sum_{i=1}^{k} f_i$ is the total number of observations.

The *variance* of a sample $X_1, X_2, \ldots, X_n$ is given by:

$$s^2 = \frac{\sum_{i=1}^{n} (X_i - \bar{X})^2}{n-1} \tag{2.2}$$

For computational purposes, the following form is used:

$$s^2 = \frac{1}{n-1} \left[\sum_{i=1}^{n} X_i^2 - n\bar{X}^2 \right]$$

If the data are in the form of a frequency distribution, the sample variance is calculated by:

$$s_y^2 = \frac{1}{n-1} \sum_{i=1}^{k} f_i (Y_i - \bar{Y})^2$$

Computationally, a simpler formula is:

$$s_y^2 = \frac{1}{n-1} \left[\sum_{i=1}^{k} f_i Y_i^2 - n\bar{Y}^2 \right] \tag{2.3}$$

The square root of the variance is called the *standard deviation*.

Example 2.4: For the random sample of five creatinine levels (micrograms per 100 milliliters) from the urine of mentally retarded children, an experimenter found the following values:

$$23, \ 141, \ 70, \ 63, \ 43$$

$$\bar{X} = \frac{23 + 141 + 70 + 63 + 43}{5} = 68$$

The sample variance is given by:

$$s^2 = \tfrac{1}{4}[(23)^2 + (141)^2 + (70)^2 + (63)^2 + (43)^2 - 5(68)^2]$$
$$= \tfrac{1}{4}[31,128 - 23,120] = 2,002$$

The sample standard deviation is given by:

$$s = \sqrt{2002} = 44.74$$

Example 2.5: Using the frequency distribution of Example 2.1, we have:

Midpoints	Frequency
181	5
202	5
223	10
244	19
265	8
286	21
307	4
328	0
349	2
370	1

So:

$$\bar{Y} = \tfrac{1}{75}[(181 \times 5) + (202 \times 5) + \cdots + (370 \times 1)]$$

$$= \frac{19,203}{75} = 256.04$$

The sample variance is:

$$s_y^2 = \tfrac{1}{74}[5(181)^2 + 5(202)^2 + \cdots + 1(370)^2 - 75(256.04)^2]$$
$$= \tfrac{1}{74}[5,033,313 - 4,916,735]$$
$$= 1575.36$$

The standard deviation is:

$$s_y = 39.69$$

Coding Data

Coding simplifies data computations. Large numbers are replaced by smaller numbers by the following procedure. From each number subtract a constant and then divide by a constant. Suppose these constants are a and b. Let Y_i be coded as Z_i with the following:

$$Z_i = \frac{Y_i - a}{b}$$

Then,

$$\bar{Z} = \frac{\bar{Y} - a}{b}$$

and

$$s_z^2 = \frac{s_y^2}{b^2}$$

or

$$s_z = \frac{s_y}{b}$$

Consider the following codes for the Example 2.5 (code 286 as 0, 265 as −1, and so on). We subtract 286 from each Y_i and then divide by 21 so that:

$$Z_i = \frac{Y_i - 286}{21}$$

The coded data are:

Midpoints (Y_i)	Codes (Z_i)	Frequency (f_i)
181	−5	5
202	−4	5
223	−3	10
244	−2	19
265	−1	8
286	0	21
307	1	4
328	2	0
349	3	2
370	4	1

The choice of 286 is arbitrary, but the number should be chosen somewhere around the middle. We divide by 21, since 21 is the length of each interval. Now we can easily compute $\bar{Z}$ and s_z^2.

$$\bar{Z} = \tfrac{1}{75}[(-5 \times 5) + (-4 \times 5) + \cdots + (4 \times 1)]$$
$$= -\tfrac{107}{75} = -1.427$$

Similarly,

$$s_z^2 = \frac{1}{75}\left[(-5^2 \times 5) + (-4^2 \times 5) + \cdots + (4^2 \times 1) - \frac{(107)^2}{75}\right] = 3.525$$

$$\bar{Y} = 286 + 21(-1.43) = 256$$

and

$$s_y^2 = (21)^2 \times 3.525 = 1{,}554.52$$

Notice that these computed values agree with the values obtained in Example 2.5, except in decimal places.

Other Statistics

Many other statistics are frequently used in data analysis. Suppose the observations are ordered from the smallest to the largest. The middle observation, in case the number of observations are odd, or the average of the two middle values, in case the observations are even, is called the *sample median*. The median divides the frequency distribution into two equal parts. For example, the median of the sample in Example 2.4 is obtained by arranging the values as 23, 43, 63, 70, 141. The median is 63, the third observation.

Suppose we have one more observation, say 157. Then the median of 6 is obtained as the average of 63 and 70:

$$\text{median} = \frac{63 + 70}{2} = 66.5$$

In case the data are given by a frequency distribution, the median can be located with the help of cumulative relative frequency distribution. In Example 2.1, we note that the median is in the interval 233.5–254.5, since it is the 38th observation. It can be approximately given as:

$$\text{median} = 233.5 + 21 \times \frac{38 - 20}{19} = 253.4$$

In general, if the median is the mth observation and is in the interval (a, b) with frequency f, we have

$$\text{median} = a + \frac{(b - a)(m - g)}{f} \tag{2.4}$$

where g is the cumulative frequency in the interval previous to the interval (a, b).

Sample Mode

The sample value with the highest frequency is called the *mode*. In a frequency distribution we have no unique mode. The interval with the highest frequency is called the *modal interval*. In Example 2.1, the modal interval is 275.5–296.5. The mode for a sample of ten given by 2, 3, 2, 3, 2, 5, 7, 8, 9, 2 is 2.

Sample Range

The *range* is the difference between the largest and the smallest observation.

It is used as a measure of dispersion of the sample in place of standard deviation.

Sample Quantiles

An important quantity in a frequency distribution is a *percentile* or a *quantile*. It is the number below which a given fraction of the sample data lies. For example, 0.1 quantile is a number below which 10% of the sample lies and above which 90% of the sample lies. 0.1 quantile is also the 10th percentile. We define *αth quantile*, denoted by $Q(\alpha)$, to be a number that divides the data into two groups, so that αth fraction of the data falls below $Q(\alpha)$ and $(1 - \alpha)$th fraction falls above $Q(\alpha)$. $Q(0.25)$ is called *lower quartile* and $Q(0.75)$ is called the *upper quartile*.

Sample Interquartile Range

The range of the distribution, which has the middle half of the data, is called the *interquartile range*. That is:

$$\text{Interquartile range} = Q(0.75) - Q(0.25)$$

For grouped data as expressed by a frequency distribution, the cumulative relative frequency polygon readily provides the value of any sample quantile. For example, finding $Q(0.70)$ requires drawing a line parallel to x-axis at 0.70 on the cumulative relative frequency polygon graph and then reading off the corresponding x-value.

For ungrouped data, when the number of observations is small, say 10, each observation when an ordered sample is taken gives 10th, 20th, 30th, and so on, percentiles. However, other percentiles or quantiles are to be obtained by interpolation. For example, if we want $Q(0.35)$, we obtain it as an average of $Q(0.3)$ and $Q(0.4)$. In general, to find $Q(p)$, let p_i be the discrete value of p for which the data provide the quantile. Suppose p is a fraction g away from p_i to p_{i+1}. We define $Q(p)$ as:

$$Q(p) = (1 - g)Q(p_i) + gQ(p_{i+1})$$

Example 2.6: Consider the data for waiting times in Exercise 1. The lower quartile is the 10th observation and the upper quartile is the 30th observation, since the total number of observations is 40. Using the stem-and-leaf plot of the data in Table 2.5, we find that:

$$Q(0.25) = 13$$
$$Q(0.75) = 28$$
$$\text{Interquartile range} = 28 - 13 = 15$$

Other quantiles can be obtained using the above formula. For instance,

$$Q(0.87) = (1 - .02)Q(0.85) + 0.02Q(0.875)$$
$$= 0.98 \times 34 + 0.02 \times 35$$
$$= 34.02$$

where $Q(0.85)$ is the 34th observation and $Q(0.875)$ is the 35th observation.
A *quantile plot*, of all the quantiles is sometimes useful.

Exercises

14. For the following sample, find the median, mode, range, lower and upper quartiles, and interquartile range.

$$28, \ 27, \ 35, \ 16, \ 28, \ 29, \ 33, \ 23$$

15. Find the median and modal interval for the frequency distribution in Example 2.1.
16. Find the median, range, and interquartile range of the data from Exercise 2.
17. Annual incomes (in thousands of dollars) of a group of 40 persons are given below. Find the median, range, upper and lower quartiles, interquartile range, $Q(.90)$, and $Q(0.10)$ for the data:

17.0	17.8	21.1	9.0
39.2	18.9	19.5	18.4
33.0	44.9	34.2	14.1
9.5	13.7	46.5	20.8
48.5	18.1	43.7	13.6
35.2	34.2	36.6	42.0
26.9	31.3	26.7	51.1
30.6	43.8	11.0	41.3
55.4	29.2	58.8	36.8
21.0	10.8	49.9	17.7

Box Plots

Another interesting method of representing a frequency distribution has recently been developed by John Tukey. The data plotted form a box, with its upper value representing the upper quartile and the lower value representing the lower quartile with the median as the line in the middle. Dashed lines are drawn to *adjacent values* as defined below. The data values lying outside the range of adjacent values are also plotted.

The *upper adjacent value* is defined as the largest observation that is less than or equal to the upper quartile plus 1.5 times the interquartile range. The lower adjacent value is similarly defined to be the smallest observation that is greater than or equal to the lower quartile minus 1.5 times the interquartile range. If any data values fall outside the two adjacent values, they are called *outside values* and are plotted as individual points. The box plots point out the symmetry of the distribution and show if there are any "outliers" in the data. Several variations of box plots, also called box-and-whisker plots, are used in practice see *Graphical Methods for Data Analysis* by Chambers, Cleveland, Kleiner, and Tukey for further details.

Example 2.7: Consider the data in Exercise 1. The upper adjacent value is the largest observation smaller than or equal to:

$$28 + (1.5 \times 15) = 50.5$$

Hence, the upper adjacent value is 45. Similarly, the lower adjacent value

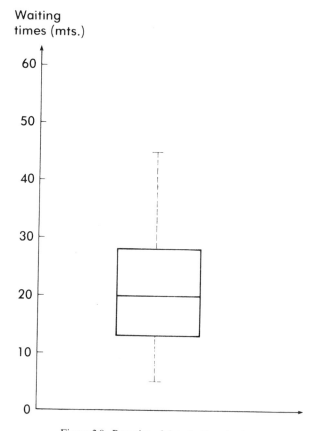

Figure 2.8. Box plot of data in Exercise 1.

Ozone
concentrations

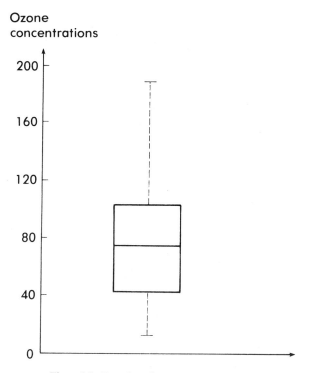

Figure 2.9. Box plot of ozone concentrations.

is 5. There are no outside values in these data. Using the quartiles and median as obtained earlier in Example 2.6, we get the box plot in Figure 2.8.

Example 2.8: For the data in Exercise 2, the lower and upper quartiles are given by the 20th and 60th observations. Using the stem-and-leaf plot of this data in Example 2.3, we have:

$$\text{lower quartile} = 41$$
$$\text{upper quartile} = 103$$
$$\text{interquartile range} = 103-41 = 62$$

Since $103 + (1.5 \times 62) = 196$, the upper adjacent value $= 188$, the largest observation. Similarly, the lower adjacent value $= 12$. The median $= 71$. The box plot is given in Figure 2.9.

Exercises

18. Give a box plot of data in Exercise 17.

19. Compare the histogram and box plot of the data set:

65	91	42	108
48	49	79	14
11	33	27	101
76	10	103	81
74	55	39	20
80	27	95	47
12	52	37	15
43	140	132	12
56	60	195	38
35	147	135	14

20. Sketch a box plot of the data in Exercise 13.

Chapter Exercises

1. The weights of the football players for the Dallas Cowboys 1980 team are given below:

180	184	183	175	180	241	250
261	214	181	225	220	251	246
216	189	286	244	214	185	199
183	258	189	179	184	226	237
180	178	222	201	192	189	197
236	256	194	256	210	190	231
193	173	236	171	204	230	176
	189	257	295	224	209	240
	275	195	241	231	173	191
	241	282	201	227	221	211

   (a) Give a frequency distribution and the histogram of the data
   (b) Obtain median, range, and both quartiles
   (c) Draw a stem-and-leaf and box plot of the data
   (d) Provide the frequency polygon and cumulative relative frequency distribution
   (e) Calculate the quantiles $Q(0.90)$ and $Q(0.10)$
   (f) Find the number of observations that fall between the interval of sample mean ± 2 standard deviation

2. For the football players of the Seattle Seahawks 1980 team, we give their heights (in inches, 6 feet $+$).

6	5	0	1
5	4	4	5
1	0	−1	2
0	0	2	1
0	2	5	−1
5	4	0	−2
−1	1	−1	
1	−2	1	
2	6	1	
3	0	6	
0	6	0	
6	1	6	
3	2	−1	
−2	1	4	
0	1	0	
6	4	−1	
6	1	−1	
1	−1	1	
2	0	4	
3	2	4	

(a) For this data, give the frequency distribution and the histogram, using discrete values

(b) Provide the mean and standard deviation of the actual data. (Note that the observations given are deviations from 72 inches)

(c) Give the median and range and compare them with mean and standard deviation

(d) Give a box plot of the data

3. The weights of frozen boxes of okra are found to be (in ounces):

14.6	14.3	18.0	15.5	15.2
14.0	16.8	14.1	17.8	16.7
18.9	15.2	15.9	13.2	16.9
17.4	16.7	17.2	18.6	17.6
16.6	12.6	12.8	15.5	18.8

(a) Find the sample mean, standard deviation, and range

(b) Find the quartiles, median, and $Q(.80)$, $Q(.20)$ and $Q(.99)$

(c) Give a box plot

(d) Give a stem-and-leaf plot

4. The systolic blood pressure of a sample of 100 males residing in Menlo Park, California, is given in the frequency table:

Systolic Blood Pressure
(in millimeters of mercury)

Class Interval	Frequency
90–99	2
100–109	2
110–119	15
120–129	31
130–139	35
140–149	8
150–159	4
160–169	3
Total	100

(a) Draw a histogram of the data
(b) Draw a frequency polygon and cumulative frequency distribution
(c) Find the number below which 20% of the sample falls
(d) Find the interquartile range

5. Data on blood urea nitrogen (BUN) of 90 pediatric patients admitted to a hospital are given below:

13	16	8	13	9	13	12	17	16
12	9	41	21	19	14	11	12	42
14	13	19	21	7	13	22	47	14
17	14	76	22	9	12	25	9	25
13	19	11	7	9	10	24	16	40
15	10	7	12	13	17	15	18	25
8	59	9	13	7	24	9	13	16
11	14	12	7	7	16	9	11	9
22	17	10	12	10	55	30	7	13
9	12	10	19	10	25	10	12	22

Reference: Okada, M., A method for estimating renal function based on the radio isotope renogram, *Computers and Biomedical Research*, 1983, *16*, 218–33.

(a) Find the mean, median, and quartiles for the above data
(b) Give a frequency distribution using ten classes
(c) Provide a stem-and-leaf plot of the data
(d) Give a box plot of the data, showing outliers, if any

6. The price per pound of chicken in the month of June 1983 as found in 20 major U.S. cities was:

0.49	0.75
0.78	0.49
0.45	0.45
0.43	0.69
0.49	0.56
0.39	0.49
0.65	0.49
0.59	0.79
0.46	0.75
0.39	1.39

(a) Find the median, mode, and mean of the prices
(b) Find the lower and upper quartiles
(c) Draw a box plot

7. Data on survival in days of orthotopic liver allografts in 44 inbred rats were obtained by Zimmerman, F. A.; Knoll, P. P.; Davies, H. ff. S.; Gokel, J. M.; and Schmid, T., the fate of orthotopic liver allografts in different rat strain combinations, *Transplantation Proceedings*, 1983, *15*, 1272–75.

191	29	90	179	114	175	116	152	17
218	31	171	198	147	198	124	178	17
232	32	211	216	195	201	136	226	18
288	39	281	222	236	209	198	251	19
631	32	297	305	291	298	231	—	20

(a) Give its frequency distribution using six classes
(b) Provide a stem-and-leaf plot of the data
(c) Give a box plot of the data

Summary

The information available in a sample can be easily digested by looking at a *frequency distribution*. The *relative frequencies* also provide an estimate of the chance that an observation falls in a given interval. Moderate-size data can be organized into a *stem-and-leaf* plot, which retains the value of an individual observation in addition to providing a histogram like figure. *One-dimensional scatter plots* of the data provide a visual form in which the concentration of observations or maverick observations *outliers* can be easily observed. The *histogram, frequency polygon,* and *cumulative frequency polygon* are useful visual representations of the frequency distribution. Many descriptive statistics, including *sample average, sample variance, standard deviation, sample median, mode,* and *range,* are useful. *Percentiles* and *quantiles* provide the points below which a given fraction of the sample falls. A *box plot* is a graphical representation of data.

References

Chambers, John M.; Cleveland, William S.; Kleiner, Beat; and Tukey, Paul A., *Graphical Methods for Data Analysis,* Belmont, Calif.: Wadsworth International Group, 1983.

Dixon, Wilfrid, J., and Massey, Frank J., Jr. *Introduction to Statistical Analysis.* New York: McGraw-Hill, 1969.

Dunn O. J. *Basic Statistics.* New York: John Wiley & Sons, 1964.

Huntsberger, David V., and Billingsley, Patrick. *Elements of Statistical Inference.* Boston: Allyn and Bacon, 1981.

Madson, Richard W., and Moeschberger, Melvin L. *Statistical Concepts with Applications to Business and Economics.* Englewood Cliffs, N. J.: Prentice-Hall, 1980.

Wainer, Howard. "On Multivariate Displays," in *Recent Advances in Statistics,* Edited by Rizvi, M. H.; Rustagi, J. S.; and Siegmund, D. New York: Academic Press, 1983.

Walker, H. M., and Lev, J. *Statistical Inference.* New York: Holt, Rinehart and Winston, 1953.

Wallis, W. A., and Roberts, H. V. *Statistics: A New Approach.* New York: The Free Press of Glencoe, 1956.

chapter three

Probability

Elements of chance are present in all aspects of real-life phenomena. Probability theory has been developed to provide a measure for chance occurrences. When we toss a coin, we do not know if it will fall heads or tails; when we toss a die, the outcome may be any one of the six numbers 1, 2, ... 6; and when a patient is examined by a physician, the diagnosis may be one of several possibilities. The sex of an unborn child is uncertain and so is the survival of a cancer patient. It is also uncertain if it will rain tomorrow or if the Dow Jones index of stocks will go up.

The amount of uncertainty of occurrence in each of these phenomena is different. The purpose of probability theory is to develop a method to measure the uncertainty and provide a framework for its use in applications. If we assume, for example, that the outcomes in the toss of a coin are equally likely, then the probability of a head can be taken to be one-half, since there are two outcomes and one of them is a head. Similarly, the probability of a tail is one-half. In the case of a toss of a fair die, implicitly we assume that the outcomes are equally likely. Thus the probability that we get a two is 1/6, since there are six possible outcomes. However, it is not easy to assign a number for probability of a head in the toss of an unfair coin. In such situations, we may have to depend on the relative frequency of a head in several tosses of the same coin to provide an *estimate* of the probability.

The mathematical theory of probability is a well-developed discipline. The early development of probability theory came through questions of gambling in the seventeenth century, and names of several famous mathematicians are associated with it. Statistics uses probability theory extensively, and many statistical concepts cannot be understood without an understanding of probability.

Sample Space and Events

Scientists perform laboratory experiments to study various phenomena or observe occurrences of phenomena outside the laboratory. We regard both of these methods as experimentation. All the possible outcomes of an experiment should be known before we can study a phenomenon. The set of all possible outcomes of an experiment is called a *sample space*.

Example 3.1: A coin is tossed. There are two possible outcomes, heads (H) and tails (T). The sample space of this experiment consists of only two elements $\{H, T\}$.

Example 3.2: A die is tossed. There are six outcomes denoted by the number of dots on various faces. The sample space consists of six elements $\{1, 2, 3, 4, 5, 6\}$.

Example 3.3: A coin is tossed twice. The sample space now has four elements $\{HH, HT, TH, TT\}$.

Example 3.4: A person is examined by a medical doctor. Assume that he has one of two diseases, D_1 or D_2, or he is healthy (H). The possible outcomes are $\{D_1, D_2, H\}$.

Example 3.5: The oral temperature of a person is being observed. The outcomes are possible values of temperatures of, say, between $95°$ and $110°$. The sample space is the set of all numbers $\{95 \leq x \leq 100\}$.

Example 3.6: A population is being screened so as to obtain the first case of AIDS denoted by A. The first person examined may have A, or the first person is normal but the second has A, (NA), or the first two are normal and the third has A, (NNA), ... and so on. The sample space here is not finite and consists of an infinity of such outcomes. It is $\{A, NA, NNA, NNNA, \ldots\}$.

Example 3.7: A die is tossed twice. Let an outcome be denoted by (a, b), where a is the outcome of the first toss and b is the outcome of the second toss. The sample space consists of the following outcomes marked also as $S_1, S_2, \ldots, S_{36}$ for later use:

$$
\begin{array}{llllll}
S_1 = (1,1) & S_7 = (2,1) & S_{13} = (3,1) & S_{19} = (4,1) & S_{25} = (5,1) & S_{31} = (6,1) \\
S_2 = (1,2) & S_8 = (2,2) & S_{14} = (3,2) & S_{20} = (4,2) & S_{26} = (5,2) & S_{32} = (6,2) \\
S_3 = (1,3) & S_9 = (2,3) & S_{15} = (3,3) & S_{21} = (4,3) & S_{27} = (5,3) & S_{33} = (6,3) \\
S_4 = (1,4) & S_{10} = (2,4) & S_{16} = (3,4) & S_{22} = (4,4) & S_{28} = (5,4) & S_{34} = (6,4) \\
S_5 = (1,5) & S_{11} = (2,5) & S_{17} = (3,5) & S_{23} = (4,5) & S_{29} = (5,5) & S_{35} = (6,5) \\
S_6 = (1,6) & S_{12} = (2,6) & S_{18} = (3,6) & S_{24} = (4,6) & S_{30} = (5,6) & S_{36} = (6,6)
\end{array}
$$

Example 3.8: From an urn containing two black balls (B_1, B_2) and three red balls (R_1, R_2, R_3), one ball is drawn at random. The sample space is $\{B_1, B_2, R_1, R_2, R_3\}$. If we draw another ball, after replacing the first ball, we have the following 25 outcomes in the sample space:

$$
\begin{array}{lllll}
B_1, B_1 & B_2, B_1 & R_1, B_1 & R_2, B_1 & R_3, R_1 \\
B_1, B_2 & B_2, B_2 & R_1, B_2 & R_2, B_2 & R_3, B_2
\end{array}
$$

B_1, R_1	B_2, R_1	R_1, R_1	R_2, R_1	R_3, R_1
B_1, R_2	B_2, R_2	R_1, R_2	R_2, R_2	R_3, R_2
B_1, R_3	B_3, R_3	R_1, R_3	R_2, R_3	R_3, R_3

The notion of a set is fundamental in the study of probability. A set is a collection of objects that have some common property. We may talk of a set of patients, a set of students, a set of books, or a set of doses of a drug. Other terms like *class, space, aggregate,* and *collection* are also used for the term *set.*

All possible outcomes of an experiment form the fundamental set, which has been called a sample space. In a given investigation we are interested in certain subsets, which we designate *events.* That is, an event is a subset of a sample space.

Example 3.9: Suppose we are interested in the outcome having an even number of dots in the toss of a die. Then the event $\{2, 4, 6\}$ is a subset of the sample space $\{1, 2, 3, 4, 5, 6\}$.

Example 3.10: Suppose the oral temperature of an individual is between 95° and 110° F. The event of obtaining an oral temperature of less than 99° is the event gives by the set $\{x: 95° \leq x < 99°\}$.

Example 3.11: In example 3.3, the event, A, of getting exactly one head in the toss of a coin twice consists of the outcomes HT and TH. Hence, $A = \{HT, TH\}$.

Example 3.12: In dealing a card from a deck of 52 playing cards, the sample space consists of 52 elements, 13 for each color with designations of ace, king, queen, jack, 10, 9, ... 3, 2. The event, A, of getting an ace is:

$$A = \{\text{Heart ace, Spade ace, Diamond ace, Club ace}\}$$

Other events: If A and B are two events, we can have events that are derived from their simultaneous occurrence, or from the occurrence of only one of them, or as the occurrence of none of them. We shall use the notation of set unions, intersections, and complements to describe these newly derived events.

Definition: $A \cap B$ is the event where A occurs, or B occurs, or both occur. This is represented by the shaded area of the Venn diagram in Figure 3.1.

Definition: $A \cap B$ denotes the simultaneous occurrence of both events A and B. The shaded area in Figure 3.2 describes the event $A \cap B$.

Example 3.13: In the toss of a die, let:

$$A = \text{Event of getting an even number}$$
$$B = \text{Event of getting a number less than or equal to 3}$$
$$A = \{2, 4, 6\}$$
$$B = \{1, 2, 3\}$$

Then

$$A \cup B = \{1, 2, 3, 4, 6\}$$
$$A \cap B = \{2\}$$

AUB

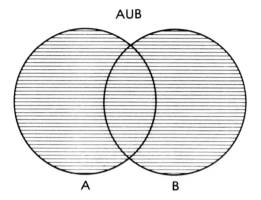

Figure 3.1. Event $A \cup B$.

A∩B

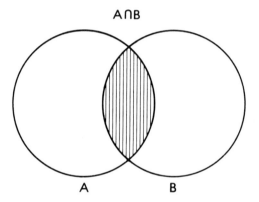

Figure 3.2. Event $A \cap B$.

Example 3.14: In the toss of a die twice in Example 3.7, let:

$$A = \text{Event of getting 2 or 12}$$
$$B = \text{Event of getting the same number on both tosses}$$

Then

$$A = \{S_1, S_{36}\}$$
$$B = \{S_1, S_8, S_{15}, S_{22}, S_{29}, S_{36}\}$$

$$A \cup B = \{S_1, S_8, S_{15}, S_{22}, S_{29}, S_{36}\}$$
$$A \cap B = \{S_1, S_{36}\}$$

Here event B contains the event A, or the occurrence or event A implies the occurrence of B.

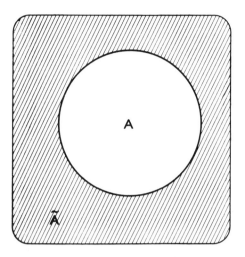

Figure 3.3. Nonoccurrence of event *A*.

Definition: Two events *A* and *B* are called mutually exclusive if the joint occurrence of both *A* and *B* is impossible. That is $A \cap B = \phi$, where ϕ denotes the impossible event (empty set).

Definition: The event $\tilde{A}$ denotes the nonoccurrence of event *A*. That is, the set of elements in $\tilde{A}$ is the set of all those elements of the sample space that are not in *A*. The shaded area in Figure 3.3 gives the event $\tilde{A}$.

Example 3.15: Let *A* be the event that an individual is sick and *B* be the event that he is overweight:

$$\tilde{A} = \text{Event that the individual is not sick}$$

$$\tilde{B} = \text{Event that the individual is not overweight}$$

$$\tilde{A} \cap \tilde{B} = \text{Event that the person is neither sick nor overweight}$$

Example 3.16: In an emergency room of a hospital, several injured persons were brought in over a period of time. Of these, 40 had face injuries, 50 had arm injuries, and 60 had leg injuries. However, 15 injured both their face and arm, 20 both arms and legs, and 25 both face and legs; 5 injured all three. Let the events *A*, *B*, and *C* describe having injuries of the face, arms, and legs, respectively. The numbers in Figure 3.4 represent the total number of elements in a given event.

Suppose we want to know the number of patients with exactly two injuries. That is:

$$A \cap B \cap \tilde{C} = \text{event of exactly two injuries on face}$$

$$\text{and arms, but not on legs}$$

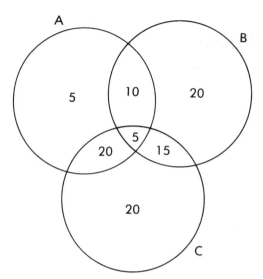

Figure 3.4. Injuries as given in Example 3.16.

Similarly,

$$A \cap C \cap \tilde{B} = \text{event of exactly two injuries on face}$$
$$\text{and legs, but not on arms}$$

$$B \cap C \cap \tilde{A} = \text{event of exactly two injuries on legs}$$
$$\text{and arms, but not on face}$$

Therefore, the total number of elements in the set gives the answer.

$$(A \cap C \cap \tilde{B}) \cup (B \cap C \cap \tilde{A}) \cup (A \cap B \cap \tilde{C})$$

Figure 3.4 provides this total as $= 10 + 20 + 15 = 45$. Similarly, the number of patients with at least two injuries is given by $45 + 5 = 50$.

Exercises

1. Write the sample space when a card is drawn from a deck of 52 playing cards.
2. A medical doctor examines two persons for an ailment. Each person may be sick (S) or healthy (H). What is the sample space?
3. What is the sample space of temperature when the person being tested may have a temperature range between 97 and 105?
4. Describe the sample space when a coin is tossed three times.
5. Let $A = \{1, 2, 3, 4\}$, $B = \{3, 4, 5, 7\}$, $C = \{3, 5, 7, 9\}$ be subsets of integers from 1 through 20. Find the events:

(a) $\tilde{A}, \tilde{B}, \tilde{C}$
(b) $A \cap B$
(c) $A \cup B$
(d) $A \cap B \cap C$
(e) $(\widetilde{A \cup B})$
(f) $\tilde{A} \cap \tilde{B}$

6. A die is tossed twice. The sample space is given in Example 3.7. Give the events:

   $A =$ Event of getting a total of 7 or 11
   $B =$ Event of getting 5
   $C =$ Event of getting 5, 7, or 11

7. In Example 3.16, find the number of patients with no injuries if it is assumed that 150 persons were brought to the hospital emergency room. Find the number of patients having injuries only on the face and arms (but not on the legs).

8. An urn has two black and two red balls. Two balls are drawn with replacement.

   (a) How many elements are there in the sample space?
   (b) What is the event A, where a black ball and a red ball is drawn? (Give elements of A.)
   (c) What is the event B, where no black ball is drawn?
   (d) Find event C, where both balls drawn are black.

9. In Exercise 8, give the events:

   (a) $A \cup B$ (b) $\tilde{A}$ (c) $A \cap B$
   (d) $B \cap C$ (e) $A \cap \tilde{B}$ (f) $A \cap B \cap C$

10. Are A and C mutually exclusive in Exercise 8?
11. Using the sample space of Example 3.7, show that $(\widetilde{A \cap B}) = \tilde{A} \cup \tilde{B}$ by finding the elements in events of both sides, where

    $A =$ Event of getting 7 or 11
    $B =$ Event of getting same number on both tosses

Probability of an Event

Probability is a numerical value associated with an outcome of an experiment representing the degree of certainty in its occurrence. In the case of equally likely outcomes, this can be understood in terms of the relative frequency of the occurrence of the experimental outcome. When a coin is tossed several times and the outcomes of heads and tails are equally frequent, then we may associate half as the probability of heads and half as the probability of tails. The probability that a person has disease D can be similarly ascertained from the frequency with which disease D occurs in the population. If we find that 50 persons have disease D in a population of 1,000, we can associate .05 as the probability of D and .95 as the probability of not having disease D.

The examples show that the probabilities are fractions and the total of these fractions for all possible outcomes adds to one.

Let S be the sample space with elements $S_1, S_2, S_3, \ldots$.

We assign nonnegative numbers to the elements in such a way that they add to one. Let $p_1, p_2, \ldots$ be assigned to $S_1, S_2, \ldots$, respectively, with:

$$p_1 \geq 0, p_2 \geq 0, \ldots \quad \text{and} \quad p_1 + p_2 + \cdots = 1$$

The number p_i is called the *probability* of the outcome S_i, $i = 1, 2, \ldots$. We shall denote probability of S_i by $P(S_i)$. We have just defined $p_1 = P(S_1)$, $p_2 = P(S_2)$, $p_3 = P(S_3)$, and so on.

Since an event consists of a few of the outcomes, the probability of an event can be taken to be the sum of the probabilities of all outcomes in *that* event. For example, if:

$$A = \{S_1, S_2, S_3\}$$
$$P(A) = P(S_1) + P(S_2) + P(S_3)$$

or

$$P(A) = p_1 + p_2 + p_3$$

Notice that $P(A)$ is also a fraction and

$$P(S) = P(S_1) + P(S_2) + \cdots$$
$$= p_1 + p_2 + \cdots$$
$$= 1$$

That is, the probability of the sure event is one. Let A and B be two mutually exclusive events with, say,

$$A = \{S_1, S_2, S_3\} \quad \text{and} \quad B = \{S_4, S_5\}$$

Then

$$A \cup B = \{S_1, S_2, S_3, S_4, S_5\}$$

and

$$P(A \cup B) = P(S_1) + P(S_2) + P(S_3) + P(S_4) + P(S_5)$$
$$= P(A) + P(B)$$

Then the above assignment provides that the probability of event A satisfies:

(i) $0 \leq P(A) \leq 1$
(ii) $P(S) = 1$
(iii) If $A \cap B = \phi$, then $P(A \cup B) = P(A) + P(B)$

In general, the probability of an event is defined with the axioms:

(i) $0 \leq P(A) \leq 1$
(ii) $P(S) = 1$
(iii) If $A_1, A_2, \ldots$ is the sequence of mutually exclusive subsets, then
$P(A_1 \cup A_2 \cup \cdots) = P(A_1) + P(A_2) + \cdots$

Example 3.17: A fair coin is tossed three times. The sample space has eight elements:

$$
\begin{aligned}
S_1 &= HHH & S_5 &= THH \\
S_2 &= HHT & S_6 &= THT \\
S_3 &= HTH & S_7 &= TTH \\
S_4 &= HTT & S_8 &= TTT
\end{aligned}
$$

Assume $\frac{1}{8}$ as the probability of each outcome. Let $A =$ event of obtaining exactly one head $= \{S_4, S_6, S_7\}$ so that

$$P(A) = P(S_4) + S(P_6) + P(S_7) = \tfrac{1}{8} + \tfrac{1}{8} + \tfrac{1}{8} = \tfrac{3}{8}$$

Similarly, let $B =$ event of getting one tail

$$= \{S_2, S_3, S_5\}$$

Thus,

$$P(B) = P(S_2) + P(S_3) + P(S_5) = \tfrac{3}{8}$$

Let $C =$ event of getting two heads. Then

$$C = \{S_2, S_3, S_5\} \quad \text{and} \quad P(C) = \tfrac{3}{8}$$

Example 3.18: Let X denote the outcome that a subject has blood pressure below 80, Y denote the outcome that it is between 80 and 100, and Z the outcome that it is above 100. Suppose two subjects are measured. The sample space has nine elements.

$$
\begin{aligned}
S_1 &= XX & S_4 &= YX & S_7 &= ZX \\
S_2 &= XY & S_5 &= YY & S_8 &= ZY \\
S_3 &= XZ & S_6 &= YZ & S_9 &= ZZ
\end{aligned}
$$

Suppose we assign probabilities:

$$
\begin{aligned}
p_1 &= \tfrac{1}{27} & p_4 &= \tfrac{1}{9} & p_7 &= \tfrac{5}{27} \\
p_2 &= \tfrac{1}{27} & p_5 &= \tfrac{1}{9} & p_8 &= \tfrac{5}{27} \\
p_3 &= \tfrac{1}{27} & p_6 &= \tfrac{1}{9} & p_9 &= \tfrac{5}{27}
\end{aligned}
$$

Define $A =$ event that blood pressure is above 100 for one or both subjects

$$= \{S_3, S_6, S_7, S_8, S_9\}$$
$$P(A) = P(S_3) + P(S_6) + P(S_7) + P(S_8) + P(S_9)$$
$$= \tfrac{1}{27} + \tfrac{1}{9} + \tfrac{5}{27} + \tfrac{5}{27} + \tfrac{5}{27} = \tfrac{19}{27}$$

Example 3.19: Assign the probability of $\frac{1}{36}$ with each outcome in

Example 3.7. Let:

$$A = \text{getting 7 or 11}$$
$$= \{S_6, S_{11}, S_{16}, S_{21}, S_{26}, S_{31}, S_{30}, S_{35}\}$$
$$P(A) = \tfrac{8}{36} = \tfrac{2}{9}$$
$$B = \text{same number on both dice}$$
$$= \{S_1, S_8, S_{15}, S_{22}, S_{29}, S_{36}\}$$
$$P(B) = \tfrac{6}{36} = \tfrac{1}{6}$$

Example 3.20: Three thumbtacks are tossed. They fall either vertically (*V*) or on a slant (*S*). The sample space has eight points:

$$S_1 = VVV \qquad S_5 = SVV$$
$$S_2 = VVS \qquad S_6 = SVS$$
$$S_3 = VSV \qquad S_7 = SSV$$
$$S_4 = VSS \qquad S_8 = SSS$$

Let the following assignment of probabilities be made:

$$p_1 = \tfrac{1}{27} \qquad p_5 = \tfrac{2}{27}$$
$$p_2 = \tfrac{2}{27} \qquad p_6 = \tfrac{4}{27}$$
$$p_3 = \tfrac{2}{27} \qquad p_7 = \tfrac{4}{27}$$
$$p_4 = \tfrac{4}{27} \qquad p_8 = \tfrac{8}{27}$$

Let:

$$A = \text{event of getting 2 vertical and 1 slant}$$

or

$$A = \{S_2, S_3, S_5\}$$
$$P(A) = \tfrac{2}{27} + \tfrac{2}{27} + \tfrac{2}{27} = \tfrac{6}{27} = \tfrac{2}{9}$$

Similarly, if *B* is the event of getting only one vertical,

$$P(B) = \tfrac{4}{9}$$

Example 3.21: A machine makes a certain number of defective items (*D*); the rest are nondefective (*N*). Suppose 4 items are examined. The sample space has 16 elements. The corresponding assignment of probabilities is also given.

$S_1 = DDDD$	$\tfrac{1}{81}$	$S_9 = NDDD$	$\tfrac{2}{81}$
$S_2 = DDDN$	$\tfrac{2}{81}$	$S_{10} = NDDN$	$\tfrac{4}{81}$
$S_3 = DDND$	$\tfrac{2}{81}$	$S_{11} = NDND$	$\tfrac{4}{81}$
$S_4 = DDNN$	$\tfrac{4}{81}$	$S_{12} = NDNN$	$\tfrac{8}{81}$
$S_5 = DNDD$	$\tfrac{2}{81}$	$S_{13} = NNDD$	$\tfrac{4}{81}$
$S_6 = DNDN$	$\tfrac{4}{81}$	$S_{14} = NNDN$	$\tfrac{8}{81}$
$S_7 = DNND$	$\tfrac{4}{81}$	$S_{15} = NNND$	$\tfrac{8}{81}$
$S_8 = DNNN$	$\tfrac{8}{81}$	$S_{16} = NNNN$	$\tfrac{16}{81}$

Let A = event of exactly two defectives. Then

$$A = \{S_4, S_6, S_7, S_{10}, S_{11}, S_{13}\}$$

and

$$P(A) = \tfrac{4}{81} + \tfrac{4}{81} + \tfrac{4}{81} + \tfrac{4}{81} + \tfrac{4}{81} + \tfrac{4}{81} = \tfrac{24}{81} = \tfrac{8}{27}$$

Let B = event of getting three defectives. Then

$$B = \{S_2, S_3, S_5, S_9\}$$

and

$$P(B) = \tfrac{8}{81}$$

Now, $A \cup B$ = event of getting two or three defectives. Then

$$P(A \cup B) = P(A) + P(B) \quad \text{since } A \text{ and } B \text{ are mutually exclusive}$$
$$= \tfrac{24}{81} + \tfrac{8}{81} = \tfrac{32}{81}$$

Exercises

12. Find the probability of an event that one subject's blood pressure is less than 80 in Example 3.18.

13. Consider the sample space in Exercise 8 and assume that the outcomes are equally likely. Find the probability of events A, B, and C.

14. Assuming that each outcome in Exercise 5 has a probability of $\tfrac{1}{20}$, find the probability of events A, B, $\tilde{A} \cap \tilde{B}$, $A \cup B$, $\widetilde{A \cup B}$.

15. Mrs. Smith encounters three traffic lights on her way to the office. Each of these lights could be red (R), green (G), or yellow (Y). Write the sample space. Assume that the outcomes are equally likely and find the probability that she finds:

    (a) all lights green
    (b) all lights red
    (c) only one light red

16. A person has a penny (P), a nickle (N), and a dime (D) in his pocket. He selects two coins at random. Write the sample space. Assume that the outcomes are equally likely and find the probability of events:

    (a) A = event of getting 15 cents
    (b) B = event of getting 6 cents
    (c) C = event of getting 20 cents
    (d) $A \cup B$
    (e) $B \cap C$

17. A card is chosen from a deck of 52 cards. Assign equal probabilities to all the outcomes and find the probability of events:

    (a) A = event of getting an ace
    (b) B = event of getting a face card

(c) $C =$ event of getting either a jack or a queen
(d) $A \cup B$
(e) $B \cup C$
(f) $B \cap C$

18. A four-sided die, with faces marked with 1, 2, 3, and 4 dots, is tossed twice. Give the sample space and assume equally likely outcomes. Find the probability of events:

(a) $A =$ event that the total is 5
(b) $B =$ event that the total is 7
(c) $C =$ event that the same number appears on both
(d) $B \cup C$
(e) $B \cap C$
(f) $C \cap \tilde{B}$
(g) $\tilde{A}$

19. In the Ohio State Lottery, all possible combinations of six digits are used and a winner gets a million dollars prize. A winner must match all the digits. What is the probability of winning a million dollars? What is the probability of not winning?

20. What is the sample space in a two-digit lottery? Suppose that any one getting at least one zero wins the lottery. What is the probability of winning the lottery?

Relations Among Probabilities of Events

In this section, a few important relations involving probabilities of events are given. These relations can be derived with the help of the axioms for the definition of probability of an event, but for simplicity, we use Venn diagrams to illustrate these results.

I. $$P(\tilde{A}) = 1 - P(A)$$

The sure event, S, the whole sample space, is composed of two mutually exclusive events A and $\tilde{A}$ (see Figure 3.5) so that:

$$P(S) = P(A) + P(\tilde{A})$$

or

$$P(A) + P(\tilde{A}) = 1$$

Therefore,

$$P(\tilde{A}) = 1 - P(A) \tag{3.1}$$

II. If A is a subset of B, that is, the occurrence of A implies the occurrence of B, then:

$$P(A) \le P(B)$$

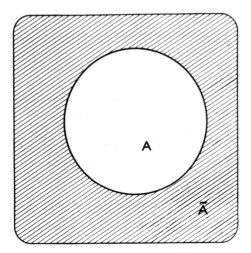

Figure 3.5. Events A and $\tilde{A}$.

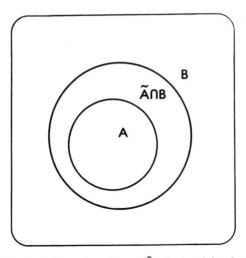

Figure 3.6. Event B as union of A and $\tilde{A} \cap B$ when A is subset of **B**.

Figure 3.6 shows that the event B is the union of two mutually exclusive events A and $\tilde{A} \cap B$. So that:

$$P(B) = P(A) + P(A \cap \tilde{B})$$

Since $P(\tilde{A} \cap B) \geq 0$, we have:

$$P(B) \geq P(A) \qquad (3.2)$$

III. $\qquad P(A \cup B) = P(A) + P(B) - P(A \cap B) \qquad (3.3)$

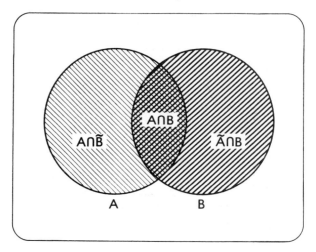

Figure 3.7. The event $A \cup B$ as a union of mutually exclusive events $A \cap \tilde{B}, A \cap B$ and $\tilde{A} \cap B$.

The shaded area in Figure 3.7 represents the event $A \cup B$. Note that the event A is the union of mutually exclusive events $A \cap B$ and $A \cap \tilde{B}$, so that:

$$P(A) = P(A \cap B) + P(A \cap \tilde{B}) \qquad (3.4)$$

Similarly,

$$P(B) = P(B \cap A) + P(B \cap \tilde{A}) \qquad (3.5)$$

Now

$$P(A \cup B) = P(A \cap \tilde{B}) + P(B \cap \tilde{A}) + P(A \cap B) \qquad (3.6)$$

Using Equations (3.4) and (3.5) and substituting their values in (3.6), we get the result in Equation (3.3).

Of course, when A and B are mutually exclusive events,

$$P(A \cap B) = 0 \quad \text{and} \quad (3.3)$$

becomes

$$P(A \cup B) = P(A) + P(B)$$

Example 3.22: Consider the injuries in Example 3.16 and assume that they are equally likely. We notice from Figure 3.4 that:

$$P(A) = \tfrac{40}{95}$$
$$P(B) = \tfrac{50}{95}$$
$$P(A \cap B) = \tfrac{15}{95}$$

Hence, $P(A \cup B) = \tfrac{40}{95} + \tfrac{50}{95} - \tfrac{15}{95} = \tfrac{15}{19}$

That is, the probability of having injuries of the arms or legs or both is $\tfrac{15}{19}$.

Similarly,

$$P(\tilde{A}) = 1 - P(A) = 1 - \frac{40}{95} = \frac{55}{95} = \frac{11}{19}$$

Example 3.23: A card is chosen at random from a deck of 52 cards. Let A = event of getting a spade and B = event of getting an ace. The event of getting an ace or a spade or the ace of spades is given by $A \cup B$. To find $P(A \cup B)$, we need:

$$P(A) = \frac{13}{52} = \frac{1}{4}$$
$$P(B) = \frac{4}{52} = \frac{1}{13}$$

and

$$P(A \cap B) = \frac{1}{52}$$

So that:

$$P(A \cup B) = \frac{1}{4} + \frac{1}{13} - \frac{1}{52} = \frac{16}{52} = \frac{4}{13}$$

The probability of not getting an ace is $P(\tilde{B})$,

$$= 1 - \frac{1}{13} = \frac{12}{13}$$

Example 3.24: In Exercise 18, the event $B \cup C$ is the event of getting the total of six or having the same number on both tosses or both of these occurrences. Now:

$$P(B) = \frac{3}{16}$$
$$P(C) = \frac{4}{16}$$
$$P(B \cap C) = \frac{1}{16}$$

Hence,

$$P(B \cup C) = \frac{3}{16} + \frac{4}{16} - \frac{1}{16} = \frac{3}{8}$$

Exercises

21. Assume equally likely outcomes in the toss of a six-sided die twice, and find the probability of getting 2, 5, or 12.

22. In diagnosing a common cold, it was noticed that the probability of the occurrence of watery nasal discharge (A) was .58, the probability of sneezing (B) was .71, and the probability of both was .39. Find the probability of:

    (a) Occurrence of either watery nasal discharge or sneezing or both
    (b) No sneezing
    (c) Neither sneezing nor watery discharge

23. Suppose the probability of death from cancer of the lung is .001 and from cancer of the stomach is .002. Assume that these events are mutually exclusive. Find the probability of:

(a) Death from either of these causes
(b) Death from a cause other than cancer of the lung

24. Given that $P(\tilde{A}) = .5$, $P(\tilde{B}) = .3$ and $P(A \cap B) = .3$, what is $P(A \cup B)$?
25. When A is a subset of B, with $P(A) = .5$, $P(B) = .7$, what is the probability of $A \cap B$?

Conditional Probability and Independence

Conditional probability gives the probability of an event when it is known that another related event has already occurred. There are situations where the occurrence of an event is affected by the presence of certain conditions. For example, we may be interested in the probability of a person who is a smoker being ill with a specific disease. Another situation is considered in drawing balls from an urn containing two red and four white balls. Suppose two balls are drawn without replacement. The probability of a red ball on the second draw depends on the outcome of the first draw. If the first ball drawn is white, the probability that the ball on the second draw is red is $\dfrac{2}{(2+3)} = \dfrac{2}{5}$. But if the first ball drawn is red, the probability of a red ball on the second draw is $\dfrac{1}{(1+4)} = \dfrac{1}{5}$.

The *conditional probability* of an event A, given the event B, is denoted by $P(A|B)$.

Example 3.25: Out of 100 patients, 30 suffer from a common cold. There are 35 patients with red noses and 20 of them have a common cold, as shown in the following:

	Common cold (A)	Other ($\tilde{A}$)
Red nose (B)	20	15
No red nose ($\tilde{B}$)	10	55

A patient is chosen at random and his red nose is noticed. What is the probability that he has a common cold? That is, we want $P(A|B)$. This is $20/(20+15) = \frac{4}{7}$.

Notice that:

$$\tfrac{20}{100} = \text{the probability of having both a}$$
$$\text{common cold and a red nose} = P(A \cap B)$$

and

$$\tfrac{35}{100} = \text{the probability of having only a red nose} = P(B)$$

Hence,

$$P(A|B) = \frac{20}{35} = \frac{20/100}{35/100} = \frac{P(A \cap B)}{P(B)}$$

Motivated by this example, the conditional probability can be defined.

Definition: The conditional probability of event A given that event B has occurred,

$$P(A|B) = \frac{P(A \cap B)}{P(B)}, \qquad \text{provided } P(B) \neq 0 \qquad (3.6)$$

Example 3.26: Consider families with two children. The sample space is

$$S_1 = bb \qquad S_3 = gb$$
$$S_2 = bg \qquad S_4 = gg$$

where $b = $ boy, $g = $ girl. Assume equally likely outcomes:

$B = $ event of having one of the children a boy $= \{S_1, S_2, S_3\}$

$A = $ event that both children are boys $= \{S_1\}$

The probability that the other child is a boy, given that one of the children is a boy:

$$= P(A|B) = (\tfrac{1}{4})/(\tfrac{3}{4}) = \tfrac{1}{3}$$

When the conditional probability of event A, given B, is known, we can find the probability of the joint occurrence of both events A and B. Multiply both sides of (3.6) by $P(B)$:

$$P(A \cap B) = P(A|B)P(B) \qquad (3.7)$$

Similarly, using the definition of $P(B|A)$:

$$P(A \cap B) = P(B|A)P(A) \qquad (3.8)$$

Further, the probability of an event A can be written in terms of $A \cap B$ and $A \cap \tilde{B}$, since $A = (A \cap B) \cup (A \cap \tilde{B})$. That is, $P(A) = P(A \cap B) + P(A \cap \tilde{B})$. Using Equation (3.7), we get:

$$P(A) = P(A|B)P(B) + P(A|\tilde{B})P(\tilde{B}) \qquad (3.9)$$

Example 3.27: In a university, 4% of the students are in engineering, and 75% of the engineering students and 50% of the remaining nonengineering students smoke cigarettes. A student is chosen at random and found to be a smoker. What is the probability that he is an engineering student? That is, we want $P(A|B)$, where

$A = $ event of being an engineering student

$B = $ event of being a smoker

It is given that:

$$P(B|A) = .75, \qquad P(A) = .04$$

and

$$P(B|\tilde{A}) = .50, \qquad P(\tilde{A}) = 1 - P(A) = .96$$

Using Equation (3.9), we have, similarly,

$$P(B) = P(B|A)P(A) + P(B|\tilde{A})P(\tilde{A})$$
$$= (.75 \times .04) + (.50 \times .96) = .51$$

and

$$P(A \cap B) = .75 \times .04 = .03$$

Hence,

$$P(A|B) = .03/.51 = .059$$

Independence

The concept of independence is central in probability theory. If the conditional probability of the event A, given event B, does not depend on event B, we call the events A and B independent. That is:

$$P(A|B) = P(A)$$

Using definition of $P(A|B)$ from Equation (3.6), we get

$$P(A \cap B) = P(A)P(B) \qquad (3.10)$$

as the condition of independence of two events A and B. This concept can be extended to more than two events.

Sometimes mutually exclusive events are confused with independent events. If A and B are mutually exclusive, $A \cap B = \phi$ and hence $P(A \cap B) = 0$. Thus we are not interested in the joint occurrence of two events when they are mutually exclusive. But this is needed to consider independence.

Example 3.28: Suppose the probability that a person recovers from a disease is p. Three persons are treated. Let A be the event that the first person recovers from the disease, B the event that the second person recovers, and C the event that the third person recovers. If A, B, and C are independent, then the event that all the three persons recover is $A \cap B \cap C$, and its probability is given by:

$$P(A \cap B \cap C) = P(A)P(B)P(C)$$
$$= p \times p \times p = p^3$$

Example 3.29: The probability that Smith gets an A in a statistics course is .3 and that Jones gets an A is .8. The assumption of independence of these two events can be made. The probability that both will get an A in Statistics is $.3 \times .8 = .24$.

The probability that either Smith or Jones or both will get an A is given by $.8 + .3 - (.8)(.3) = .86$, using the expression for the probability of union of two events.

Example 3.30: The probability that a person wins any prize in the Ohio Lottery is .8. What is the probability that two persons will win the prize, if the winning of a prize by one is independent of winning the prize by the other? This probability is $.8 \times .8 = .64$.

Exercises

26. Show that if A and B are two independent events, then:

    (a) $\tilde{A}$ and B are also independent
    (b) $\tilde{A}$ and $\tilde{B}$ are also independent

27. In a university, the distribution of students is such that 5% are in medicine, 10% are in law, 15% are in engineering, and 70% are in the remaining fields. Of those who wear ties, 90% are medical students, 70% are law students, 10% are engineering students, and 5% are the remaining students. A student is chosen at random and is found to be wearing a tie. What is the probability of his being:

    (a) a medical student?
    (b) a law student?
    (c) an engineering student?

28. The conditional probability of diabetes, given that a person takes insulin, is much higher than for the general population. Should one conclude that insulin causes diabetes?

29. The following table gives the probability of death in a population group in the United States. The probabilities of death in each age interval of ten years are given. Note that they are unconditional probabilities, and death can occur in an interval only if a person has survived the previous intervals.

Age interval	Probability of death
0–10	.064
–20	.013
–30	.022
–40	.032
–50	.064
–60	.127
–70	.211
–80	.269
–90	.170
–100	.028

Find the probability of:

(a) a person in this group reaching age 50
(b) a death occurring in ages 50–60
(c) a person living after age 90

30. Let the probability of a male birth be 0.514. Consider families having two children. Given that one of their two children is a male, what is the probability that:

(a) the other is a male
(b) the other is a female

31. Of patients having open-heart surgery, 20% die during the operation. Of those that survive, 10% die from after effects. What is the overall proportion of cases dying from one or the other of these causes?

Bayes Formula

Many applications require conditional probability of event B given event A in terms of the conditional probability of A given B. Such situations may arise in differential diagnosis. For instance, a physician wants to know the probability of a disease (B) given a symptom (A) in terms of the known probability of the symptom (A) given the disease (B) as part of medical knowledge. We have seen earlier in equations (3.7) and (3.8) that

$$P(A \cap B) = P(A|B)P(B) = P(B|A)P(A)$$

so that

$$P(B|A) = \frac{P(A|B)P(B)}{P(A)}$$

Using Equation (3.9), we have:

$$P(B|A) = \frac{P(A|B)P(B)}{P(A|B)P(B) + P(A|\tilde{B})P(\tilde{B})} \tag{3.10}$$

Formula (3.10) is a special case of Bayes Formula. The general form of the formula is given next. $P(B)$ can be regarded as a *prior* probability of B, while $P(B|A)$ is known as the *posterior* probability.

Consider event B, which occurs when any one of the mutually exclusive k events $B_1, B_2, \ldots, B_k$ occur. That is:

$$B = B_1 \cup B_2 \cup \cdots \cup B_k$$

Then, we have:

$$P(B_i|A) = \frac{P(A|B_i)P(B_i)}{P(A|B_1)P(B_1) + \cdots + P(A|B_k)P(B_k)} \tag{3.11}$$

Bayes formula is named after Reverend Thomas Bayes, who developed it in 1764–65. The importance of Bayes Formula arises from its applications in business decision making, among other major applications, by statisticians who themselves are called Bayesians. A considerable theory of Bayesian statistics has been developed. For an introduction, see D. V. Lindley's *Bayesian Statistics: A Review*, published by The Society of Industrial and Applied Mathematics (Philadelphia, 1971).

In the following we use Bayes Formula for a problem of differential diagnosis.

Example 3.31: Let the condition of a heart patient be described in terms of the following events:

$$A = \text{normal condition}$$
$$B = \text{ventrical septal defect}$$
$$C = \text{arterial septal defect}$$

Let the presence of symptoms be given by:

$$X = \text{easy fatigue}$$
$$Y = \text{chest pain}$$
$$Z = \text{dyspnea}$$
$$U = \text{pulsatile liver}$$

In the actual study conducted by Warner, Toronto, Veasy, and Stephenson in 1961 for congenital heart diseases, there were thirty-three diseases and fifty symptoms.

The conditional probabilities of a symptom under the known disease are given in the symptom-disease matrix:

Symptom-Disease Matrix

Symptoms	Diseases		
	A	B	C
X	.5	.8	.1
Y	.1	.1	.8
Z	.8	.1	.4
U	.6	.2	.3

Assume further that the incidence of diseases A, B, and C are given, say by $P(A) = .1$, $P(B) = .4$, and $P(C) = .2$, reflecting the proportion of patients with these diseases in the populations.

Let the absence of a symptom X be denoted by $\tilde{X}$, and assume that the symptoms are independent given *any* disease. Suppose we want to find the probability of a disease, say B, given symptoms, say, X, $\tilde{Y}$, Z, and U. Let $V = X \cap \tilde{Y} \cap Z \cap U$. Then by Bayes Formula, we have:

$$P(B|V) = \frac{P(V|B)P(B)}{P(V|A)P(A) + P(V|B)P(B) + P(V|C)P(C)}$$

Now, since the symptoms are independent.

$$P(V|A) = P(X \cap \tilde{Y} \cap Z \cap U|A)$$
$$= P(X|A)P(\tilde{Y}|A)P(Z|A)P(U|A)$$

Similarly $P(V|B)$ and $P(V|C)$ can be expressed in terms of $P(X|B)$, $P(Y|B)$, and so forth. Using numbers from the symptom-disease matrix, we have:

$$P(V|B)P(B) = .8 \times (1 - .1) \times .1 \times .2 \times .4$$
$$= .00576$$

Similarly,

$$\cdot\, P(V|A)P(A) = .02160$$
$$P(V|C)P(C) = .00048$$

giving

$$P(B|V) = \frac{.00576}{.02160 + .00576 + .00048}$$
$$= 0.207$$

Similarly,

$$P(A|V) = .776$$
$$P(C|V) = .017$$

The above probabilities show that the most probable disease is A, which means that the person most likely is normal.

Example 3.32: A new model car is manufactured by General Motors in plants at cities X and Y. The city X plant makes 70% of the total cars, and 30% are made by the city Y plant. At city X plant 10% of the cars are blue (A), and 5% of the cars made by city Y are blue. Mr. Smith bought a blue car. What is the probability that it came form city X plant (B); that is, what is $P(B|A)$?

Here $P(A|B) = .10$, $P(B) = .70$, $P(A|\tilde{B}) = .05$. Then using Equation (3.10), we know:

$$P(B|A) = \frac{.10 \times .70}{(.10 \times .70) + (.05 \times .30)}$$
$$= .824$$

Exercises

32. Find the following probabilities for Example 3.31:

   (a) $P(A|\tilde{X} \cap \tilde{Y} \cap Z \cap U)$
   (b) $P(A|\tilde{X} \cap \tilde{Y} \cap \tilde{Z} \cap U)$
   (c) $P(A|X \cap Y \cap \tilde{Z} \cap \tilde{U})$

33. Given the occurrence of easy fatigue, chest pain, and dyspnea, but no pulsative liver condition in Example 3.31, find the most likely diagnosis.

34. Consider families with three children. Assume that the birth of a boy or a girl is equally likely. Given that one of the children is a boy, what is the probability that the remaining children will be:

   (a) both boys?
   (b) both girls?
   (c) one boy and one girl?

35. In a diagnostic test for a disease, let D be the event that the person tested has the disease and A be the event that the test says that the person tested has the disease. It is given that $P(A|D) = .9$ and $P(\tilde{A}|\tilde{D}) = .95$. Find the probability that a person tested has the disease, given that the test says that he has the disease. That is, find $P(D|A)$. It is given that $P(D) = .005$.

36. There are two urns. Urn I contains five red and seven white balls, and urn II contains three red and four white balls. An urn is selected at random—that is, the probability of selecting I is $\frac{1}{2}$—and a ball is drawn from it. The color of this ball is red. What is the probability that it came from urn II?

Combinatorial Problems in Probability

We have seen earlier that the probability of an event, in the case of equally likely outcomes, can be obtained as the ratio of the total number of elements in the event and the total number of outcomes in the sample space. Counting various possibilities forms an important part of combinatorial mathematics. The following provides a few elementary methods of counting that are useful in probability.

Multiplication Rule

Suppose the number of elements in set A is m and the number of elements in set B is n. An element chosen from set A and an element chosen from set B form a pair. The total number of possible pairs is mn. In general, let sets $A_1, A_2, \ldots, A_k$ have $n_1, n_2, \ldots, n_k$ elements, respectively. The number of k-tuples formed by selecting one element from each set is given by the multiplication rule:

$$N = n_1, n_2, \ldots, n_k \tag{3.12}$$

Formula (3.12) is a fundamental result of combinatorial analysis.

Example 3.33: Suppose Mary has five skirts, six blouses, and three pairs of shoes. Any combination can be worn. The total number of different outfits she can wear is:

$$5 \times 6 \times 3 = 90$$

Example 3.34: Suppose a sample of two balls is drawn without replacement from an urn containing M distinguishable balls. The number of possible samples drawn is:

$$M(M - 1)$$

Example 3.35: A fast-food restaurant serves different relishes with its hamburgers. One can order onion or no onion, tomato or no tomato, ketchup or no ketchup, and mustard or no mustard. The total number of different relishes one can order with hamburgers is:

$$2 \times 2 \times 2 \times 2 = 16$$

Permutations or Arrangements

Consider arranging three books (a, b, c) on a shelf. Here are six different arrangements:

$$abc \quad acb \quad bca \quad bac \quad cab \quad cba$$

If two of the three books are to be arranged, we have:

$$ab \quad ba \quad ca \quad ac \quad bc \quad cb$$

When the number of items is large, such arrangements are not easy to write down for enumeration. However, we can determine the number of such permutations without too much difficulty.

Let $(n)_r$ be the number of permutations of n distinct items taken r at a time:

$$(n)_r = n(n - 1)(n - 2) \ldots (n - r + 1) \tag{3.13}$$

The argument to show Equation (3.12) is straightforward. The problem is to fill r places with n items. The first place can be filled with any one of n items. Suppose the first place has been filled. There are $n - 1$ items left, any one of which can be placed on the second place.

Corresponding to each way of filling up the first place, there are $n-1$ ways of filling up the second place. Hence there are $n(n-1)$ ways of filling up the first two places together. Arguing in the same way, we have:

$$(n)_r = n(n-1)\ldots r \text{ factors}$$
$$= n(n-1)\ldots(n-(r-1))$$
$$= n(n-1)\ldots(n-r+1)$$

By $n!$, known as n *factorial*, we mean the product of all integers 1 through n. That is:

$$n! = n(n-1)(n-2)\ldots2.1$$

Then $(n)_r$ can be written in terms of factorials:

$$(n)_r = \frac{n!}{(n-r)!}$$

Example 3.36: The number of possible words with four different letters using the English alphabet is

$$(26)_4 = 26 \times 25 \times 24 \times 23 = 358{,}800$$

Example 3.37: The number of ways in which three distinguishable test tubes can be arranged in a row is:

$$(3)_3 = 3 \times 2 \times 1 = 6$$

Example 3.38: The number of arrangements of four guests around a round table is $3! = 6$. If the guests are a, b, c, d, then note that circular arrangements of the type $abcd$ and $bcda$, etc., are the same arrangements. Hence, we fix one person and then arrange the others among themselves. In general, n persons can be arranged at a round table in $(n-1)!$ ways.

Example 3.39: Suppose there is a couple among the four persons to be seated at a round table. If all arrangements are equally likely, what is the probability that the couple is seated together?

The number of arrangements when the couple is seated together are obtained as follows. Consider the couple as one individual. Then we arrange only 3 around a round table in 2! ways. However, the couple can be arranged among themselves in 2! ways. Therefore, there are $2! \times 2!$ ways of arrangements. The probability is:

$$\frac{2 \times 2}{3 \times 2 \times 1} = \frac{2}{3}$$

Example 3.40: Four college students want to choose four different specialties. If they choose them at random, what is the probability that they will choose all different specialties. There are $4 \times 4 \times 4 \times 4 = 4^4$ total ways of choosing the four specialties. However, the number of choosing four different ones is 4! Hence the probability is:

$$\frac{4!}{4^4} = \frac{3}{32}$$

Combinations or Selections

When we want to find the number of ways of selecting r items out of n, the order of arrangements is of no significance. The number of *selections* (*combinations*) of n items taken r at a time is denoted by:

$$\binom{n}{r} = \frac{n!}{r!(n-r)!} \tag{3.14}$$

To show the result of Equation (3.14), we argue as follows. Let $\binom{n}{r} = y$. Choose one of the selections y and arrange r items in it. This can be done in $r!$ ways. Therefore, the total number of arrangements is $y(r!)$. But they are $\dfrac{n!}{(n-r)!}$. Therefore,

$$r! \, y = \frac{n!}{(n-r)!}$$

or

$$y = \frac{n!}{r!(n-r)!}$$

Example 3.41: The number of ways of selecting a bridge hand, that is, selecting 13 cards out of 52, is:

$$\binom{52}{13} = \frac{52!}{13!\,39!} \doteq 6.35 \times 10^{12}$$

Example 3.42: From a set of 40 players, a team of 11 players is to be chosen. The total number of possibilities is $\binom{40}{11} = \dfrac{40!}{11!\,29!}$.

Example 3.43: A committee of 3 persons is to be selected from a group of 10. The total number of committees is $\binom{10}{3} = \dfrac{10!}{3!\,7!} = 120$.

Example 3.44: In a poker hand, we have two aces and three other cards. What is the probability of this event? The total number of poker hands is obtained by selecting 5 cards out of 52. This is $\binom{52}{5}$. Similarly, getting

2 aces out of 4 is $\binom{4}{2}$, and getting 3 out of the remaining 48 cards is $\binom{48}{3}$. The total number of such combinations is $\binom{4}{2}\binom{48}{3}$. Hence, the probability of getting 2 aces and 3 other cards in a poker hand is:

$$\frac{\binom{4}{2}\binom{48}{3}}{\binom{52}{5}} = .40$$

Since the number of ways of selecting r things out of n is the same as selecting $n - r$ things out of n:

$$\binom{n}{r} = \binom{n}{n-r} \tag{3.15}$$

Also the number of ways of selecting all of n out of n items is 1, so that:

$$\binom{n}{n} = 1 \tag{3.16}$$

Since

$$1 = \binom{n}{n} = \frac{n!}{n!(n-n)!} = \frac{n!}{n!0!} = \frac{1}{0!}$$

$$0! = 1 \tag{3.17}$$

The results of Equations (3.15), (3.16), and (3.17) are often used in computations.

Example 3.45: The number of ways of distributing cards among four players in the game of bridge is $\binom{52}{13}\binom{39}{13}\binom{26}{13}\binom{13}{13} = \frac{(52)!}{(13!)^4}$.

Example 3.46: The probability that in a bridge hand there is a complete suit is given by:

$$\frac{\binom{4}{1}}{\binom{52}{13}}$$

since the total number of bridge hands is $\binom{52}{13}$ and the number of ways of selecting a complete suit is $\binom{4}{1}$.

Exercises

37. Find:

    (i) $(20)_3$, $(5)_3$, $(5)_5$

    (ii) $\dbinom{7}{3}$, $\dbinom{8}{2}$, $\dbinom{52}{4}$, $\dbinom{52}{13}$

38. In how many ways can we select a committee of three out of five physicians, including two cardiologists, two surgeons, and an anesthesiologist, with the condition that only one cardiologist and one surgeon are on the committee?

39. Suppose six guests, including a couple, are to be seated at a round table. What is the probability that the couple is not seated together, assuming that all outcomes are equally likely?

40. Two balls are drawn without replacement from an urn containing five red balls and three black balls. What is the probability that the sample contains one black ball and one red ball?

41. A restaurant serves three kinds of sandwiches and four kinds of soups. How many different kinds of soup-and-sandwich combinations can it serve?

42. Eight persons are seated at a round table, including three close friends. The three friends want to be seated together. What is the total number of ways of seating them?

43. Five cards are selected at random for a poker hand. What is the probability of getting:

    (a) all five cards of the same suit?
    (b) two kings and three aces?

Chapter Exercises

1. Find the number of licence plates that have:

    (a) six possible digits
    (b) six positions, with the first three involving letters and the last three with digits
    (c) six positions, with the first two with letters and the last four with digits

2. What is the probability that a licence plate with your initials is assigned to you by chance in the case of Exercise 1(b), where all outcomes are equally likely?

3. $P(A) = .04$ and $P(B) = 0.5$. Find $P(A \cap B)$ if

    (a) A and B are mutually exclusive
    (b) A and B are independent

4. A psychologist is studying the performance of certain learning strategies by having rats pass through various kinds of mazes. A rat is trained to choose one of the three mazes at random. The probability that a rat can cross a maze is given by 0.8, 0.5, 02. Suppose the rat has crossed a maze. What is the probability that it crossed:

   (a) the first maze?
   (b) the second maze?

5. A box contains 5 rotten apples among a total of 24. A sample of 10 apples is examined. What is the probability that it contains no rotten apples? One rotten apple? Two rotten apples? Three rotten apples? Four rotten apples? Five rotten apples?

6. On a grocery-store counter there are 120 loaves of bread stocked, ten of which are a day old. What is the probability that the two loaves you pick at random are not fresh? One loaf is fresh? Two loaves are fresh?

7. Suppose the birthday of a person is equally likely to be on any one of 365 days in the year. Of five persons in a room, what is the probability that they all have different birthdays?

8. In Exercise 7, what is the probability that there are at least two persons with the same birthday?

9. In a tennis tournament, a player is involved in three games, and his probability of winning any one of them is 0.6. These games are independent. What is the probability that he wins all three games? Only two games? Only one game? None of the games? At least two games?

10. Three tank guns are employed in a battle with the probability of successful hits of 0.8, 0.7, and 0.5. A gun is selected at random and its shot hits the target. What is the probability that gun I was chosen? That gun II was chosen?

11. A box has three drawers. The top drawer contains two gold coins, the middle drawer contains two silver coins, and the bottom drawer contains one silver and one gold coin. A drawer is selected at random and you know that one of the coins is gold. What is the probability that the

    (a) second coin is silver?
    (b) second coin is gold?

Summary

All possible outcomes of an *experiment* form a *sample space*. A subset of the sample space is an *event*. An axiomatic definition of probability of an event requires that the probability of any two *mutually exclusive* events be the sum of their probabilities, probability for the whole sample space be one and be nonnegative for any event. The *conditional probability* of an event given another event is also a probability and is defined by the ratio of the probability of the joint occurrence of both events and the probability of the given event. *Bayes Formula* provides the formula for *posterior* probability of an event in terms of its *prior* probability when another event is known to have occurred. *Permutations or arrangements* of *n* things taken *r* at a time and *combinations or selections* of *n* things taken *r* at a time form important combinatorial results of use in computing probabilites, especially in problems with equally likely outcomes.

References

Cornfield, Jerome. The Bayesian outlook and its applications, *Biometrics*, 1969, *25*, 617–642.

Feller, W. *Introduction to Probability Theory and Its Applications*. New York: John Wiley & Sons, 1960.

Lindley, Dennis V. *Bayesian Statistics: A Review*. Philadelphia: Society of Industrial and Applied Mathematics, 1971.

Parzen, Emanuel. *Modern Probability Theory and Its Applications*. New York: John Wiley & Sons, 1960.

Warner, H. R.; Toronto, A. F.; Veasy, L. G., and Stephenson, R. A mathematical approach to medical diagnosis, *J. American Medical Association*, 1961, 177–183.

chapter four

Random Variables and Discrete Probability Distributions

The concept of the probability of an event was introduced in the last chapter. When we describe the probability for all the possible outcomes of an experiment, the sample space may be quite large. Often we are interested in certain specific characteristics of the sample space. For example, when we toss a coin four times, we have sixteen outcomes. For practical purposes, however, we may be interested only in the number of heads or the number of tails in each outcome. The set of interest now for this experiment is reduced to only five numbers, the possible number of heads 0, 1, 2, 3, and 4. This simplification is obtained by introducing a function on the sample space, such as the counting function, in the case of tossing a coin. Similarly, when a patient is being diagnosed, there are two possiblities—disease (D) or normal (N). When two patients are screened, the number of possibilities is four, and when ten patients are screened, the number of possibilities increases to $2^{10} = 1,024$. The interest here may be simply in the total number of diseased among the ten persons screened, which can have values 0, 1,..., 10.

Random Variables

Random variables are rules (functions) that assign a number to each outcome of the sample space.

The concept of random variable is fundamental in probability. The random variable is associated with the probability that has been defined on the sample space, and it condenses information on probability from the sample space.

When the random variable assumes a finite or countably infinite number of values, we call it a *discrete random variable*. When the random variable assumes real values, it is called a *continuous random variable*. Examples of a continuous random variable are the weight or height of persons, the time it takes a drug to have an effect, cholesterol levels, and so forth. The probability distributions for continuous random variables will be discussed in Chapter 5.

Example 4.1: From a population of 1,000 items, 5 items are selected without replacement. The sample space here has $\binom{1000}{5}$ elements. Each item may be good or bad. Let the random variable count the number of good items in each sample. Then the possible outcomes are only 0, 1, 2, 3, 4, or 5, and the random variable is discrete.

Example 4.2: Suppose we select individuals from a population until we find the first case of AIDS. The number of persons selected is a discrete random variable having values 1, 2, 3,... .

Example 4.3: The number of colonies of bacteria on a petrie dish is a discrete random variable having values 0, 1, 2,... .

Example 4.4: The number of accidents in a given city is a discrete random variable having values 0, 1, 2,... .

Example 4.5: A number is chosen at random from an interval (0, 1). Then the variable is continuous.

Notation: A random variable is usually denoted by a capital letter and the values it takes by a lower-case letter. The probability that random variable X has value x gives the *discrete probability distribution* of random variable X and is denoted by

$$P\{X = x\} = p_X(x) \tag{4.1}$$

Notice that $p_X(x) \geq 0$ for all x and $\sum_x p_X(x) = 1$.

An event in the sample space on which a random variable has been defined can be described in terms of the random variable. The corresponding probability of the event thus can be obtained from that of the event in the original sample space. Consider the following example.

Example 4.6: A cage contains mice, 40% of them white and the rest black. A sample of three is selected with replacement. Let X denote the number of white mice in the sample. The possible values are $x = 0, 1, 2, 3$. Let the sample space be given by:

$$
\begin{aligned}
S_1 &= WWW & S_5 &= WWB \\
S_2 &= WBW & S_6 &= WBB \\
S_3 &= BWW & S_7 &= BWB \\
S_4 &= BBW & S_8 &= BBB
\end{aligned}
$$

$X = 2$ describes the event $A = \{S_2, S_3, S_5\}$. Similarty, $X = 3$ describes the event $B = \{S_1\}$, and so on. These probabilities can be found as before. For example:

$$P\{X = 2\} = P(A) = P\{S_2\} + P\{S_3\} + P\{S_5\}$$
$$= (.4)^2(.6) + (.4)^2(.6) + (.4)^2(.6)$$
$$= 3(.4)^2(.6)$$

The probabilities listed below give the probability distribution of X:

x	$p_X(x)$
0	$(.6)^3$
1	$3(.4)(.6)^2$
2	$3(.4)^2(.6)$
3	$(.4)^3$

These values can be expressed in one formula:

$$p_X(x) = \binom{3}{x}(.4)^x(.6)^{3-x} \quad x = 0, 1, 2, 3$$

Example 4.7: Suppose a sample of two balls is drawn without replacement from an urn containing three white balls and two red balls. The number of white balls in the sample is a random variable X having possible values $x = 0, 1, 2$. The probability that $X = 0$ is obtained by the ratio of the number of outcomes with no white and two red balls in the sample and the total number of selections of two balls from the urn. The number of ways of getting no white and two red balls is $\binom{3}{0}\binom{2}{2} = 1$ and the total number of samples is $\binom{5}{2} = 10$. Therefore:

$$P(X = 0) = \tfrac{1}{10}$$

Similarly,

$$P(X = 1) = \frac{\binom{3}{1}\binom{2}{1}}{\binom{5}{2}} = \frac{6}{10}$$

$$P(X = 2) = \frac{\binom{3}{2}\binom{2}{0}}{\binom{5}{2}} = \frac{3}{10}$$

The probability distribution of x is given by:

x	0	1	2
$p_X(x)$	$\frac{1}{10}$	$\frac{6}{10}$	$\frac{3}{10}$

Example 4.8: A fair die is tossed. Let X be the number of dots observed with possible values $x = 1, 2, \ldots, 6$. The probability distribution of X is given by:

x	1	2	3	4	5	6
$p_X(x)$	$\frac{1}{6}$	$\frac{1}{6}$	$\frac{1}{6}$	$\frac{1}{6}$	$\frac{1}{6}$	$\frac{1}{6}$

Exercises

1. A biased coin is tossed three times. Assume the probability of a head is .6. Let X be the number of heads in a given toss. Give the probability distribution of X.
2. A fair die is tossed twice and the total number of dots on both faces is observed. Notice that it will be 2, 3, 4,... 12. Give the probability distribution of the total number of dots.
3. From a group containing four doctors and two nurses, three persons are selected. The number of doctors in the sample is a random variable X having values 0, 1, 2, 3, 4. Find the probability distribution of X, assuming that the selection is with replacement.

Binomial Distribution

Many trials result in two possible outcomes, such as in tossing a coin, in examining a patient, or in an opinion poll. Trials having two possible outcomes are called *Bernoulli trials*. Suppose we denote them by success S, and failure F. The binomial probability distribution is concerned with the random variable that counts the number of successes in a fixed number of independent Bernoulli trials. Let p be the probability of success in a trial and let $q = 1 - p$. Suppose there are n trials, and let X be the number of successes. Then the probability distribution of X is given by:

$$p_X(x) = \binom{n}{x} p^x (1 - p)^{n - x}$$

or

$$p_X(x) = \frac{n!}{x!(n - x)!} p^x q^{n - x} \qquad (4.2)$$

Some examples are given in Table 4.1.

Table 4.1
Binomial Probabilities

n	x	$p_X(x)$
1	0	q
	1	p
2	0	q^2
	1	$2pq$
	2	p^2
3	0	q^3
	1	$3q^2p$
	2	$3qp^2$
	3	p^3
4	0	q^4
	1	$4q^3p$
	2	$6q^2p^2$
	3	$4qp^3$
	4	p^4
5	0	q^5
	1	$5q^4p$
	2	$10q^3p^3$
	3	$10q^2p^3$
	4	$5qp^4$
	5	p^5

Binomial probabilities are so named because they occur in binomial expansion. Binomial theorem gives the expansion of the sum of two terms to a positive integral power. The following equation is well known and can be shown by finite induction:

$$(a+b)^n = \binom{n}{0}a^n + \binom{n}{1}a^{n-1}b + \cdots + \binom{n}{x}a^{n-x}b^x + \cdots + \binom{n}{n}b^n$$

$$= \sum_{x=0}^{n} \binom{n}{x}b^x a^{n-x}$$

Note that

$$(a+b)^2 = a^2 + 2ab + b^2 = \binom{2}{0}a^2 + \binom{2}{1}ab + \binom{2}{2}b^2$$

$$(a+b)^3 = a^3 + 3a^2b + 3ab^2 + b^3$$

$$= \binom{3}{0}a^3 + \binom{3}{1}a^2b + \binom{3}{2}ab^2 + \binom{3}{3}b^3$$

and so on.

Let $a = p$ and $b = 1 - p = q$, then $a + b = 1$. The general term in the binomial theorem provides the probabilities in the binomial distribution. Furthermore:

$$\sum_{x=0}^{n} P_X(x) = \sum_{x=0}^{n} \binom{n}{x} p^x (1-p)^{n-x}$$

$$= (p + 1 - p)^n = 1$$

The basic property of the probability distribution given in Equation (4.1) is verified. Binomial probability distribution plays a central role in probability theory. For various values of n, p, and x, the probabilities are tabulated extensively. For example, tables published by the National Bureau of Standards, U.S. Government Printing Office (1950), include one for n from 2 to 49 and p from .01 to .5. Table I in the Appendix gives the binomial probabilities for use in this book

Figures 4.1 (a) and (b) give the binomial probabilities for $n = 10$, $p = .5$, and $n = 10$, $p = .2$.

Example 4.9: Suppose the probability of survival from a given disease is .05. Among ten patients who have the same probability of survival, the probability that two survive from the disease is given by the binomial probability distribution with $n = 10$, $x = 2$, and $p = .05$. From Equation (4.2) it is:

$$\binom{10}{2}(.05)^2(.95)^8$$

$$= .0746 \text{ from Table I}$$

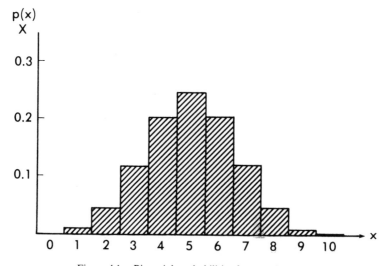

Figure 4.1a. Binomial probabilities for $n = 10$, $p = 5$.

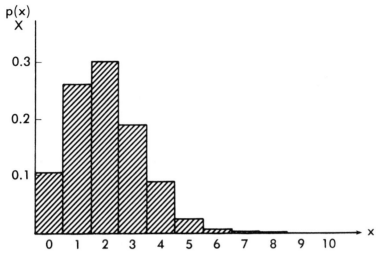

Figure 4.1b. Binomial probabilities for $n = 10$, $p = .2$.

Example 4.10: There are 100 defectives in a lot of 1,000 items. A sample of ten with a replacement is chosen and one defective is found. Here $n = 10$, $p = .1$, $x = 1$, and the probability is:

$$\binom{10}{1}(.1)^1(.9)^9$$

$$= .3874 \text{ from Table I}$$

The probability that there is at most one defective in the sample is:

$$p_X(0) + p_X(1)$$

$$= \binom{10}{0}(.1)^0(.9)^{10} + \binom{10}{1}(.1)^1(.9)^9$$

$$= .3487 + .3874$$

$$= 0.7361 \text{ from Table I}$$

Example 4.11: Let the probability of death from an auto accident be .001. The event that at least one death occurs is the complement of the event that no death occurs. Let $n = 15$, $p = .001$, and $x = 0$:

$$P\{x = 0\} = \binom{15}{0}(.001)^0(.999)^{15}$$

$$= .985$$

$$P\{X \geq 1\} = 1 - P\{X = 0\} = 1 - .985 = .015$$

Exercises

4. Let X be the number of successes in n Bernoulli trials, with p as the probability of success. Find:

   (a) $P\{X = 7\}$, when $n = 15$, $p = .2$
   (b) $P\{X = 8\}$, when $n = 10$, $p = .4$
   (c) $P\{1 \leq X \leq 5\}$ when $n = 5$, $p = .05$
   (d) $P\{X \geq 11\}$, when $n = 15$, $p = 0.2$

5. The probability that a drug is effective is .8. What is the probability that amont ten patients on whom this drug is used five or fewer will show any effect?

6. The probability that a student is absent from a class is .1. What is the probability that in a class of ten students

   (a) no one will be present?
   (b) only four will be present?
   (c) all the students will be present?

7. For $n = 25$, use the binomial tables to find the two values of p that satisfy the equation $P\{X = 8\} = .075$.

8. A fair die is tossed ten times. Find the probability of getting a six four times.

9. The probability that a laboratory sample becomes contaminated is .2. Find the probability that one out of five samples is contaminated.

10. The shots fired from a gun have a probability of .8 of scoring a hit, and they are independent. Find the probability that eight out of ten times the gun will score a hit.

11. In a mating with genotypes Aa and aa, what is the probability of one Aa, given that there are three offspring? Note that the offspring of this cross are only Aa and aa, and each has a probability of .5.

12. In Exercise 11, if it is known that one of the children is of genotype Aa, find the conditional probability that at least one of the remaining children is Aa?

Expectation of a Random Variable

The expected value or mean of a random variable gives the central value of its probability distribution. Just as the center of gravity of a body gives a point through which the total mass of the body seems to act, in the same way the expected value gives a reference point for the probability distribution. For a discrete random variable, the expectation of X is defined as the weighted average of the values of X, the weights being the corresponding probabilities. If

X has probability distribution $p_X(x)$, the *expected value* of X, denoted by $E(X)$, is given by:

$$E(X) = \sum_{all\,x} x p_X(x) \tag{4.3}$$

The expected value is also called the *mean* and is usually denoted by Greek the letter *mu* μ_X. That is:

$$\mu_X = E(X) \tag{4.4}$$

Example 4.12: The number of injuries a person incurs is random variable X having the probability distribution:

x	0	1	2
$p_X(x)$	$\frac{1}{3}$	$\frac{1}{3}$	$\frac{1}{3}$

The expected number of injuries is given by $E(X) = \left(0 \times \frac{1}{3}\right) + \left(1 \times \frac{1}{3}\right) + \left(2 \times \frac{1}{3}\right) = 1$. That is, on the average, a person will receive one injury.

Example 4.13: A lottery ticket costs one dollar. The probability of winning a $101 prize is .001. The gain of the player is a random variable with the distribution:

x	-1	100
$p_X(x)$	.999	.001

The expected gain in this case, then, is:

$$E(X) = (-1 \times .999) + (100 \times .001)$$
$$= -.899$$

That is, the player is expected to lose on the average 89.9 cents.

Example 4.14: A number is chosen at random from 1, 2, 3, 4, and 5. The probability distribution is:

x	1	2	3	4	5
$p_X(x)$	$\frac{1}{5}$	$\frac{1}{5}$	$\frac{1}{5}$	$\frac{1}{5}$	$\frac{1}{5}$

The expected value of X is:

$$E(X) = \left(1 \times \frac{1}{5}\right) + \left(2 \times \frac{1}{5}\right) + \left(3 \times \frac{1}{5}\right) + \left(4 \times \frac{1}{5}\right) + \left(5 \times \frac{1}{5}\right)$$
$$= 3$$

Example 4.15: The number of colonies of bacteria on a plate in a certain experiment has the probability distribution:

x	0	1	2	3
$p_X(x)$	$\frac{1}{8}$	$\frac{1}{8}$	$\frac{1}{2}$	$\frac{1}{4}$

The mean of X is:

$$E(X) = \left(0 \times \frac{1}{8}\right) + \left(1 \times \frac{1}{8}\right) + \left(2 \times \frac{1}{2}\right) + \left(3 \times \frac{1}{4}\right)$$
$$= 1\frac{7}{8}$$

Example 4.16: For $n = 3$, let X be the number of successes with probability p of a success. Table 4.1 gives the probabilities:

$$E(X) = (0 \times q^3) + (1 \times 3pq^3) + (2 \times 3p^2q) + (3 \times p^3)$$
$$= 3p(q^2 + 2pq + p^2)$$
$$= 3p(p + q)^2 = 3p$$

In general, the expected value of a binomial random variable is:

$$E(X) = np \qquad (4.5)$$

Example 4.17: In 100 tosses of a fair coin, the number of heads (X) has a binomial distribution with $n = 100$ and $p = \frac{1}{2}$. The expected number of heads is:

$$100 \times \frac{1}{2} = 50$$

Example 4.18: Suppose that the probability that an auto will be involved in an accident is .001. If in a city there are 15,000 autos, then the number of accidents is a binomial random variable with $n = 15,000$ and $p = .001$. The expected number of accidents in the city is;

$$15,000 \times .001 = 15$$

Although the mean of a random variable is an important point of its probability distribution, several different probability distributions may lead to the same mean. For example, if X has the distribution

x	-1	0	1
$p_X(x)$	$\frac{1}{3}$	$\frac{1}{3}$	$\frac{1}{3}$

then

$$E(X) = \left(-1 \times \frac{1}{3}\right) + \left(0 \times \frac{1}{3}\right) + \left(1 \times \frac{1}{3}\right) = 0$$

If X has the distribution

x	-2	-1	0	1	2
$p_X(x)$	$\frac{1}{5}$	$\frac{1}{5}$	$\frac{1}{5}$	$\frac{1}{5}$	$\frac{1}{5}$

the expected value is again zero, since

$$E(X) = \left(-2 \times \frac{1}{5}\right) + \left(-1 \times \frac{1}{5}\right) + \left(0 \times \frac{1}{5}\right) + \left(1 \times \frac{1}{5}\right) + \left(2 \times \frac{1}{5}\right) = 0$$

Another characteristic of a probability distribution is given by *variance*, which is a measure of scatter of the values of X around the mean value.

Definition: The *variance* of random variable X is the expected value of the squared deviation from its mean. That is:

$$\text{Variance of } X = \text{Var}(X) = E(X - \mu)^2 \qquad (4.6)$$

This expression can be simplified to give:

$$\text{Var}(X) = E(X^2) - \mu^2 \qquad (4.7)$$

The variance of X is also denoted by σ_X^2 (small Greek letter *sigma* squared). The *standard deviation* is the square root of the variance.

There are many other measures of scatter, such as $E|X - \mu|$, known as the absolute deviation from the mean, but variance is the most commonly used. Note that if X has values close to the mean, $\text{Var}(X)$ will be small. For the binomial random variable, the variance is:

$$\text{Var}(X) = npq \qquad (4.8)$$

Example 4.19: For the distribution of X:

x	-1	0	1
$p_X(x)$	$\frac{1}{3}$	$\frac{1}{3}$	$\frac{1}{3}$

$$E(X^2) = \left((-1)^2 \times \frac{1}{3}\right) + \left(0 \times \frac{1}{3}\right) + \left(1^2 \times \frac{1}{3}\right) = \frac{2}{3}$$

$$\text{Var}(X) = \frac{2}{3} - 0 = \frac{2}{3}$$

For the distribution of a random variable X,

x	-2	-1	0	1	2
$p_X(x)$	$\frac{1}{5}$	$\frac{1}{5}$	$\frac{1}{5}$	$\frac{1}{5}$	$\frac{1}{5}$

$$\text{Var}(X) = E(X^2) - 0 = \left(-2^2 \times \frac{1}{5}\right) + \left(-1^2 \times \frac{1}{5}\right) + \left(0 \times \frac{1}{5}\right)$$
$$+ \left(1^2 \times \frac{1}{5}\right) + \left(2^2 \times \frac{1}{5}\right)$$
$$= 2$$

Example 4.20: In Example 4.17, the variance of the number of heads in 100 tosses is:

$$\mathrm{Var}(X) = 100 \times \frac{1}{2} \times \frac{1}{2} = 25$$

Example 4.21: In Example 4.18, the variance of the number of accidents is:

$$\mathrm{Var}(X) = 15,000 \times .001 \times .999$$
$$= 14.985$$

Example 4.22: Suppose a drug cures a disease with a probability of .2. One hundred patients take this drug. The expected number of patients cured is $100 \times .2 = 20$. The variance is $100 \times .2 \times .8 = 16$.

Example 4.23: In one game of the Illinois State Lottery, the value of the ticket is one dollar, and there are prizes of $11, $101, and $1001 with probabilities of winning .05, .002, and .00001, respectively. The probability distribution of the gain, X, is:

x	-1	10	100	1,000
$p_X(x)$	.94799	.05	.002	.000,01

$$E(X) = (-1 \times .94799) + (10 \times 0.05) + (100 \times .002) + (1,000 \times .000,01)$$
$$= -0.23799$$
$$\mathrm{Var}(X) = (-1^2 \times .94799) + (10^2 \times .05) + (100^2 \times .002)$$
$$+ (1,000^2 \times .000,01) - (-.23799)^2$$
$$= 35.89$$

Exercises

13. Find the mean and variance for the following distribution:

(a)

x	1	2	3
$p_X(x)$	$\frac{1}{3}$	$\frac{1}{3}$	$\frac{1}{3}$

(b) $p_X(x) = \binom{5}{3}\left(\frac{1}{3}\right)^x \left(\frac{2}{3}\right)^{5-x}$, $\quad x = 0, 1, 2, 3, 4, 5$

(c)

x	-1	0	1
$p_X(x)$	.25	.5	.25

(d)

x	1	2	3	4	5
$p_X(x)$	.2	.2	.2	.2	.2

(e)

x	1	2	3	4	5
$p_X(x)$	.4	.08	.04	.08	.4

14. The probability of winning a game is 1/3. What is the expected gain for a player if the player gets $10 when he wins but pays one dollar when he loses?

15. In a weekly Ohio State Lottery game, the cost of the ticket is 50 cents. Suppose the probabilities of winning prizes of $10 and $1,000 are .02 and .000,02, respectively. What is the probability distribution of the gain to the player? Find its mean and variance.

16. Find the mean and variance for the following binomial distributions:

   (a) $n = 20$, $p = .01$
   (b) $n = 50$, $p = .20$
   (c) $n = 100$ $p = .30$
   (d) $n = 1,000$ $p = .01$
   (e) $n = 100,000$ $p = .001$

17. A machine makes defectives with a probability of .02. In a sample of 50 items produced, what is the expected number of defectives? What is the variance?

18. A student has probability .8 that he will be able to answer any question in a test correctly. A test contains 15 questions. What is the expected number of questions that he will be able to answer correctly? What is the variance?

19. A printing machine makes an error with probablity .01 per page. For a book of 1,000 pages, find the expected number of errors and their variance.

20. A cookie company makes cookies with raisins with the following distribution of raisins:

x	0	1	2	3	4
$p_X(x)$	.2	.3	.4	.05	.05

Find the expected number of raisins in a cookie and its variance.

Poisson Probability Distribution

Many practical applications of binomial distribution involve large number of trials with a small probability of success. That is, n is large and p is small. In such cases, the binomial probabilities can be approximated by Poisson probabilities. Consider an auto insurance company with 10,000 policyholders. The probability that an individual insured has an accident is small, say .001. The binomial probability for the number of persons involved in an accident can then be approximated. Similarly, consider a chemist who is interested in the probability distribution of inert particles in a solution. Here the solution can be regarded as consisting of hundreds of drops, and in each drop there are only a few inert particles. Again, the number of trials (drops) is large and the probability of success (inert particles) is small.

To define Poisson probabilities, we need exponential function e^x, which occurs frequently in applications. The number e is the base of natural logarithms and an approximate value of e correct to 7 decimal places is:

$$e = 2.7182818$$

The Poisson distribution is given by

$$p_X(x) = \frac{e^{-\lambda}\lambda^x}{x!}, \quad x = 0, 1, 2, \ldots \tag{4.9}$$

and provides an approximation to binomial probabilities when n is large and p is small (n tends to infinity and p tends to zero), but $np = \lambda$ (Greek letter *lambda*). Poisson distribution can also be derived with the help of axioms and does not use the above approximation.

The mean and variance of Poisson distribution are given by:

$$\mu = \lambda \tag{4.10}$$
$$\sigma^2 = \lambda \tag{4.11}$$

Poisson probabilities are tabulated in Table II in the Appendix.

Example 4.22: Suppose misprints on a page in a book have a Poisson distribution with $\lambda = 5$. The probability that a reader finds three misprints on a page is:

$$\frac{e^{-5}5^3}{3!} = .1404$$

The probability of finding fewer than three misprints on a page is:

$$= P(\text{no misprint}) + P(\text{one misprint}) + P(\text{two misprints})$$
$$= .0067 + .0337 + .0842$$
$$= .1246$$

Example 4.23: The number of bacteria in equal areas of a microscope field has a Poisson distribution with $\lambda = 2$. The probability that we find four bacteria in a given field is:

$$\frac{e^{-2}2^4}{4!} = .0902$$

$$P\{3 \le X \le 5\} = P\{X = 3\} + P\{X = 4\} + P\{X = 5\}$$
$$= .1804 + .0902 + .0361$$
$$= .3067$$

Exercises

21. The probability that a car has an accident at an intersection is .001. Find the probability that among 10,000 cars, there will be three accidents. Assume that the number of accidents has a Poisson distribution with $\lambda = 10{,}000 \times .001 = 10$.

22. The number of bacteria seen in equal areas of a microscopic field has a Poisson distribution with $\lambda = 3$. Find the probability that:

    (a) there are two bacteria in a given field
    (b) there is none
    (c) there are more than two bacteria

23. Given that the number of vacant beds on one day in a hospital has a Poisson distribution with $\lambda = 5$, what is the probability that the number of vacant beds on a given day is:

    (a) between three and five, including both three and five
    (b) more than or equal to five
    (c) less than or equal to three
    (d) either less than three or more than five

24. Suppose the number of vacancies for justices of the U.S. Supreme Court in any given presidential term has a Poisson distribution with $\lambda = 3$. What is the probability that President Reagan will be required to fill two vacancies in his term of office?

25. In an auto plant, the absenteeism has been on the average .02 as a result of illness or other factors. In a work force of 250 persons, how many absentees would you expect? Given that the number of absentees has a Poisson distribution, find the probability that:

    (a) the number of absentees is five
    (b) more than four are absent
    (c) the number of absentees is fewer than three
    (d) no one is absent

Geometric Distribution

Suppose a married couple desperately wants a daughter and continues to have children until a girl is born. The number of children, then, is an important random variable. In the treatment of leukemia, we want to know the number of weeks a patient is in remission before he gets his first relapse, or a medical team may need to know the number of persons to be screened before they discover a case of AIDS. These are instances of a *geometric random variable.*

Suppose a sequence of successes (*S*) and failures (*F*) is being observed until we get the first successes. Let $P\{\text{Success}\} = p$ and $P\{\text{Failure}\} = 1 - p = q$. Let X be the total number of trials needed. There are $X - 1$ failures before having one success. Since the trials are independent, the equation

$$p_X(x) = q^{x-1}p, \quad x = 1, 2, \ldots \tag{4.12}$$

defines the *geometric probability distribution*. The *mean* and *variance* of this distribution are:

$$\mu = \frac{1}{p} \tag{4.13}$$

$$\sigma^2 = \frac{q}{p^2} \tag{4.14}$$

Example 4.24: Given that the probability of a female child is $\frac{1}{2}$, the expected number of children the above couple will have before having a girl is 2, from Equation (4,13).

Suppose a couple wants to know how many children to have, such that the probability is at least .95 that they have one daughter. The probability of having x children is:

$$\left(\frac{1}{2}\right)^{x-1}\frac{1}{2} = \left(\frac{1}{2}\right)^x$$

The probability of having m children or fewer is obtained by adding the probabilities:

$$P\{m \text{ children or fewer}\} = \sum_{x=1}^{m}\left(\frac{1}{2}\right)^x$$

These probabilities are tabulated for a few values of m:

m	1	2	3	4	5	6
probability	.5	.75	.87	.94	.97	.99

Hence, five or fewer children would give a probability of .97.

Example 4.25: The probability of having a certain disease is .001 in a given population. The expected number of cases to be screened before we get a case of this disease is the mean of the geometric distribution with $p = .001$. That is:

$$\mu = 1,000$$

The variance is 999,000 and the standard deviation is approximately 1,000.

A generalization of the geometric probability distribution occurs when the sequence of Bernoulli trials is stopped at the point where a preassigned number of successes is obtained. Let X be the total number of trials required to obtain r successes. Then X is said to have a *negative binomial distribution*. This probability distribution is given by:

$$p_X(x) = \binom{x-1}{r-1} p^r q^{x-r} \tag{4.15}$$

$$x = r, r+1, r+2, \ldots$$

When $r = 1$, the negative binomial distribution is the same as the geometric distribution.

The mean and variance of the negative binomial distribution are:

$$\mu = \frac{r}{p} \tag{4.16}$$

and

$$\sigma^2 = \frac{rq}{p^2} \tag{4.17}$$

Example 4.26: Suppose a couple wants to have children until they have two daughters. Then the distribution of the number of children is a negative binomial, the probabilities being given by Equation (4.15) with $r = 2$:

$$\binom{x-1}{1} p^2 q^{x-2}, \quad x = 2, 3, 4, \ldots$$

The probability distribution for $p = \frac{1}{2}$ is:

$$(x-1)(\tfrac{1}{2})^x, \quad x = 2, 3, 4, \ldots$$

The expected number is four and variance is also four.

Exercises

26. The probability of a sunny day in the Santa Clara Valley of California is .7. Suppose sunny and cloudy days occur independently. Find the probability that a visitor to the valley finds that there are six sunny days before a cloudy day.

27. Suppose the probability of a male birth is .52, and births are independent. Find the probability that a couple who wants to have children until they get a son would have three daughters before having a son.

28. A machine produces defective items with a probability of .05. A sampling inspector of the quality-control department wants to stop the operation if he finds a second defective in the sequence he is observing. What is the probability distribution of X, the total number of items examined before the operation is stopped? What is its mean and variance?

Other Discrete Distributions

Several distributions other than binomial, Poisson, and geometric occur in practical problems. We discuss two of them here: the discrete uniform distribution, which is connected with random sampling, and the hypergeometric distribution, which is connected with a sampling without replacement. Their probability distributions and expressions for mean and variance are given.

Discrete Uniform Distribution

When an item is selected "at random" from a given set of items, we imply that each item has the same probability of selection. Suppose we have n items with values $x_1, x_2, \ldots, x_n$ and one is selected at random. Let x be the value selected. Then the discrete uniform probability distribution gives:

$$p_X(x) = \frac{1}{n}, \quad x = x_1, x_2, \ldots, x_n \tag{4.18}$$

The mean and variance are:

$$\mu = \frac{1}{n}(x_1 + x_2 + \cdots x_n) \tag{4.19}$$

and

$$\sigma^2 = \frac{1}{n}(x_1^2 + x_2^2 + \cdots + x_n^2) - \mu^2 \tag{4.20}$$

Example 4.27: In a country fair, the prizes in a game are marked on slips of paper chosen by the winner. The prizes are $10, $15, $20, $25, and $30. The probability distribution of prizes is given by:

x	10	15	20	25	30
$p_X(x)$	$\frac{1}{5}$	$\frac{1}{5}$	$\frac{1}{5}$	$\frac{1}{5}$	$\frac{1}{5}$

The expected value of the prize and its variance are:

$$\mu = \tfrac{1}{5}(10 + 15 + 20 + 25 + 30) = 20$$
$$\sigma^2 = \tfrac{1}{5}(10^2 + 15^2 + 20^2 + 25^2 + 30^2) - 20^2$$
$$= 50$$

Hypergeometric Distribution

Let X be the number of red balls in a sample of r balls selected without replacement from an urn containing m red and $n - m$ white balls. The probability distribution of X is called the *hypergeometric distribution*, given by:

$$p_X(x) = \frac{\binom{m}{x}\binom{n-m}{r-x}}{\binom{n}{r}}$$

$$x = 0, 1, 2, \ldots \min(m, r) \qquad (4.21)$$

The mean and variance of X are given by:

$$E(X) = \frac{rm}{n} \qquad (4.22)$$

$$V(X) = \frac{(n-r)rm}{(n-1)n}\left(1 - \frac{m}{n}\right) \qquad (4.23)$$

Example 4.28: In a sampling inspection of a lot with 100 items, a sample of 10 items is taken. Suppose the lot has 5 defective items. The probability that the sample has one defective item is obtained from Equation (4.21), using

$$n = 100 \quad m = 5 \quad r = 10 \quad \text{and } X = 1$$

and this probability is given by

$$\frac{\binom{5}{1}\binom{95}{9}}{\binom{100}{10}} = .339$$

The mean of the number of defectives in the sample of size 10 is $10 \times \dfrac{5}{100} = .5$ and the variance is .43.

Exercises

29. Five balls are drawn from an urn containing seven red and eight white balls without replacement. What is the probability distribution of the number of red balls in the sample? Find the mean and variance.

30. A sample of 5 is selected from group of 50 women having 10 ERA supporters, without replacement. Find the probability that there are two ERA supporters in the sample. What is the probability that there are none? What is the probability that all of them are ERA supporters?

31. In Example 4.27, suppose you win a prize ten times. What is the total amount of the prize expected?

Joint Distributions

Many times, two or more random variables are observed simultaneously. For example, we may obtain both the systolic and diastolic blood pressure of a patient, measure the height and weight of a subject, or get the hardness and carbon content of a steel rod. Consider two random variables X and Y. The discussion will extend to several random variables. The *joint probability distribution* of two discrete random variables X and Y is given by:

$$p_{X,Y}(x, y) = P(X = x, Y = y) \qquad (4.24)$$

Example 4.29: Suppose X is the number of injuries received and Y the number of days of work missed by a factory worker. Let X have two values, 0 and 1, and Y have three values, 0, 1, and 2, and let:

$$P(X = 0, Y = 0) = .5 \qquad P(X = 1, Y = 0) = .1$$
$$P(X = 0, Y = 1) = .1 \qquad P(X = 1, Y = 1) = .1$$
$$P(X = 0, Y = 2) = .1 \qquad P(X = 1, Y = 2) = .1$$

The joint distribution $p_{X,Y}(x, y)$, can be written in the form of the table:

x \ y	0	1	2
0	.5	.1	.1
1	.1	.1	.1

Example 4.30: A fair coin is tossed two times. Let X be the number of heads and Y be the number of runs of heads and tails in an outcome. The sample space is:

$$\text{HH} \qquad \text{HT} \qquad \text{TH} \qquad \text{TT}$$

Now X has three possible values (0, 1, and 2), whereas Y has two values (1 and 2). The joint probability distribution is:

x \ y	1	2
0	$\frac{1}{4}$	0
1	0	$\frac{1}{2}$
2	$\frac{1}{4}$	0

The probability distribution of one of the random variables X or Y is called the *marginal probability distribution*. The marginal distribution of X, for example, is obtained by adding all the probabilities for various values of Y (in the margin). Similarly, the marginal probability distribution of Y is obtained by adding all the probabilities for X. For example, if $p_{X,Y}(x, y)$ is given by:

x \ y	0	1	2	Marginal distribution $p_X(x)$
1	.2	.1	.1	.4
2	.1	.1	.1	.3
3	.1	.1	.1	.3
Marginal distribution $p_Y(y)$	.4	.3	.3	1.0

Similar to the definition of conditional probability of event A given B—that is, $P(A|B) = P(A \cap B)/P(B)$—we have a *conditional probability distribution* of one random variable given the other. The conditional probability distribution of X given that $Y = y$ will be denoted by $p_{X|Y}(x, y)$ and is given by

$$p_{X|Y}(x|y) = \frac{p_{X|Y}(x, y)}{p_Y(y)} \tag{4.25}$$

when $p_Y(y) \neq 0$.

Example 4.31: The conditional probability that a factory worker in Example 4.29 has one injury, given that he is absent for two days, is $p_{X|Y}(1|2)$, which is

$$\frac{p_{X|Y}(1, 2)}{p_Y(2)} = \frac{.1}{.2} = \frac{1}{2}$$

The concept of independence of two random variables X and Y follows from that of events. We say that random variables X and Y are *independent* if:

$$p_{X|Y}(x, y) = p_X(x)p_Y(y) \tag{4.26}$$

for all x and y.

Example 4.32: The random variables X, Y with the following distribution are independent.

x \ y	0	1
0	$\frac{1}{4}$	$\frac{1}{4}$
1	$\frac{1}{4}$	$\frac{1}{4}$

Their marginals are:

$$p_X(x) = \tfrac{1}{2}, \quad x = 0, 1$$

and

$$p_Y = \tfrac{1}{2}, \quad y = 0, 1$$

Thus Equation (4.26) is satisfied:

Example 4.33: For the distribution in Example 4.29, X and Y are not independent, since $p_{X,Y}(0,0) = .5$, while $p_X(0) = .7$ and $p_Y(0) = .6$ and $.7 \times .6 = .42$, which does not equal .5.

A measure of dependence between the random variables X and Y is given by *covariance*. Just as the variance of a random variable is defined as the expected value of the squared deviation of itself from the mean, the covariance is the expected value of the product of the deviations of each random variable from their respective means. That is, the *covariance* between X and Y is given by

$$\sigma_{X,Y} = \text{Cov}(X, Y) = E[(X - \mu_X)(Y - \mu_Y)] \tag{4.27}$$

where μ_X and μ_Y represent the expected values of X and Y, respectively. Note that $\text{Cov}(X, X) = \text{Var } X$. For computational convenience:

$$\sigma_{XY} = E(XY) - \mu_X\mu_Y \tag{4.28}$$

This measure of covariance has units of both X and Y. If X is in pounds and Y is in inches, then the units of covariance will be pounds-inches. To make it dimensionless, another quantity has been developed, called the *correlation coefficient*. Since the standard deviation of X has units of X and standard deviation of Y has units of Y, the covariance between X and Y is divided by the product of the standard deviations of X and Y to define correlation coefficient. That is, the *correlation coefficient* between X and Y defines a dimensionless quantity given by ρ, the Greek letter *rho*,

$$\rho = \frac{\sigma_{XY}}{\sigma_X\sigma_X} \tag{4.29}$$

When $p = 0$, we call X and Y *uncorrelated*.

Example 4.34: For Example 4.29, the marginal distributions are:

x	0	1
$p_X(x)$	.7	.3

y	0	1	2
$p_Y(y)$	.6	.2	.2

First we obtain means and variances as follows:

$$\mu_X = (0 \times .7) + (1 \times .3) = .3$$
$$\mu_Y = (0 \times .6) + (1 \times .2) + (2 \times .2) = .6$$
$$\sigma_X^2 = (0^2 \times .7) + (1^2 \times .3) - (.3)^2 = 0.21$$
$$\sigma_Y^2 = (0^2 \times .6) + (1^2 \times .2) + (2^2 \times .2) - (.6)^2 = .64$$

Now
$$E(XY) = (0 \times 0 \times .5) + (0 \times 1 \times .1) + (0 \times 2 \times .1)$$
$$+ (1 \times 0 \times .1) + (1 \times 1 \times .1) + (1 \times 2 \times .1)$$
$$= .3$$

So that
$$\sigma_{XY} = .3 - (.3 \times .6) = .12$$

and
$$\rho = \frac{.12}{\sqrt{.21}\sqrt{.64}} = 0.327$$

Example 4.35: In Example 4.32, the marginal distributions are:

x	0	1
$p_X(x)$	$\frac{1}{2}$	$\frac{1}{2}$

y	0	1
$p_Y(z)$	$\frac{1}{2}$	$\frac{1}{2}$

So that
$$\mu_X = \tfrac{1}{2}, \mu_Y = \tfrac{1}{2}$$
$$\sigma_X^2 = \tfrac{1}{4}, \sigma_Y^2 = \tfrac{1}{4}$$

$$E(XY) = \left(0 \times 0 \times \frac{1}{4}\right) + \left(0 \times 1 \times \frac{1}{4}\right) + \left(1 \times 0 \times \frac{1}{4}\right) + \left(1 \times 1 \times \frac{1}{4}\right)$$
$$= \frac{1}{4}$$

so that
$$\sigma_{XY} = \frac{1}{4} - \frac{1}{2} \times \frac{1}{2} = 0$$

Hence, $\rho = 0$.

This example demonstrates that if two random variables are independent, they are uncorrelated. However, the converse is not always true, and this will be discussed later.

Exercises

32. For the joint distribution $p_{XY}(x, y)$ given by the following, find μ_X, μ_Y, σ_X^2, σ_Y^2, σ_{XY}, and ρ.

(a)

x \ y	1	2	3
0	$\frac{1}{9}$	$\frac{1}{18}$	$\frac{1}{18}$
1	$\frac{1}{9}$	$\frac{2}{9}$	0
2	$\frac{2}{9}$	0	$\frac{2}{9}$

(b)

x \ y	1	2	3
0	$\frac{1}{9}$	$\frac{1}{9}$	$\frac{1}{9}$
1	$\frac{1}{9}$	$\frac{1}{9}$	$\frac{1}{9}$
2	$\frac{1}{9}$	$\frac{1}{9}$	$\frac{1}{9}$

33. Are X and Y independent in (a) and (b) in Exercise 32?

Chapter Exercises

1. A player for the Yankees has a batting average of .300. In any game, he has a chance of batting five times. Find the probability that het gets:

    (a) one hit
    (b) no hit
    (c) between two and four hits
    (d) first hit after four times at bat (use geometric distribution)

2. The number of seeds not germinating has a Poisson distribution with $\lambda = 3$. Find the probability that two seeds do not germinate.

3. An all-star team is to be selected at random (without replacement) from a collection of 40 players, including 5 quarterbacks. What is the probability that in a team of 11 selected at random there will be:

    (a) no quarterback?
    (b) one quarterback?
    (c) all quarterbacks?

4. In a population of 100 persons, it is known that there are five cases of hepatitis. Ten persons are selected at random with replacement. Find the probability that one has hepatitis.

Summary

The *random variable* is a function defined on sample space. It is *discrete* when the values taken by the random variable are finite or countably finite; otherwise, it is *continuous*. The probabilities associated with a random variable are given by its *probability distribution*.

The following probability distributions of discrete random variables have been discussed in this chapter.

(i) *Binomial distribution* gives the probabilities of the number of successes in n trials with only two outcomes, success and failure, with p as the probability of a success. The *binomial probability distribution* is

$$\binom{n}{x}p^x q^{n-x} \quad x = 0, 1, 2, \ldots, n$$

The mean is np and the variance is npq.

(ii) *Poisson distribution* gives the probabilities of a random variable X having countable values $0, 1, 2, \ldots$ with the probability distribution:

$$\frac{e^{-\lambda}\lambda^x}{x!}$$

The mean and the variance are λ.

(iii) *Geometric distribution* gives the probabilities of the number of trials needed to obtain one success in independent Bernoulli trials, given by:

$$q^{x-1}p \quad x = 1, 2, \ldots$$

The mean is $1/p$ and the variance is q/p^2.

(iv) *Negative binomial distribution* is the generalization of geometric distribution when r successes are needed. The probability distribution is given

Table 4.2
Discrete Probability Distributions

Name	Probability distribution	Range of x	Mean	Variance
Binomial	$\binom{n}{x}p^x q^{n-x}$	$x = 0, 1, 2, \ldots, n$	np	npq
Poisson	$\dfrac{e^{-\lambda}\lambda^x}{x!}$	$x = 0, 1, 2, \ldots$	λ	λ
Geometric	$q^{x-1}p$	$x = 1, 2, 3, \ldots$	$\dfrac{1}{p}$	$\dfrac{q}{p^2}$
Discrete Uniform	$\dfrac{1}{n}$	$x = x_1, x_2, \ldots, x_n$	$\dfrac{\sum x_i}{n}$	$\dfrac{\sum x_i^2}{n} - \dfrac{(\sum x_i)^2}{n^2}$
Hypergeometric	$\dfrac{\binom{m}{x}\binom{n-m}{r-x}}{\binom{n}{r}}$	$x = r, r+1, \ldots$	$r\dfrac{m}{n}$	$\dfrac{n-r}{n-1}r\dfrac{m}{n}\left(1-\dfrac{m}{n}\right)$
Negative binomial	$\binom{x-1}{r-1}q^{x-r}p^r$	$x = r, r+1, \ldots$	$\dfrac{r}{p}$	$\dfrac{rq}{p}$

Source: Johnson, Norman L., and Kotz, Samuel, *Discrete Distributions*, Boston: Houghton Mifflin Company, 1969.

by:

$$\binom{x-1}{r-1} q^{x-r} p^r \quad x = r, r+1, \ldots$$

The mean is r/p and the variance is rq/p^2.

(v) The *discrete uniform distribution* gives the probability distribution when a selection is at random and gives equal probabilities to all items selected.

(vi) The *hypergeometric probability distribution* provides probabilities in a sampling without replacement. When a sample of r balls is selected without replacement from an urn containing m white and $n-m$ red balls, the probability distribution of the number of white balls in the sample gives the hypergeometric probability distribution:

$$\frac{\binom{m}{x}\binom{n-m}{r-x}}{\binom{n}{r}}, \quad x = 0, 1, 2, \ldots, \min(r, m)$$

The mean is $r(m/n)$ and the variance is $\dfrac{n-r}{n-1} r \dfrac{m}{n} \left(1 - \dfrac{m}{n}\right)$. For a sampling with replacement, use binomial distribution.

chapter five

Continuous Probability Distributions

Probability distributions for discrete random variables were discussed in the last chapter. There are many practical situations where random variables are not discrete, such as the oral temperature of a subject, the lifetime of an electric tube, the daily milk yield of a cow, the amount of triglycerides in a blood sample, or the height of a college student. These are continuous random variables, and we need ways to describe their probability distributions. In the case of discrete random variables, the number of possibilities is finite or countably infinite and, hence, probabilities can be defined for every value of the random variable. However, this is not possible for continuous random variables, since one has to define probabilities for an uncountable set of quantities. The total probability will exceed one if one assigns any probability, however small, to each of the uncountable outcomes; therefore, the probability of any single occurrence is taken to be zero. To give a probability distribution of a continuous random variable, we require the help of a frequency curve.

Example 5.1: A train arrives at random at a station anytime between 8 A.M. and 8.15 A.M. The assumption of random arrival means that the probability of arrival between 8:05 and 8:10 is the same as the probability of arrival between 8:07 and 8:12. Also, the probability that the train arrives in the interval 8:07 and 8:15 is larger than the probability that it arrives between 8:07 and 8:12. That is, the probability of arrival is dependent on the length of the interval. The probability that the train will arrive between 8:00 and 8:15 is one. These probabilities can be expressed with the help of areas under a curve that is drawn in such a way that the total area under it, above the horizontal axis, and bounded by 8:00 and 8:15 is one. Also note that the probability that the train

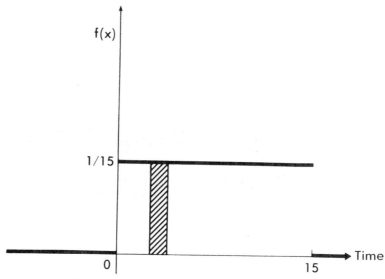

Figure 5.1. Probability density function of arrival time.

arrives exactly at 8:10 is zero, since it is impossible to point out the exact instance, however accurate our time measurements. The only thing we can measure is the interval of time. When we say that it is 8:10 A.M., we mean that we are within a second or microsecond of 8:10 A.M., depending upon the accuracy of the watch being employed to measure time.

Figure 5.1 gives the curve for the time of arrival. This curve is known as a *frequency curve* or *probability density function* and will be discussed in more detail in the next section.

Exercise

1. Are the following random variables continous or discrete? Why?

    (a) The number of cars in a county in California
    (b) Real numbers chosen at random from [0, 1]
    (c) The weight of an apple
    (d) The number of telephone calls made by an individual
    (e) The length of an individual telephone call
    (f) The number of birds seen by a bird-watcher before he sees one cardinal

The Probability Density Function

A *frequency curve* or a *probability density function* gives the probabilities of a continuous random variable. Figure 5.1 gives an example of a frequency curve

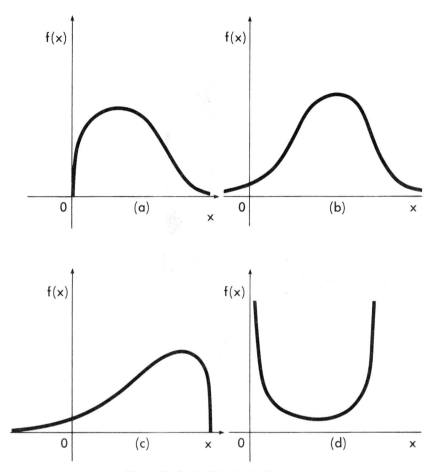

Figure 5.2. Probability density functions.

for the train-arrival time. The probabilities for random variable X, giving the arrival time of the train can be obtained by areas under this curve.

In general, a probability density function of random variable X is any function $f(x)$ that is non negative and such that the area under it bounded by the horizontal x-axis and the range of the random variable is one Several frequency curves are sketched in Figure 5.2.

Example 5.2: The probability density function of random variable X is given in Figure 5.3. The line makes a triangle of height 2 over the base of length 1; hence, the total area is one. The range of the random variable is $(0, 1)$. The probability that X is between .3 and .6 is the area of the trapezium. That is:

$$P(.3 \leq X \leq .6) = \tfrac{1}{2}(.6 + 1.2) \times (.6 - .3)$$
$$= .27$$

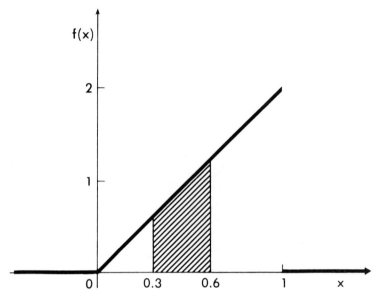

Figure 5.3. Probability density function.

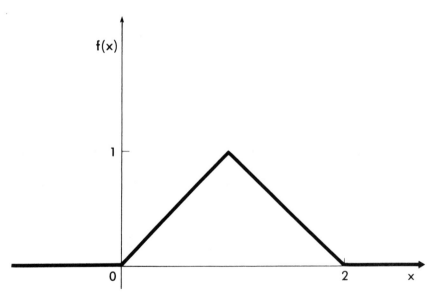

Figure 5.4. Triangular probability density function.

This probability density function can be algebraically defined to be a line between (0, 1) and 0 elsewhere:

$$f(x) = 2x, 0 \leq x \leq 1$$
$$= 0, \text{ otherwise} \qquad (5.0)$$

Example 5.3: Let the distribution of random variable X be the triangular distribution with a range of (0, 2) having attitude one and equal sides. The probability density function is given in Figure 5.4. Using areas, we can find, for example, the probabilities:

$$P(1 \leq X \leq 2) = \tfrac{1}{2}$$
$$P(.8 \leq X \leq 1) = \tfrac{1}{2}(0.8 + 1) \times (.2) = .18$$
$$P(1 \leq X \leq 1.4) = \tfrac{1}{2}(1 + .6) \times (0.4) = .32$$
$$P(.8 \leq X \leq 1.4) = .18 + .32 = .5$$
$$P(X = 1) = 0$$

The definition of a probability density function of a continuous random variable by a curve can also be seen from a frequency interpretation of probabilities. The curve can be regarded either as a smoothed version of a histogram or as a frequency polygon of a frequency distribution. In general, a probability density is specified by a function involving constants, for example Equation (5.0).

Expectation

In the case of a discrete random variable, the expected value has been defined as the weighted average of the values of the random variable with probabilities acting as weights. However, since there are no probabilities specifically for each value of the random variable in the continuous case, it is not possible to define the expected value with the help of a weighted average. The extension of the concept of a weighted average for the continuous case requires the concepts of integral calculus and is beyond the scope of this book. We shall regard the expected value of the random variable as the centre of gravity of a body if the probabilities were replaced by masses. Essentially the expected value at which the probability distribution can be balanced. The variance can also be thought of as the extension of the weighted average of the squared deviation of the value from the mean. Those well versed in the concepts of calculus will find the following definitions of mean and variance of X to be straightforward:

$$\mu_X = E(X) = \int x f(x)\, dx, \quad \sigma_X^2 = \text{Var } X = \int x^2 f(x)\, dx - \mu_X^2$$

Exercises

2. Find the following probabilities for the continuous random variable X having the probability density function in Figure 5.5:

   (a) $P(X \leq 1)$
   (b) $P(X \geq 1.5)$
   (c) $P(1 \leq X \leq 1.5)$
   (d) $P(.2 \leq X \leq 1.2)$
   (e) $P(X \geq 3)$
   (f) $P(X = 1)$

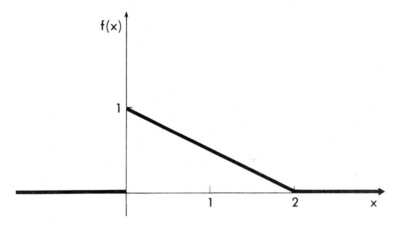

Figure 5.5. Exercise 2

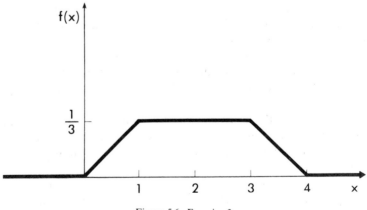

Figure 5.6. Exercise 3

3. For the probability density function in Figure 5.6, find the probabilities:

   (a) $P(X \geq .5)$
   (b) $P(X \leq 2.5)$
   (c) $P(.5 \leq X \leq 2)$
   (d) $P(1.5 \leq X \leq 2.5)$
   (e) $P(3 \leq X \leq 5)$
   (f) $P(|X| \geq 1)$

4. Verify if the curves given in Figures 5.7a, b, and c are probability density functions.

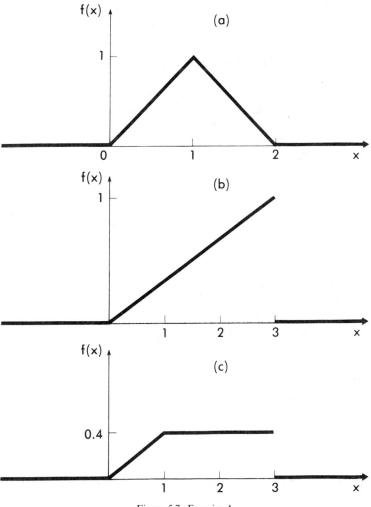

Figure 5.7. Exercise 4

5. What is the probability density function of X, a point chosen at random from interval $(1, 2)$?

(a) Graph the probability density function

$$f(x) = \begin{cases} 2(1 - x), & 0 \le x \le 1 \\ 0, & \text{otherwise} \end{cases}$$

and find $P(.5 \le X \le .75)$ and $P(-1 \le X \le .5)$.

(b) For the triangular probability density function in Example 5.3, find:

(i) $P(0 \le X \le 1)$
(ii) $P(X \le 1.5)$
(iii) $P(X \ge .5)$

The Cumulative Distribution Function

The probability density function gives probabilities with the help of areas, and this may not be convenient in many cases. A function that gives probabilities by its ordinates—that is, by the value of the function—is the *cumulative distribution function*. This function gives the cumulative probability that a random variable X is less than or equal to x. That is, $F(x)$, given by $F(x) = P(X \le x)$, is called the cumulative distribution function of X.

Example 5.4: Consider the following probability distribution of a discrete random variable X.

x	-2	0	2
$p_X(x)$	$\frac{1}{3}$	$\frac{1}{3}$	$\frac{1}{3}$

Then $F(x)$ can be obtained as follows. Since $P(X < x) = 0$, for $x < -2$, we have $F(x) = 0$ for all $x < -2$. At $x = -2$, the probability for X is $\frac{1}{3}$ and therefore $F(x) = \frac{1}{3}$ when $x = -2$. When x is between -2 and 0, there is no probability for X; hence, $F(x) = \frac{1}{3}$ for x between -2 and 0, but not including 0. That is:

$$F(x) = \tfrac{1}{3}, \quad -2 \le x < 0$$

When $x = 0$, $P(X = 0) = \frac{1}{3}$ and hence,

$$P(X \le 0) = P(X = -2) + P(X = 0) = \tfrac{1}{3} + \tfrac{1}{3} = \tfrac{2}{3}$$

so that

$$F(x) = \tfrac{2}{3}, \quad \text{at } x = 0$$

Also, there is no probability added until $x = 2$, so that

$$F(x) = \tfrac{2}{3}, \quad 0 \le x < 2$$

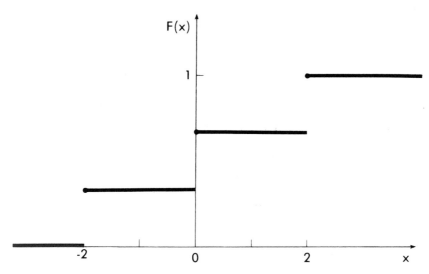

Figure 5.8. Cumulative distribution function.

Similarly,

$$F(x) = 1, \quad x \geq 2$$

The cumulative distribution function can then be graphed in Figure 5.8 as a staircase function. Notice that the function has jumps of height $\frac{1}{3}$ at $-2, 0$, and 2 giving the probabilities at these points, and the function is between 0 and 1.

Example 5.5: A fair die is tossed twice. The number of dots has the probability distribution:

x	2	3	4	5	6	7	8	9	10	11	12
$p_X(x)$	$\frac{1}{36}$	$\frac{2}{36}$	$\frac{3}{36}$	$\frac{4}{36}$	$\frac{5}{36}$	$\frac{6}{36}$	$\frac{5}{36}$	$\frac{4}{36}$	$\frac{3}{36}$	$\frac{2}{26}$	$\frac{1}{36}$

The cumulative distribution function cumulates probabilities and gives the distribution $F(x)$, as given in Figure 5.9.

When X is a continuous random variable, the cumulative distribution function is a continuous function having a similar shape as in the discrete case; however, there are no jumps. A typical cumulative distribution function gives a smooth increasing curve between 0 and 1 and provides probabilities for any interval (a, b).

$$P(a < X \leq b) = P(X \leq b) - P(X \leq a)$$
$$= F(b) - F(a)$$

Example 5.6: The uniform cumulative distribution function on $(0, 1)$ is given in Figure 5.10 by a straight line making a $45°$ angle at the origin. Suppose we want $P(.3 < X \leq .5)$. This is $F(.5) - F(.3)$ and therefore equals $.5 - .3 = .2$.

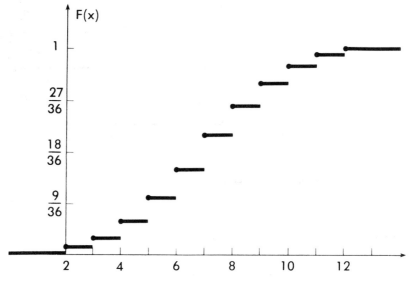

Figure 5.9. Cumulative distribution function of Example 5.5.

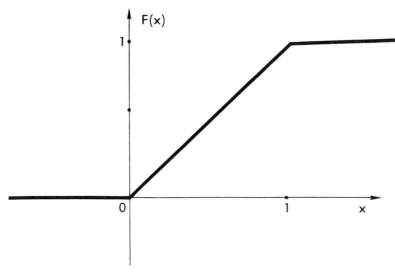

Figure 5.10. Cumulative distribution function of Uniform.

Example 5.7: The exponential density $f(x) = e^{-x}$ has the cumulative distribution function:

$$F(x) = 1 - e^{-x}, \quad x \geq 0$$
$$= 0 \text{ otherwise}$$

Figure 5.11 gives its graph.

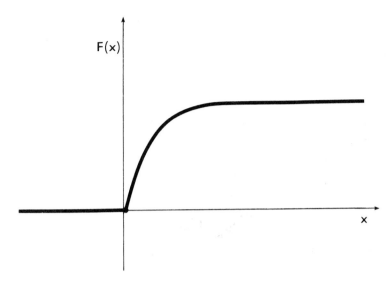

Figure 5.11. Cumulative distribution function of exponential.

Most of the tabulated probability distributions are given in terms of the cumulative distribution functions. The cumulative distribution function has the same relation to the probability density function as the cumulative relative frequency distribution has to the relative frequency distribution.

Exercises

6. Draw the cumulative distribution function for the following probability distribution of discrete random variables.

(a)

x	0	1
$p_X(x)$	.5	.5

(b)

x	− 2	− 1	0	− 1	2
$p_X(x)$	.2	.2	.2	.2	.2

(c)

x	1	2	3	4	5	6
$p_X(x)$	.1	.1	.3	.3	.1	.1

7. For the distribution function

$$F(x) = \begin{cases} 0, & x < 0 \\ x^2, & 0 \le x \le 1 \\ 1, & x \ge 1 \end{cases}$$

find the probabilities:

(a) $P(.5 \le X \le 1)$
(b) $P(0 \le X \le .5)$
(c) $P(.2 \le X \le .7)$

8. Plot the cumulative distribution function:

$$F(x) = \begin{cases} 0, & x < 0 \\ 1 - e^{-2x}, & x \ge 0 \end{cases}$$

Find the probabilities:

(a) $P(2 \le X \le 3)$
(b) $P(X \ge 5)$
(c) $P(X \le 2)$

Normal Distribution

One of the most frequently used probability distributions is the normal, also called Gaussian. An important use of normal distribution is in approximating the binomial probabilities when the number of trials is large. Another use is when the sample size is large and the sample average has normal distribution.

The probability density function of the normal is:

$$f(x) = \frac{1}{\sqrt{2\pi}\,\sigma} e^{-(x-\mu)^2/2\sigma^2}, \quad -\infty < x < \infty \tag{5.1}$$

The Greek letters μ and σ are constants involved in the distribution and provide various normal frequency curves. Such constants involved in the probability density function are called *parameters* of the distribution. For example, in binomial distribution, n and p are parameters; in Poisson distribution, λ is the parameter.

A few typical frequency curves for the normal are given in Figure 5.12. The curves are symmetric about μ. For small value of σ, the peak of the curve is high, while the curve is flat for large values of σ. The mean and variance of the distribution are:

$$E(X) = \mu \tag{5.2}$$
$$\mathrm{Var}(X) = \sigma^2 \tag{5.3}$$

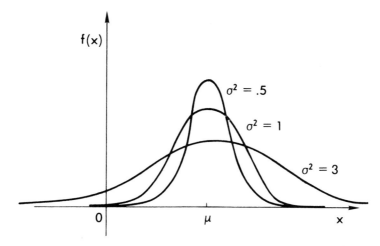

Figure 5.12. Normal density functions.

Standard Normal Distribution

When the mean of a random variable is zero and the variance is one, the random variable is said to be *standardized*. If X is a random variable with a mean of μ and a variance of σ^2, then

$$Z = \frac{X - \mu}{\sigma} \tag{5.4}$$

gives the standardized random variable Z. The mean of Z is zero and the variance is one. The probability density function of a standardized normal random variable is:

$$\phi(z) = \frac{1}{\sqrt{2\pi}} e^{-z^2/2} \tag{5.5}$$

The cumulative distribution function of the normal is given by:

$$\Phi(z) = P(Z \le z)$$

Figure 5.13 gives the density function and Figure 5.14 gives the cumulative distribution function of the standard normal random variable. The values of $\Phi(z)$ are tabulated in Table III in the Appendix.

To find the probabilities from the table, we proceed as follows. Suppose we want $P(1 \le Z \le 2)$; that is, the shaded area in Figure 5.15. Then:

$$P(1 \le Z \le 2) = P(Z \le 2) - P(Z \le 1)$$
$$= \Phi(2) - \Phi(1) = .9772 - .8413 \text{ from Table III}$$
$$= .1359$$

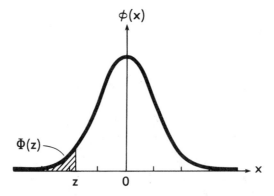

Figure 5.13. Standard normal density.

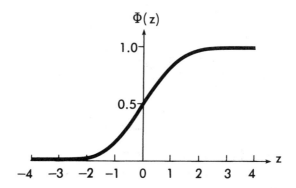

Figure 5.14. Cumulative distribution function of standard normal.

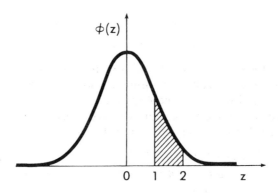

Figure 5.15. Probability of Z between 1 and 2.

Since the standard normal distribution is symmetric about zero, the probability in the lower tail, say $P(Z \leq -a)$, is the same as the corresponding probability in the upper tail, that is, $P(Z \geq a)$. Since $P(Z \leq -a) = \Phi(-a)$, we have:

$$\Phi(-a) = 1 - \Phi(a) \tag{5.6}$$

This is used frequently in calculating normal probabilities when negative numbers are involved.

Example 5.8: Find the probability that Z is in the interval -1 and 2. That is:

$$\begin{aligned}
P(-1 \leq Z \leq 2) &= \Phi(2) - \Phi(-1) \\
&= \Phi(2) - (1 - \Phi(1)) \\
&= \Phi(2) + \Phi(1) - 1 \\
&= .9772 + .8413 - 1 = .8185
\end{aligned}$$

For a random variable X having a normal distribution with a mean of μ and a variance of σ^2, we standardize the variable using Equation (5.4) and then use the table of the standardized normal random variable.

Example 5.9: Let X have a normal distribution with $\mu = 5$ and $\sigma = 7$. Then:

$$\begin{aligned}
P(12 \leq X \leq 19) &= P\left(\frac{12-5}{7} \leq \frac{X-5}{7} \leq \frac{19-5}{7}\right) \\
&= P(1 \leq Z \leq 2) \\
&= \Phi(2) - \Phi(1) = .1359
\end{aligned}$$

Example 5.10: Let the oral temperature of a group have a normal distribution with a mean of $98°$ F and a standard deviation of $.4°$ F. The probability that a randomly selected subject has a temperature between 97.2 and 98.8 is:

$$\begin{aligned}
P(97.2 \leq X \leq 98.8) &= P\left(\frac{97.2-98}{.4} \leq \frac{X-98}{.4} \leq \frac{98.8-98}{.4}\right) \\
&= P(-2 \leq Z \leq 2) \\
&= \Phi(2) - \Phi(-2) = \Phi(2) - [1 - \Phi(2)] \\
&= 2\Phi(2) - 1 = 2(.9772) - 1 = .9544
\end{aligned}$$

Example 5.11: To find the number k such that the standard random variable Z has probability .95 that $Z \leq k$, we proceed as follows:

$$P(Z \leq k) = .95$$

From tables, we know that:

$$P(Z \leq 1.645) = .95$$

Hence, $k = 1.645$. This is the 95th percentile of z.

Example 5.12: To find k, such that

$$P(Z \geq k) = .90, \text{ we use}$$
$$P(Z \geq -1.282) = .90,$$

so that $k = -1.282$.

Example 5.13: Find k, such that:

$$P(|Z| \leq k) = .90$$

Again, since $P(|Z| \leq k) = P(-k \leq Z \leq k)$, we find that:

$$P(-1.645 \leq Z \leq 1.645) = .90$$

Hence, $k = 1.645$.

Example 5.14: Let X have a normal distribution with a mean of 5 and a standard deviation of 2. What is the value of k such that $P(X \geq k) = .95$?

$$P(X \geq k) = P\left(\frac{X-5}{2} \geq \frac{k-5}{2}\right) = .95$$

Since $P(Z \geq 1.645) = .95$, we get

$$\frac{k-5}{2} = 1.645$$

$$\text{or} \quad k = 5 + (2 \times 1.645) = 8.29$$

Example 5.15: Find k, such that

$$P(|X - 10| \leq k) = .90,$$

where X has a normal distribution with mean 10 and variance 4:

$$P(|X - 10| \leq k) = P(10 - k \leq X \leq 10 + k)$$

$$= P\left(-\frac{k}{2} \leq \frac{X-10}{2} \leq \frac{k}{2}\right)$$

$$= P\left(-\frac{k}{2} \leq Z \leq \frac{k}{2}\right)$$

Since $P(-1.645 \leq Z \leq 1.645) = .90$, we get:

$$\frac{k}{2} = 1.645$$

$$\text{or}$$

$$k = 3.29$$

Example 5.16: Suppose that the scores in an examination are normally distributed with a mean of 65 and a standard deviation of 10. Let three persons take this examination. What is the probability that two will get an A if the person getting above 80 gets an A?

The probability of getting an A is:

$$P(X \geq 80), \quad \text{where } X \text{ is the score}$$

Now

$$P(X \geq 80) - P\left(\frac{X - 65}{10} \geq \frac{80 - 65}{10}\right) - P(Z \geq 1.5) = .0668$$

Getting an A is similar to getting a success in a binomial distribution with probability of success, $p = .0668$. To find the probability that two out of three students taking the examination get an A, we use the binomial probabilities with $n = 3$, $p = .0668$, and $x = 2$. Thus the probability is $\binom{3}{2}(.0668)^2(.9332) = .0125$.

Normal Approximation to the Binomial

We have seen that an approximation to the binomial probabilities when n is large but p is small is given by Poisson distribution. When n is large but p is not necessarily small, binomial probabilities can be approximated by normal probabilities. Figure 5.16 shows binomial probability histograms for $n = 10$, 20, 25 and $p = 0.4$ with a corresponding normal distribution superimposed on each. We note that as n increases, the binomial probabilities given by the histogram and the corresponding probabilities given by the normal frequency curve tend to be closer.

The mean of the approximating normal is $\mu = np$ and variance $\sigma^2 = npq$. There is a rule of thumb that when n is more than 30 and p is not near 0 or 1, the approximation is quite good. Since binomial distribution is the distribution of a discrete random variable and the probabilities are given by rectangles, whereas the probabilities by the normal curve are areas under the curve, so the probabilities are always approximately equal. Furthermore, note that if X is a binomial variable, $P(X = 2)$ means that the area of the rectangle between 1.5 and 2.5 is to be obtained. Similarly, for finding $P(1 \leq X \leq 3)$, which provides the range for obtaining normal approximation. That is, .5 is subtracted from the lower value and .5 is added to the upper value of the range of X when finding normal approximate probabilities. So that $P(1 \leq X \leq 3) \doteq P(.5 \leq X \leq 3.5)$.

Example 5.17: Let X have a binomial distribution with $n = 50$ and $p = .2$. Then $P(X \leq 10) \doteq P(X \leq 10.5)$ from the normal distribution with mean $\mu = 50 \times .2 = 10$ and variance $\sigma^2 = 50 \times .2 \times .8 = 8$.

Example 5.18: Suppose the probability that grass seeds will germinate is

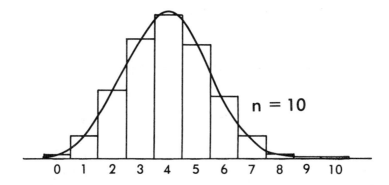

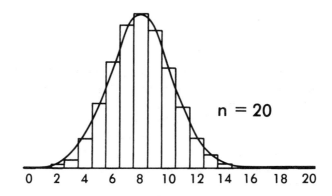

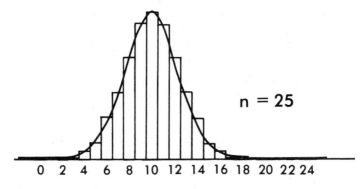

Figure 5.16. Normal approximation to binomial, n = 10, n = 20, n = 25.

0.8. The probability that out of 100 seeds planted, more than 90 will germinate is

$$P(X \geq 90) \doteq P(X \geq 89.5)$$

for the normal distribution with mean $\mu = 80$ and variance $\sigma^2 = 16$. So that:

$$P(X \geq 89.5) = P\left(Z \geq \frac{89.5 - 80}{4}\right)$$

$$= P(Z \geq 2.378) = 1 - \Phi(2.378)$$

$$= 1 - .9913 = .008$$

Exercises

9. For the standard normal distribution, find:

(a) $P(Z \leq 2.3)$	(k) $P(Z \leq -3)$
(b) $P(Z \geq 1.3)$	(l) $P(Z \geq 3)$
(c) $P(-1.21 \leq Z \leq 2.35)$	(m) $P(Z \leq 1.645)$
(d) $P(Z \leq -2.5)$	(n) $P(Z \leq -1.645)$
(e) $P(-1.21 \leq Z \leq -0.5)$	(o) $P(Z \geq 1.645)$
(f) $P(1 \leq Z \leq 2.35)$	(p) $P(Z \geq -1.645)$
(g) $P(Z \leq -3.1)$	(q) $P(Z \geq 1.96)$
(h) $P(Z \geq -10.5)$	(r) $P(Z \geq -1.96)$
(i) $P(-3 \leq Z \leq 3)$	(s) $P(-2.38 \leq Z \leq 2.38)$
(j) $P(-1.96 \leq Z \leq 1.96)$	(t) $P(-1.645 \leq Z \leq 1.645)$

10. For the standard normal distribution, find k, such that:

(a) $P(Z - 2 \leq k) = .95$
(b) $P(Z - 2 \geq k) = .95$
(c) $P(|Z| \leq k) = .95$
(d) $P(|Z| \geq k) = .90$

11. In Example 5.10, suppose independent measurements of temperatures were taken on three subjects. Find the probability that:

(a) none of them had a temperature above 98
(b) two of them had a temperature above 98.4
(c) all three had temperatures between 98.2 and 98.4

12. For the normal distribution with mean 15 and standard deviation 3, find k, such that:

(a) $P(X \leq k) = .05$
(b) $P(X \geq k) = .01$
(c) $P(|X - 15| \geq k) = .05$
(d) $P(|X - 15| \leq k) = .90$

13. Let the scores of an examination be normally distributed with $\mu = 65$ and $\sigma^2 = 100$. Find the range of scores that will be used to determine the grades. It is assumed that grades A, B, C, D, and F will be given to the upper 15%, 25%, 40%, 10%, and 10% of students, respectively. What are the grades if the actual scores obtained by four students are 73, 85, 22, and 35?

14. Highly precise components are needed in an assembly and the components are acceptable if the diameter is at most 5 cm. Suppose a machine makes these components with diameters normally distributed with $\mu = 4.9$ and $\sigma = .04$ cm. If 10 components are randomly examined, what is the probability that all of the components are acceptable?

Some Other Important Continuous Probability Distributions

There are many other distributions besides the normal distribution that we encounter in applications. In this section, we discuss two such distributions, the uniform and the exponential.

Uniform Distribution (Rectangular Distribution)

The probability density function of a uniform distribution where the random variable takes values over the interval (a, b) is given by:

$$f(x) = \frac{1}{b - a}, \quad a \le x \le b$$

$$= 0 \quad \text{otherwise} \tag{5.8}$$

The graph is given in Figure 5.17.

The mean and variance of the uniform distribution are:

$$E(X) = \frac{a + b}{2} \tag{5.9}$$

$$\text{Var } X = (b - a)^2/12 \tag{5.10}$$

Example 5.19: A pointer is rotated in a circle at random. The probability distribution of the angle X is uniform, and X takes values between 0 and 360°. So the probability density of X is:

$$f(x) = \frac{1}{360}, \quad 0 \le x \le 360$$

$$= 0 \quad \text{otherwise}$$

Suppose we want the probability that the pointer is in the shaded area of

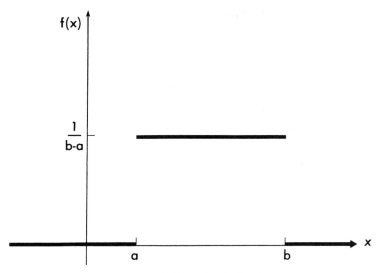

Figure 5.17. Uniform density function.

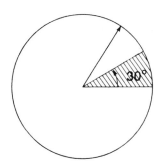

Figure 5.18. Circular pointer.

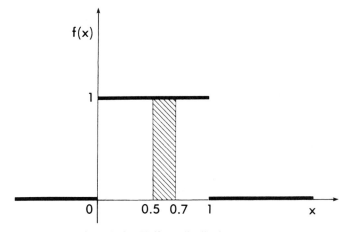

Figure 5.18a. Uniform distribution on $(0, 1)$.

Figure 5.18. The probability is:

$$\frac{30}{360} = \frac{1}{12}$$

Example 5.20: A number is chosen at random between 0 and 1. The probability that the number chosen is between .5 and .7 is 0.2, as seen in the the area of rectangle needed is over the range $.5 \leq x \leq 3.5$, shaded area of Figure 5.18a. The conditional probability that X is between .5 and .7 given that X is less than .8 is given by:

$$P(.5 \leq X \leq .7 | X \leq .8) = \frac{P(.5 \leq X \leq .7)}{P(X \leq .8)}$$

$$= \frac{.2}{.8} = \frac{1}{4}$$

Exponential Probability Distribution

An important model of a random phenomenon in reliability and mortality studies is that of an exponential distribution. The probability density function for the exponential is given by:

$$f(x) = \theta e^{-\theta x}, \quad x \geq 0$$
$$= 0 \qquad \text{otherwise} \qquad (5.11)$$

A typical density function is given in Figure 5.19. The parameter θ (*theta*) plays

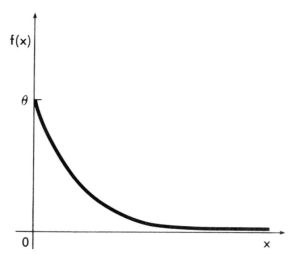

Figure 5.19. Exponential density function.

an important role in the mean and variance of the distribution given by:

$$E(X) = \frac{1}{\theta} \quad \text{and} \quad \text{Var}(X) = \frac{1}{\theta^2} \tag{5.12}$$

The cumulative distribution function of the exponential is:

$$F(x) = 1 - e^{-\theta x}, \quad x \geq 0$$
$$= 0 \qquad \text{otherwise} \tag{5.13}$$

For a given value of θ, the function can be easily calculated and the corresponding probabilities obtained.

Example 5.21: Let the life of electric light bulbs be exponentially distributed with a mean of 500 hours. Then $\theta = 1/500$. The probability that a bulb lasts 600 hours or more is:

$$P(X \geq 600), \text{ which equals}$$
$$= 1 - P(X \leq 600)$$
$$= 1 - (1 - e^{-\frac{1}{500} \times 600}), \text{ using (5.13)}$$
$$= e^{-1.2} = .3$$

Exercises

15. Guests arrive at a party between 8:00 and 8:30 P.M. at random. The host expects 10 guests. What is the probability that:

   (a) no guest will arrive before 8:15?
   (b) two guests will arrive before 8:15?
   (c) all guests will arrive before 8:20?

16. Students can finish a test at random within 45 minutes. What is the probability that a student will finish this test in more than 40 minutes? What is the probability that out of five students taking the test, two will finish in less than 30 minutes?

17. The length of a conversation on a telephone line is exponentially distributed with a mean of 7 minutes. What is the expected number of people in a city of 1,000 whose conversation lasts more than 10 minutes? What is the probability that a conversation lasts between 5 and 10 minutes?

18. Suppose customers wait at a grocery counter for a time X, which is an exponentially distributed random variable with a mean waiting time of 10 minutes. What is the probability that a customer will wait for less than 5 minutes? For more than 8 minutes? Between 7 and 10 minutes?

Sampling Distributions

Statistics studies populations with the help of samples. The sample, however, must be representative of the population. The population may be a collection of households, the weights of college students, gas station sales of gasoline, diameters of piston rings manufactured by a vendor, the total yield of a wheat farm in Iowa, and so on. Let there be a finite number of elements, N, in the population. By a sample, we mean the selection of a smaller number of items of the population for intensive study. There are $\binom{N}{n}$ possible samples of size n from the population of N items. A *random sample* gives equal probabilities to all the possible samples. That is, the probability of selecting a particular sample is $1 \big/ \binom{N}{n}$. There are various methods of obtaining random samples from a population. The most common uses random numbers. A list of a million random digits has been published by the Rand Corporation for this purpose. A sample of four pages of this list is given in Table VII in the Appendix.

Suppose we want to select a random sample of 5 observations from a population consisting of 100 individuals. This can be accomplished by taking two-digit random numbers. The items of the population are arbitrarily numbered from 0 to 99. The sequence of 5 random numbers provided by the table of random numbers gives the sequence of items to be selected. If a number is repeated, we discard the number and select another one. Proceeding in this manner, we get a random sample of 5 items.

Example 5.22: Suppose a random sample of 5 students is to be selected from a class of 35. We number the students from 0 to 34. Now we select two-digit numbers from the table of random numbers starting any page, any row, and any column. Suppose we get 09, 30, 73, 83, 17,..., and so on. Then we select student number 9, 30, 17,..., etc., ignoring numbers 73, 83, etc., which are not in our list.

In general, *random sample* of n observations means that n values of a random variable X are obtained. These values must be independent of each other, and the population should remain the same during the process of selection. That is, a random sample of size n from a population having the probability density function $f(x)$ is a set of n random variables $X_1, X_2, \ldots, X_n$, which are independent of each other and have the same distribution. A random sample is a set of independent and identically distributed random variables.

The quantities obtained from the sample are called *statistics*. Average, median, range, standard deviation, quartiles, and percentiles are examples of statistics. *Sampling distributions* refer to the probability distributions of these statistics. For using various statistics in making inferences from the sample and for any problem of decision making based on samples, sampling distributions are necessary.

Example 5.23: Consider an urn with balls marked 0 and 1. 30% of the balls are marked 0 and 70% are marked 1. The probability distribution of the number on the ball when a ball is chosen at random from this urn is:

x	0	1
$p_X(x)$	.3	.7

For this distribution, $E(X) = .7$ and $Var(X) = .21$.

Suppose a sample of size 2 is to be obtained. We do this by replacement so that independence is maintained. Denote the numbers on the balls by X_1 and X_2, respectively. The joint probability distribution of X_1, X_2 is:

x_1 \ x_2	0	1
0	.09	.21
1	.21	.49

Consider the statistic providing the total on the balls:

$$T = X_1 + X_2$$

The distribution of T is the sampling distribution of T. Let $p_T(t) = P(T = t)$. For $T = 1$, $X_1 = 0$ and $X_2 = 1$, or $X_1 = 1$ and $X_2 = 0$ so that:

$$p_T(1) = P(X_1 = 0 \text{ and } X_2 = 1) + P(X_1 = 1 \text{ and } X_2 = 0)$$
$$= .21 + .21 = .42$$

Similarly, we can find $p_T(0)$ and $p_T(2)$. The probability distribution of T is given by:

t	0	1	2
$p_T(t)$	.09	.42	.49

Now $E(T) = 1.4$ and $Var(T) = .42$. We note that:

$$E(T) = 2E(X) \quad \text{and} \quad Var(T) = 2Var(X)$$

Example 5.24: The number of cavities in children's teeth has the distribution

x	1	2
$p_X(x)$	$\frac{2}{3}$	$\frac{1}{3}$

so that $E(X) = \frac{4}{3}$ and $Var(X) = \frac{2}{9}$.

Suppose a random sample of three children is taken and their cavities observed, given by X_1, X_2, and X_3. Let:

$$\bar{X} = \frac{X_1 + X_2 + X_3}{3}$$

what is the sampling distribution of $\bar{X}$? The eight possible outcomes are:

$$
\begin{array}{ll}
1, 1, 1 & 2, 1, 1 \\
1, 1, 2 & 2, 1, 2 \\
1, 2, 1 & 2, 2, 1 \\
1, 2, 2 & 2, 2, 2
\end{array}
$$

The possible values of the sample average are $1, \frac{4}{3}, \frac{5}{3}$, and 2. Suppose we want $p_{\bar{X}}(1)$. That is, $P(\bar{X} = 1)$, which is $P(1, 1, 1)$, and this is $\frac{8}{27}$, and so on. The probability distribution of $\bar{X}$ is given by:

$\bar{x}$	1	$\frac{4}{3}$	$\frac{5}{3}$	2
$p_{\bar{X}}(\bar{x})$	$\frac{8}{27}$	$\frac{12}{27}$	$\frac{6}{27}$	$\frac{1}{27}$

$E(\bar{X}) = \frac{4}{3}$ and $Var(\bar{X}) = \frac{2}{27}$. Again we notice that:

$$E(\bar{X}) = E(X)$$
$$Var(\bar{X}) = \tfrac{1}{3}\,Var(X)$$

These results are special cases of the following fact. If $X_1, X_2, \ldots, X_n$ is a random sample and $a_1, a_2, \ldots, a_n$ are constants, then:

$$E(a_1 X_1 + a_2 X_2 + \cdots + a_n X_n) = a_1 E(X_1) + a_2 E(X_2) + \cdots + a_n E(X_n),$$
$$Var(a_1 X_1 + a_2 X_2 + \cdots + a_n X_n) = a_1^2\,Var\,X_1 + a_2^2\,Var\,X_2 + \cdots + a_n^2\,Var\,X_n$$

When $a_1 = a_2 = \cdots = a_n = 1/n$, we get a special case of the above:

$$E(\bar{X}) = E(X) \tag{5.14}$$

$$Var(\bar{X}) = \frac{1}{n}Var(X) \tag{5.15}$$

There are many continuous probability density functions for which we need the distribution of a sample average. Here we give the result for the normal distribution. If $X_1, X_2, \ldots, X_n$ is a random sample from the normal distribution with mean μ and variance σ^2, the distribution of $\bar{X}$ is also normal with mean μ and variance σ^2/n.

Since the normal distribution is extensively used in statistical applications, sampling distributions of statistics using random samples obtained from normal distributions are extremely important. Three distributions that will be used in later chapters are chi-square, t, and F, and they are extensively tabulated. We give here simple motivations that led to these distributions.

Chi-square Distribution

If Z is a standard normal random variable, it can take negative as well as positive values. When we square Z, we get only positive values. The distribution of the square of Z no longer remains symmetric and has a chi-square distribution with one *degree of freedom*. The degree of freedom essentially is an intrinsic parameter of the distribution and should reflect the meaning given in mechanics. Here only one variable is squared, so we have one degree of freedom. The sum of independent chi-square random variables also has a chi-square distribution with their degrees of freedom added. A random variable is said to have χ_n^2 (chi-square with n degrees of freedom) when it is the sum of squares of n independent standard normal random variables. It can be shown that:

$$E(\chi_n^2) = n, \quad \mathrm{Var}(\chi_n^2) = 2n \tag{5.16}$$

The chi-square distributions for various degrees of freedom are given in Figure 5.20. The cumulative distribution function is given in Table IV in the Appendix for a few degrees of freedom.

If the random variable X has a normal distribution with mean μ and

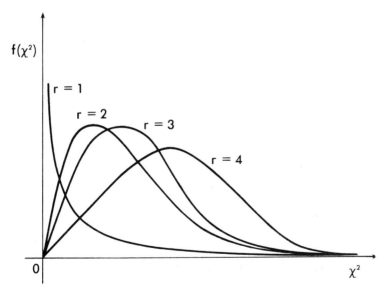

Figure 5.20. Chi-square distributions.

variance σ^2, then $\dfrac{X-\mu}{\sigma}$ is a standard normal random variable and $\dfrac{(X-\mu)^2}{\sigma^2}$ has a χ_1^2 distribution.

When $X_1, X_2, \ldots X_n$ is a random sample from the normal distribution with mean μ and variance σ^2, we have the distribution of

$$\sum_{i=1}^{n} \frac{(X_i - \mu)^2}{\sigma^2} \quad \text{as } \chi_n^2$$

However, we use the sum of squares of deviations from the sample average in calculating sample variance, so we need to know the probability distribution of $\dfrac{\sum(X_i - \bar{X})^2}{\sigma^2}$, which is also chi-square, but with $n-1$ degrees of freedom.

That is, if $X_1, X_2, \ldots, X_n$ is a random sample from a normal distribution with mean μ and variance σ^2, the distribution of:

$$\frac{\sum(X_i - X)^2}{\sigma^2} \quad \text{is } \chi_{n-1}^2 \tag{5.17}$$

One degree of freedom is reduced for using $\bar{X}$ in place of μ. The αth percentile of χ_r^2 will be denoted by $\chi_{r,\alpha}^2$. For example, $\chi_{9,.05}^2$ is a number such that the probability of a chi-square random variable with 9 degrees of freedom less than $\chi_{9,.05}^2$ is .05.

Since

$$s^2 = \frac{\sum(X_i - \bar{X})^2}{n-1}$$

$$\frac{(n-1)s^2}{\sigma^2} = \frac{\sum(X_i - \bar{X})^2}{\sigma^2}$$

Therefore, we also say:

$$\frac{(n-1)s^2}{\sigma^2} \quad \text{has } \chi_{n-1}^2$$

or

$$s^2 \text{ has } \frac{\sigma^2}{n-1}\chi_{n-1}^2$$

Example 5.25: Suppose that $\sum(X_i - \bar{X})^2$ from a random sample based on 10 observations from a normal distribution with a variance of 25 is given to be greater than 475.5. Since $\sum(X_i - \bar{X})^2/25$ has a χ_9^2 distribution, the probability that $\sum(X_i - \bar{X})^2 > 475.5$ is the same as $P(\chi_9^2 > 19.02)$, which equals .025 from Table IV in the Appendix.

t-distribution

A *t*-distribution arises from the use of the ratio of the sample average to the standard deviation of observations from a normal distribution. If Z is a standard normal random variable and χ_n^2 has a chi-square distribution with n degrees of freedom, and they are independent, then

$$t_n = \frac{Z}{\sqrt{\dfrac{\chi_n^2}{n}}} \tag{5.18}$$

has a *t*-distribution with n degrees of freedom. The *t*-distribution is tabulated in Table V in the Appendix.

It is known that the sample average and sample variance of observations from a normal distribution are independent. Also, the sample average has the normal distribution with the same mean but with a variance of $\dfrac{\sigma^2}{n}$. Let:

$$s^2 = \frac{\sum (X_i - \bar{X})^2}{n - 1} \tag{5.19}$$

Then $\dfrac{\sqrt{n}(\bar{X} - \mu)}{s}$ has a *t*-distribution with $n - 1$ degrees of freedom.

The *t*-distribution is symmetric about 0 and thus its mean is 0. The αth percentile of a *t*-distribution with n degrees of freedom will be denoted by $t_{n,\alpha}$.

Example 5.26: A random sample of nine observations of prices of a commodity provided the sample standard deviation of 2.3. Suppose the prices are normally distributed with a mean of 10. The probability that the average of these prices is greater than 11.426 can be obtained by using the *t*-distribution with 8 degrees of freedom. That is:

$$P(\bar{X} \geq 11.426) = P\left(\frac{\sqrt{9}(\bar{X} - 10)}{2.3} \geq \frac{\sqrt{9}(11.426 - 10)}{2.3} \right)$$

$$= P(t_8 \geq 1.86)$$

$$= .05$$

since $t_{8,.95} = 1.86$.

F-distribution

The distribution of the ratio of two independent chi-square random variables occurs in many applications as the ratio of variances of two independent samples from normal distribution. Let s_1^2 and s_2^2 be the sample variances of samples from two different normal populations based on n_1 and n_2 observations. Then the ratio

$$F_{n_1-1,n_2-1} = \frac{s_1^2}{s_2^2} \tag{5.20}$$

has an F-distribution with $n_1 - 1$, $n_2 - 1$ degrees of freedom. The distribution is tabulated in Table VI in the Appendix. The application of the F-distribution will be seen in later chapters. The αth percentile of F_{n_1-1,n_2-1} is denoted by $F_{n_1-1,n_2-1,\alpha}$.

   Example 5.27: From the tables, the distribution of $F_{3,7}$ gives:

$$P(F_{3,7} \geq 4.35) = .05$$
$$P(F_{3,7} \geq 5.89) = .025$$
$$P(F_{3,7} \geq 8.45) = .01$$

Similarly,
$$P(F_{7,3} \leq 8.89) = .95$$
$$P(F_{7,3} \leq 14.6) = .975$$
$$P(F_{7,3} \leq 27.7) = .99$$

and so on.

Exercises

19. Find the distribution of

   (a) $U = X + Y$
   (b) $V = XY$
   (c) $Z = \dfrac{X}{Y}$

   where X and Y are two independent observations from the following population:

x	1	2
$p_X(x)$	.5	.5

20. When $X_1, X_2, \ldots, X_{10}$ is a random sample from a normal distribution with a mean of 10 and a variance of 5, find the probability that:

   (a) $\bar{X} < 11$
   (b) $\bar{X} > 12$
   (c) $9 < \bar{X} < 10$

21. Let X_1 and X_2 be two independent observations from a normal distribution with a mean of 0 and a variance of 2. Find the probability that $X_1^2 + X_2^2 > 11.98$.

When a sample of three observations is taken from the above, what is the probability that $X_1^2 + X_2^2 + X_3^2 < 1.168$?

Chapter Exercises

1. The batting average of the San Francisco Giants in a certain week for all its players has a mean of .250 and a standard deviation (known) of .10. Suppose the batting average is normally distributed. What is the probability that:

   (a) the average of batting averages of 9 players is larger than .256?
   (b) the average of batting averages of 16 players is between .239 and .260?

2. Suppose the probability of hitting a target by a recently made gun and improved ammunition is .8. Using normal approximation, find the probability that of 100 shots fired the number of shots hitting the target are:

   (a) greater than 83
   (b) between 78 and 86
   (c) less than 90

3. In a "fun run" of 5 miles in a race, several people could not complete the race. Suppose the distance run by individuals has an exponential distribution with a mean of 4 miles. What is the probability that a person chosen at random from the participants runs:

   (a) more than 3.5 miles before quitting?
   (b) less than 2.5 miles before quitting?
   (c) completes the run (runs at least 5 miles)?
   (d) Suppose that there are 10 participants in the "fun run." What is the probability that 7 of them will complete the run?

4. (a) Select 50 two-digit random numbers from the table of random numbers and give a stem-and-leaf plot. Calculate the relative frequencies. Assuming that you are observing a continuous random variable, what distribution would you expect the random numbers have?
   (b) For a uniform distribution on the range 0–100, find the expected value and variance.

Summary

The *probability density function* of a continuous random variable is a curve that gives probabilities by areas under the curve and above the horizontal line bounded by the range of the random variable. The *uniform distribution* is a model of a random variable that gives the number chosen at random between two points. The *exponential distribution* has the density $f(x) = \theta e^{-\theta x}$, $x \geq 0$ and a mean of $1/\theta$ and a variance of $1/\theta^2$.

The *cumulative distribution function* of X gives the probability of $X \leq x$

$$P(X \leq x) = F(x)$$

and is a function that is nondecreasing and is between 0 and 1. For the discrete random variable, the cumulative distribution function has jumps at the points with positive probabilities; otherwise, it is flat. The *normal distribution* has the probability density function

$$\frac{1}{\sqrt{2\pi}\sigma} e^{-(1/2\sigma^2)(x-\mu)^2}, \quad -\infty < x < \infty$$

and has mean μ and variance σ^2.

Sampling distributions describe the probability distributions of statistics obtained from the sample. Some important sampling distributions are *chi-square*, *t*, and *F*.

References

Hogg, Robert V., and Craig, Allen T. *Introduction to Mathematical Statistics*, 3rd ed. New York: Macmillan, 1970.

Johnson, Norman L., and Kotz, Samuel. *Continuous Distribution*. Boston: Houghton Mifflin Company, 1969.

chapter six

Statistical Inference— Estimation

Probability models given by probability density functions or cumulative distribution functions were discussed in the previous chapters. These models are appropriate for making inferences from data and may involve unknown parameters. For instance, when we assume that the data on cholesterol levels of subjects may have come from a normal distribution with unknown mean μ and unknown variance σ^2, we are assuming a partially specified model to analyze data on cholesterol levels. Similarly, when dealing with data on auto accidents in a city, we may assume that accidents have a Poisson distribution with parameter λ. One of the first things we want from the data is the determination of these unknown parameters. Inference in such parametric models is called *parametric inference*. When inference is concerned with distributions that are completely unknown—that is, we cannot assume if they are normal, exponential, or Poisson—we call it *nonparametric* or *distribution-free inference* (see Chapter 8).

A function of sample observations has been called a *statistic*. If a statistic estimates a parameter, it is the *point estimate* of the parameter. The sample average, sample standard deviation, and sample median are examples of point estimates of various parameters.

Since different samples lead to different values of the same statistic estimating a certain parameter, another method of estimation has been developed through intervals, called *interval estimation*. The interval is a random interval and is so constructed that it contains the unknown parameter with some large prespecified probability. The probability assumed to construct an interval reflects the amount of confidence we want to put in this interval; hence, interval estimates are also *confidence intervals* and the probability is called the *confidence coefficient*. For example, an interval of the type $(\bar{x} - a, \bar{x} + a)$ based on the statistic $\bar{x}$, sample average, may be constructed with an appropriate "a" so as to contain the parameter μ of the normal distribution.

Another kind of statistical inference in parametric models is made by *testing hypotheses* about parameters. If we wish to test the claim that the number of auto accidents has decreased after the introduction of the 55-mile-per-hour speed limit, we have a hypothesis-testing problem. When the probability model for an auto accident is Poisson with parameter λ, we will test hypotheses about the parameter λ.

The classical statistical inference is concerned with point and interval estimation and tests of hypotheses. In this chapter we give a few methods of parametric inference about parameters in the partially specified models and discuss point and interval estimates. Testing hypotheses about the parameters will be discussed in Chapter 7.

Point Estimates

We assume that a random sample is available from a population having a known probability density function with unknown parameter (parameters) and we wish to estimate the parameter. For example, let $X_1, X_2, \ldots X_n$ be a random sample of size n from a population that is normal with mean μ and variance σ^2, both of them unknown. In general, we denote the probability density function by $f(x, \theta)$ where f is the known function and θ is an unknown parameter. A *point estimate* of θ will be denoted by $\hat{\theta}$.

The theory of point estimation is concerned with the single value estimation of parameters. It also provides criteria by which an estimate is selected from a large number of possible estimates for a parameter. Essentially we want the "best" estimate under a given set of circumstances. There are several criteria for choosing an estimate. One is that of unbiasedness.

Unbiasedness

By an *unbiased estimate* $\hat{\theta}$ of the parameter θ, we mean that the sampling distribution of $\hat{\theta}$ has the expected value θ. That is, $\hat{\theta}$ is an unbiased estimate of θ if for all θ,

$$E(\hat{\theta}) = \theta \qquad (6.1)$$

The sample average $\bar{X}$ is an unbiased estimate of the population mean μ, whatever the distribution. Similarly, $s^2 = \sum(X_i - \bar{X})^2/(n-1)$ is an unbiased estimate of the population variance σ^2.

Example 6.1: Let X be a binomial random variable with parameter p based on n trials. Then we know that $E(X) = np$. Hence $\hat{p} = \dfrac{X}{n}$ is an unbiased estimate of p, since $E(\hat{p}) = E\left(\dfrac{X}{n}\right) = \dfrac{1}{n}E(X) = p$.

The *bias* of an estimate is given by:

$$E(\hat{\theta}) - \theta \tag{6.2}$$

For an unbiased estimate, the bias is zero.

Example 6.2: Let $X_1, X_2, \ldots, X_n$ be a random sample from a population with mean μ and variance σ^2. Let:

$$\hat{\sigma}^2 = \frac{\sum (X_i - \bar{X})^2}{n}$$

Then:

$$E\left(\frac{\sum (X_i - \bar{X})^2}{n}\right) = \frac{1}{n}E(\sum (X_i - \bar{X})^2) = \frac{1}{n}(n-1)\sigma^2$$

Hence, the bias of $\hat{\sigma}^2$ is:

$$\frac{n-1}{n}\sigma^2 - \sigma^2 = -\frac{\sigma^2}{n} \tag{6.3}$$

This bias is generally removed by taking the estimate of σ^2, which is unbiased, as:

$$s^2 = \frac{\sum (X_i - \bar{X})^2}{n-1}$$

There are other ways of bias reduction for estimates. One is the *method of jackknifing* (see Efron [1982]).

The choice among several unbiased estimates can be made by comparing their variances. Let $\hat{\theta}_1$ and $\hat{\theta}_2$ be two estimates of θ. Then $\hat{\theta}_1$ is more *efficient* than $\hat{\theta}_2$ if:

$$\text{Var}(\hat{\theta}_1) < \text{Var}(\hat{\theta}_2)$$

For example, the sample average and sample median are both unbiased estimates of the mean of a normal population. However, it can be shown that the sample average has a smaller variance and hence is a more efficient estimate of the population mean.

Estimates are selected by several other criteria, such as those of consistency and sufficiency (see, for example, Hogg and Craig [1970]).

Methods of Estimation

Among the several methods of finding estimates, one of the most important is that of *maximum likelihood*. Suppose a random sample of size n with values $X_1 = x_1$, $X_2 = x_2, \ldots, X_n = x_n$, from $f(x, \theta)$ has been observed. The *likelihood* of the sample is defined as:

$$L(\theta) = f(x_1, \theta) f(x_2, \theta) \ldots f(x_n, \theta) \tag{6.4}$$

That is, the likelihood is the product of probability density functions evaluated at the observed values of the sample. When the random variable is discrete, the likelihood is the probability of obtaining the sample. The *maximum likelihood estimate* of θ is obtained by maximizing the likelihood. Ordinarily, methods of calculus are needed to maximize likelihoods, and this is beyond the scope of the book. We give some maximum likelihood estimates for important cases. Sir Ronald A. Fisher popularized the method of maximum likelihood, and it is extensively used in statistics.

Example 6.3: Let $X_1, X_2, \ldots, X_n$ be a random sample from a Poisson distribution with parameter λ. The likelihood is:

$$L(\lambda) = \frac{e^{-\lambda} \lambda^{x_1}}{x_1!} \cdot \frac{e^{-\lambda} \lambda^{x_2}}{x_2!} \cdots \frac{e^{-\lambda} \lambda^{x_n}}{x_n!}$$

$$= \frac{e^{-n\lambda} \lambda^{x_1 + x_2 + \cdots + x_n}}{x_1! x_2! \cdots x_n!}$$

The maximum of the likelihood is obtained with $\hat{\lambda} = \bar{X}$, and so $\bar{X}$ is the maximum likelihood estimate.

Example 6.4: Suppose a random sample $X_1, X_2, \ldots X_n$ is given from the exponential distribution. Then the likelihood is:

$$L(\theta) = \theta e^{-\theta x_1} \cdot \theta e^{-\theta x_2} \cdots \theta e^{-\theta x_n}$$

$$= \theta^n e^{-\theta(x_1 + x_2 + \cdots + x_n)}$$

Here the maximum likelihood estimate is given by $\hat{\theta} = \dfrac{1}{\bar{X}}$.

Another method of estimation is the *method of moments*. The sample average is the first *sample moment*. By rth *sample moment*, we mean the average of rth powers of the observations. That is:

$$m_r = \frac{1}{n} \sum_{i=1}^{n} X_i^r$$

so that

$$m_1 = \frac{1}{n} \sum_{i=1}^{n} X_i = \bar{X}$$

and

$$m_2 = \frac{1}{n} \sum_{i=1}^{n} X_i^2$$

Use m_2 to calculate the sample variance s^2.

The *method of moments* requires that the rth population moment be equated to the rth sample moment. We can equate as many sample moment as we like, depending on the number of parameters. Since the choice is there for equating any moments, the moment estimates may not be unique. However, the method of moments provides unbiased estimates.

Example 6.5: For the normal distribution, the mean μ is the population moment, whereas the sample moment is $\bar{X}$. Hence, $\hat{\mu} = \bar{X}$ gives the moment estimate of μ.

Example 6.6: Let $X_1, X_2, \ldots, X_n$ be a random sample from uniform distribution on $(0, \theta)$. The mean of this population is $\frac{\theta}{2}$. Hence, equating the moments

$$\frac{\hat{\theta}}{2} = \bar{X}$$

or

$$\hat{\theta} = 2\bar{X}$$

gives the method of moment estimate. It can be shown that the maximum likelihood estimate of θ in this case is the largest value of the sample, so the method-of-moment estimate may be quite different from the maximum likelihood estimate.

Exercises

1. Given a random sample from the uniform distribution on (a, b), find the method-of-moment estimates for a and b.
2. The times of death of 10 mice, fed on a toxic material mixed with their feed, are (in days) 10, 75, 25, 33, 44, 45, 15, 23, 27, and 12. Assuming that the times of death are exponentially distributed with mean μ, find the maximum likelihood estimates of μ as well as the method-of-moment estimate of μ.
3. The total number of accidents in a given city observed over 60 days is 40. Assuming that the accidents per day have a Poisson distribution with parameter λ, what is the maximum likelihood estimate of λ?
4. Suppose a random sample of 15 heights of teenage boys is given (cm) as:

    157,   160,   145,   161,   156,   153,   148,
    149,   157,   145,   162,   164,   180,   172,   168

Assuming that the heights are normally distributed with mean μ and variance σ^2, find the moment estimates of μ and σ^2.

5. The frequency distribution of the number of misprints per page of a typed report of 100 pages is:

Number of misprints	0	1	2	3	4	5	6
Frequency	36	18	22	16	6	1	1

Suppose the misprints have a Poisson distribution with parameter λ, Find the maximum likelihood estimate of λ.

Interval Estimates

Sometimes we are interested in estimating parameters by intervals, called *confidence intervals*. A preassigned probability is given such that the interval contains the unknown parameter. This probability is called *confidence coefficient*, and it reflects the importance of the estimation procedure. The parameter is a constant and the interval is random; therefore, we say that the interval contains the parameter or the interval covers the parameters. We do not say that the parameter lies in the interval. The interpretation of confidence coefficient can be given in terms of relative frequencies. Suppose we want to estimate a parameter θ by an interval with 95% confidence coefficient. The estimated interval in the problem will either contain θ or not contain θ. If we take samples again and again and construct intervals for the parameters, then 95% confidence implies that 95% of the time the interval is expected to contain the true parameter θ, and 5% of the time it is not. Figure 6.1 shows several intervals for the parameter θ, some of which do not cover the parameter. Similarly, when we give intervals with confidence coefficient $1 - \alpha$, it is meant that in the long run $100(1 - \alpha)\%$ of the intervals will contain θ and $100\alpha\%$ of the intervals will not contain θ.

Confidence Interval for Mean of the Normal Distribution

Let $X_1, X_2, \ldots, X_n$ be a random sample from the normal distribution with mean μ and variance σ^2. Then $\hat{\mu} = \bar{X}$. We consider two cases. In the first case, σ^2 is known.

When σ^2 is known, $\dfrac{\sqrt{n}(\bar{X} - \mu)}{\sigma}$ is the standard normal random variable. We can find two numbers from the tables that give probability $1 - \alpha$ in the middle. These two numbers are $-z_{1-\alpha/2}$ and $z_{1-\alpha/2}$ where $z_{1-\alpha/2}$ gives $\alpha/2$ probability in the upper tail of the standard normal distribution. That is, the probability

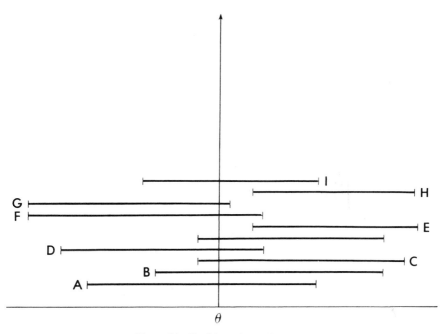

Figure 6.1. Confidence intervals for θ.

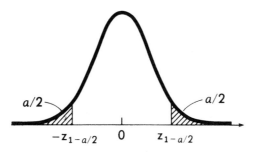

Figure 6.2. Standard normal.

in the lower tail of the normal is also $\alpha/2$, as shown in Figure 6.2. For instance, if want a 90% confidence interval, we have .90 probability in the middle and .05 probability in each tail, so that:

$$P\left\{-z_{1-\alpha/2} \leq \frac{\sqrt{n}(\bar{X} - \mu)}{\sigma} \leq z_{1-\alpha/2}\right\} = 1 - \alpha$$

The inequalities inside the probability are:

$$-z_{1-\alpha/2}\sigma/\sqrt{n} \leq \bar{X} - \mu \leq z_{1-\alpha/2}\frac{\sigma}{\sqrt{n}}$$

So that,

$$\bar{X} - z_{1-\alpha/2}\frac{\sigma}{\sqrt{n}} \leq \mu \leq \bar{X} + z_{1-\alpha/2}\frac{\sigma}{\sqrt{n}}$$

giving $(1-\alpha)$-level confidence interval for μ as

$$\left(\bar{X} - z_{1-\alpha/2}\frac{\sigma}{\sqrt{n}}, \quad \bar{X} + z_{1-\alpha/2}\frac{\sigma}{\sqrt{n}}\right) \tag{6.5}$$

Note that the probability statement is made before the data are observed. After the values of the statistic $\bar{X}$ and σ, n and z are substituted in Equation (6.5), the interval gives just two numbers, and this interval either contains the parameter or does not contain the parameter.

Example 6.7: The lengths of 16 jellyfish are given by Don McNeil in "Analysis of Jellyfish Data," *Interactive Statistics*, North-Holland Publishng Company, 1979, as (in cm):

6.5, 6, 6.5 7, 8, 7, 8, 8, 7, 8, 9, 10, 11, 12, 11, 11

Assume that the lengths are normally distributed with variance 4 (cm)^2. $\bar{X} = 8.5$ cm. Then the 90% confidence interval for the mean length is:

$$= \left(8.5 - 1.645\sqrt{\frac{4}{16}}, \quad 8.5 + 1.645\sqrt{\frac{4}{16}}\right)$$

$$= (7.68, 9.32)$$

The 99% confidence interval for the mean is:

$$= \left(8.5 - 2.58\sqrt{\frac{4}{16}}, \quad 8.5 + 2.58\sqrt{\frac{4}{16}}\right)$$

$$= (7.21, 9.79)$$

Example 6.8: The weight gain per week in pounds for 9 pigs fed on a new diet are 2.1, 3.2, 1.8, 2.5, 0.9, 7.3, 5.2, 3.5, and 1.8. Assume that the weight gains are normally distributed with a variance of 4 (pounds)^2. The confidence interval for the mean weight gain is required. In this case $\bar{X} = 3.14$, and the 95% confidence interval for the mean is:

$$= \left(3.14 - 1.96\sqrt{\frac{4}{9}}, \quad 3.14 + 1.96\sqrt{\frac{4}{9}}\right)$$

$$= (1.83, 4.45)$$

In the second case σ^2 is unknown. We estimate it by s^2 and then use the distribution of $\dfrac{\sqrt{n}(\bar{X} - \mu)}{s}$, which is t with $n-1$ degrees of freedom. Let

$t_{n-1,1-\alpha/2}$ be the number that gives the probability of $\alpha/2$ in the upper tail of the t-distribution with $n-1$ degrees of freedom. Then:

$$P\left\{-t_{n-1,1-\alpha/2} \leq \frac{\sqrt{n}(\bar{X}-\mu)}{s} \leq t_{n-1,1-\alpha/2}\right\} = 1-\alpha$$

Simplifying the inequalities, we have $(1-\alpha)$-level confidence interval for the mean as:

$$\left(\bar{X} - t_{n-1,1-\alpha/2}\frac{s}{\sqrt{n}}, \bar{X} + t_{n-1,1-\alpha/2}\frac{s}{\sqrt{n}}\right) \qquad (6.6)$$

Example 6.9: Serum cholesterol levels* of 12 subjects are (in mg/100 ml):

182, 182, 203, 187, 245, 158, 180, 284, 191, 202, 198, 191

Assuming that the population is normal with mean μ and variance σ^2, the confidence interval for the mean with a 90% confidence coefficient is:

$$= \left(\bar{X} - t_{11,.05}\frac{s}{\sqrt{n}}, \bar{X} + t_{11,.05}\frac{s}{\sqrt{n}}\right)$$

$$= \left(200.25 - 1.796\frac{33.358}{\sqrt{12}}, 200.25 + 1.796\frac{33.358}{\sqrt{12}}\right)$$

$$= (200.25 - 17.295, 200.25 + 17.295)$$

$$= (182.95, 217.55)$$

Example 6.10: The heights of a sample of 16 college students are $\bar{X} = 165$ cm and $s^2 = 25$ (cm)^2. Assuming that the heights have a normal distribution, the 95% confidence interval for the mean height is:

$$= \left(165 - \left(2.131 \times \sqrt{\frac{25}{16}}\right), 165 + \left(2.131 \times \sqrt{\frac{25}{16}}\right)\right)$$

$$= (162.3, 167.7)$$

The 98% confidence interval for the mean height is:

$$= (165 - (2.602 \times 1.25), 165 + (2.602 \times 1.25))$$

$$= (161.7, 168.3)$$

Data Source: Krempler, F.; Kostner, G. M.; Roscher, A.; Haslaner, F.; Bolzaro, K.; and Sandhoffer, F. Studies on the role of specific cell surface receptors in the removal of Lipoprotein (a) in *Man, J. Clin. Invest.*, 1983, 71, 1431–41.

Confidence Intervals for the Variance of the Normal Distribution

The variance of the normal distribution is estimated by:

$$s^2 = \sum \frac{(X_i - \bar{X})^2}{n-1}$$

Now $(n-1)s^2/\sigma^2$ has a chi-square distribution with $n-1$ degrees of freedom. Since the chi-square distribution is not symmetric, it is not proper to take equal probabilities in the tails of the distribution. However, for convenience, the usual convention allows this. We denote by $\chi^2_{n-1,1-\alpha/2}$ the point for which the upper tail probability is $\alpha/2$ and by $\chi^2_{n-1,\alpha/2}$ the point for which the probability in the lower tail is $\alpha/2$, as given in Figure 6.3. So we have:

$$P\left\{ X^2_{n-1,\alpha/2} \leq \frac{(n-1)s^2}{\sigma^2} \leq \chi^2_{n-1,1-\alpha/2} \right\} = 1 - \alpha$$

The inequalities give $(1-\alpha)$-level confidence interval for the variance σ^2 as:

$$\left(\frac{(n-1)s^2}{\chi^2_{n-1,1-\alpha/2}}, \frac{(n-1)s^2}{\chi^2_{n-1,\alpha/2}} \right) \tag{6.7}$$

Example 6.11: For the data in Example 6.9, suppose we want a 90% confidence interval for σ^2. Now $\chi^2_{11,.05} = 4.57, \chi^2_{11,.95} = 19.7$. The interval is:

$$= \left(\frac{11 \times 1112.75}{19.7}, \frac{11 \times 1112.75}{4.57} \right)$$

$$= (621.33, 2678.39)$$

Example 6.12: For the data in Example 6.10, the 95% confidence interval

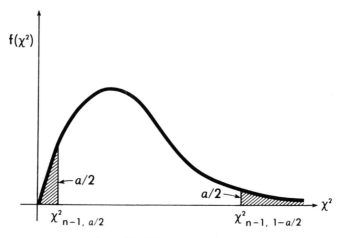

Figure 6.3. Chi-square distribution.

for the variance is given by $\chi^2_{15,.975} = 27.5, \chi^2_{15,.025} = 6.26$

$$= \left(\frac{15 \times 25}{27.5}, \frac{15 \times 25}{6.26} \right)$$

$$= (13.6, 59.9)$$

Confidence Intervals for the Proportion

The proportion, or the binomial parameter p, is estimated by X/n, where X is the number of successes in n independent Bernoulli trials, each having p as the probability of success. That is, $\hat{p} = X/n$. To determine the exact confidence intervals requires the use of binomial distribution. This is a discrete distribution, and for various values of X we cannot attain a prespecified probability. But when n is large, we can use the normal approximation to the binomial. That is, for n large, $\hat{p}$ has a normal distribution with mean p and variance $p(1 - p)/n$. Since n is large, we can assume that the variance is known and is given by $\hat{p}(1 - \hat{p})/n$. The problem is then converted into that of finding the confidence intervals for the mean of the normal mean with a known variance. Hence, $(1 - \alpha)$-level confidence interval is given by:

$$\left(\hat{p} - z_{1-\alpha/2} \sqrt{\frac{\hat{p}(1-\hat{p})}{n}}, \hat{p} + z_{1-\alpha/2} \sqrt{\frac{\hat{p}(1-\hat{p})}{n}} \right) \tag{6.8}$$

Example 6.13: The number of rotten apples in a sample of 50 is 5. The estimate $\hat{p} = \frac{5}{50} = .1$. A confidence interval of 90% for the probability of rotten apples is:

$$= \left(.1 - 1.645 \times \sqrt{\frac{.1 \times .9}{50}}, .1 + 1.645 \times \sqrt{\frac{.1 \times .9}{50}} \right)$$

$$= (.03, .17)$$

In a lot of 1,000 apples, from which this sample was taken, we would expect between 30 and 170 rotten apples with a 90% confidence coefficient.

Example 6.14: A highway patrolman measures the speed of cars on an interstate highway. In a sample of 100 cars, he found that 35 were driving beyond the speed limit. The probabilities of driving over the speed limit p is estimated by $\frac{35}{100} = .35$. A 95% confidence interval for this probability is given by:

$$= \left(.35 - 1.96 \sqrt{\frac{.35 \times .65}{100}}, .35 \times 1.96 \sqrt{\frac{.35 \times .65}{100}} \right)$$

$$= (.26, .44)$$

Confidence intervals when n is small have been obtained for various values of n through graphs, for reference see Beyer (1966), pp. 188–189.

Sample-Size Determination

Statisticians are often asked the question, "How many observations should I take?" Before this question can be answered, we must know what the problem is and what kinds of risks the experimenter is willing to take. Suppose the experimenter is interested in finding $(1 - \alpha)$ level confidence interval of the mean of a normal population whose variance is already known. Suppose further that he wants the mean to be estimated within d units of the true mean. In that case, the length of the confidence interval desired is $2d$. Hence, we have:

$$\left(\bar{x} + z_{1-\alpha/2} \frac{\sigma}{\sqrt{n}} \right) - \left(\bar{x} - z_{1-\alpha/2} \frac{\sigma}{\sqrt{n}} \right) = 2d,$$

$$2z_{1-\alpha/2} \frac{\sigma}{\sqrt{n}} = 2d,$$

or

$$n = \frac{z_{1-\alpha/2}^2 \sigma^2}{d^2} \tag{6.9}$$

Equation (6.9) gives the value of n, and a nearest integer would suffice as the sample size. Similarly, if one is willing to assume that a large sample size can be taken to estimate a proportion with $d\%$ of the true proportion with confidence level $1 - \alpha$, we have:

$$n = z_{1-\alpha/2}^2 \frac{\hat{p}(1 - \hat{p})}{d^2}$$

Since we do not know p or $\hat{p}$, we can use the largest possible value of $\hat{p}(1 - \hat{p})$ giving the sample size:

$$n = z_{1-\alpha/2}^2 / 4d^2 \tag{6.10}$$

Example 6.15: An experimenter wants to obtain a 90% confidence interval for the daily butter-fat content of milk in a breed of cows, and he wants the estimate to be within 2 pounds of the true value. Assume that the variance is 25. Then the sample size is:

$$n = (1.645)^2 \tfrac{25}{4} = 16.9$$

We therefore have a sample of size 17.

Example 6.16: An estimate of the proportion of defective items is to be obtained so that the estimate is within 2% of the true value with a 95%

confidence interval. Using Equation (6.10), we find that the sample size is:

$$n = \frac{(1.96)^2}{4(.02)^2} = 2401$$

Exercises

6. A sample of 25 systolic blood pressure of subjects gives a mean of 85. It is known that systolic blood pressure is normally distributed with variance 4. Find the 90% and 99% confidence intervals for the mean systolic blood pressure of the population.

7. The sales per day of a store (in thousands of dollars) are:

$$27.5, \ 17.2, \ 15.3, \ 8.5, \ 9.5, \ 17.8, \ 15.2$$

   Suppose the sales have a normal distribution. Find the estimate of the mean and variance and give (a) the 95% confidence interval for the mean and (b) the 90% confidence interval for the variance.

8. A sample of workers in a factory consumed the following calories in their daily diet:

$$2522, \ 2200, \ 2001, \ 2020, \ 3562, \ 4500,$$
$$4500, \ 4250, \ 4300, \ 3500$$

   Assuming that the calorie intake is normally distributed, give the 90% confidence interval for the mean caloric intake.

9. In a sample of 65 persons examined, it was found that 3 had been exposed to a disease. Provide a 99% confidence interval for the proportion of those exposed.

10. The price of gasoline is a random variable having normal distribution. The following prices of regular gas in the month of July were noticed for 10 gas stations chosen at random in the San Francisco Bay Area. Find the 95% confidence interval for the mean price:

$$\$1.19, \ 1.27, \ 1.38, \ 1.21, \ 1.23, \ 1.29, \ 1.37, \ 1.28, \ 1.18, \ 1.25$$

11. A new treatment for a disease was clinically tested on 120 patients and it was found to cure the disease completely in 37 cases. Find the 95% and 99% confidence intervals for the cure rate.

12. Fifteen boxes of a dry cereal of the same size were weighed and the following observations were obtained (in ounces):

$$15.7, \ 14.9, \ 16.2, \ 16.1, \ 15.9, \ 16.3, \ 14.7, \ 14.8$$
$$15.5, \ 16.0, \ 15.0, \ 14.8, \ 14.8, \ 15.7, \ 15.0$$

   (a) Find the confidence interval for the mean at 90% level
   (b) Find the 95% confidence interval for the variance

   Assume that the weights are normally distributed.

13. How many observations must be taken so as to estimate the proportion of drug addicts in a state within 5% of the true proportion at a confidence level of .90?

14. Suppose a confidence interval estimate of the mean weight of pigs fed on a certain diet at .95 level of confidence is to be obtained within 5 pounds of the true mean weight. Assume that the weights are normally distributed with variance 16. What is the sample size?

15. The caloric intake by individuals is assumed to be normal with a variance of 1,900. An estimate of the mean caloric intake is to be made within 100 of the true mean with confidence coefficient .95. What is the sample size?

16. The confidence interval estimate of proportion of absentees before the Thanksgiving holiday in a university is to be made with a .95 confidence level. The estimate should be within 5% of the true proportion. What is the sample size?

Confidence Intervals for the Difference of Two Means

A common problem in scientific experimentation and industrial research is concerned with the comparison of two different groups. A medical clinician may be interested in comparing the cure rate of two drugs, an agriculture scientist may want to know the relative performance of two different fertilizers, and an industrial engineer may want to see the relative performance of two different machines or two different workers. Such problems occur in several other areas. We assume here that the two populations to be considered provide independent samples. We consider the interval estimation for the difference of the means, and the difference of the proportions of two populations. Hypothesis-testing problems for the equality of means or equality of variance will be discussed in the next chapter.

Let $X_1, X_2, \ldots, X_n$ be a random sample from a population having normal distribution with mean μ_1 and variance σ_1^2 and let $Y_1, Y_2, \ldots, Y_m$ be a random sample from a population that is normal with mean μ_2 and variance σ_2^2. Let $\bar{X}$ and s_x^2 be the sample average and sample variance from the first sample and $\bar{Y}$ and s_y^2 be the sample average and sample variance from the second sample. Then the probability distribution of $\bar{X} - \bar{Y}$ is normal with mean $\mu_1 - \mu_2$ and variance $(\sigma_1^2/n) + (\sigma_2^2/m)$, so that

$$\frac{\bar{X} - \bar{Y} - (\mu_1 - \mu_2)}{\sqrt{\dfrac{\sigma_1^2}{n} + \dfrac{\sigma_2^2}{m}}}$$

has a distribution that is standard normal.

(i) σ_1^2, σ_2^2 *known*

We follow the same arguments that apply to the case of the confidence interval of the mean of a normal distribution discussed earlier. Using Equation

(6.5), we can get $(1 - \alpha)$-level confidence interval for $\mu_1 - \mu_2$ as:

$$\left(\bar{X} - \bar{Y} - z_{1-\alpha/2} \sqrt{\frac{\sigma_1^2}{n} + \frac{\sigma_2^2}{m}}, \bar{X} - \bar{Y} z_{1-\alpha/2} \sqrt{\frac{\sigma_1^2}{n} + \frac{\sigma_2^2}{m}} \right) \qquad (6.11)$$

(ii) $\sigma_1^2 = \sigma_2^2 = \sigma^2$ unknown

Since both populations are assumed to have the same variance, we estimate the common variance by pooling the variances of the two populations. Let s_p^2 be this estimate of the common variance:

$$s_p^2 = \frac{(n-1)s_x^2 + (m-1)s_y^2}{m+n-2} \qquad (6.12)$$

In this case, the probability distribution of $\bar{X} - \bar{Y}$ is normal with mean $\mu_1 - \mu_2$ and variance $\sigma^2(1/n + 1/m)$. Hence,

$$\frac{\bar{X} - \bar{Y} - (\mu_1 - \mu_2)}{s_p \sqrt{\frac{1}{m} + \frac{1}{n}}}$$

has t-distribution with $m + n - 2$ degrees of freedom. We use this fact in getting the confidence interval for $\mu_1 - \mu_2$. As in Equation (6.6), the $(1 - \alpha)$-level confidence interval for $\mu_1 - \mu_2$ is:

$$\left(\bar{X} - \bar{Y} - t_{m+n-2, 1-\alpha/2} s_p \sqrt{\frac{1}{m} + \frac{1}{n}}, \right.$$
$$\left. \bar{X} - \bar{Y} + t_{m+n-2, 1-\alpha/2} s_p \sqrt{\frac{1}{m} + \frac{1}{n}} \right) \qquad (6.13)$$

Example 6.17: It is assumed that the batting averages of players in the National League and the American League are normally distributed with variances of .0025 and .0036, respectively. A random sample of 15 National League players and a sample of 20 American League players gives, respectively:

$$\bar{X} = .274 \quad \text{and} \quad \bar{Y} = .248$$

The confidence interval for the difference of means of National League players and American League players is, from Equation (6.11), at level 90%:

$$\left(.274 - .248 - 1.645 \sqrt{\frac{.0025}{15} + \frac{.0036}{20}}, \right.$$
$$\left. .274 - .248 + 1.645 \sqrt{\frac{.0025}{15} + \frac{.0036}{20}} \right)$$

$$= (.026 - .031, .026 + .031)$$
$$= (-.005, .057)$$

Example 6.18: The weights of pigs are assumed to be normally distributed. Suppose a sample of 10 pigs fed on regular diet gave an average weight of 205 pounds and a standard deviation of 10 pounds. An experimental diet fed to 16 pigs resulted in the average of weights as 240 pounds and a standard deviation of 15 pounds. Find the 90% confidence interval of the mean weight gain, if any, as a result of the experimental diet. Here:

$$\bar{X} = 240 \qquad s_x = 15 \qquad n = 16$$
$$\bar{Y} = 205 \qquad s_y = 10 \qquad m = 10$$

Using the assumption of unknown but equal variances, Equation (6.12) yields:

$$s_p^2 = \frac{9 \times (10)^2 + 15 \times (15)^2}{9 + 15} = 178.125$$

Hence, a 90% confidence interval of the mean weight gain from Equation (6.13) is:

$$= \left(240 - 205 - 1.711 \times 13.346 \sqrt{\frac{1}{10} + \frac{1}{16}}, \right.$$

$$\left. 240 - 205 + 1.711 \times 13.346 \sqrt{\frac{1}{10} + \frac{1}{16}} \right)$$

$$= (35 - 9.2, 35 + 9.2) = (25.8, 44.2)$$

Exercises

17. For the following data, find the 90% and 95% confidence intervals for the difference of means when the variances are known. Assume normal distribution and independent samples.

	$\bar{X}$	$\bar{Y}$	σ_1^2	σ_2^2	n	m
(a)	110	120	16	36	10	15
(b)	2.4	3.5	.5	.7	30	10
(c)	10	5	2	1	15	25

18. The caloric intakes of farm workers and blue-collar workers have normal distributions with common unknown variance σ^2 and means μ_1 and μ_2, respectively. Given sample averages and variances obtained from samples of size 25 and 15 from farm workers and blue-collar workers, respectively, find a 95% confidence interval for the mean difference of caloric intake:

$$\bar{X} = 2800 \qquad s_x^2 = 250 \qquad n = 25$$
$$\bar{Y} = 2600 \qquad s_y^2 = 200 \qquad m = 15$$

19. The nicotine content in samples obtained from filter pads was estimated by two laboratories A and B. Assume that they are normally distributed with a common unknown variance. Find the 95% confidence interval for the difference of means of the two laboratories:

| Laboratory A: | .101, | .192, | .373, | .663, | .692, | .688 | |
| Laboratory B: | .157, | .193, | .366, | .660, | .688, | .671, | .701 |

20. Blood glucose levels of two groups, control and experimental, are given below. Find the 95% and 90% confidence intervals for the difference of their means. Assume normality and a common unknown variance:

| Control group: | 2.3, | 3.4, | 2.1, | 2.5, | 2.7, | 3.3 |
| Experimental group: | 3.3, | 4.2, | 2.9, | 3.6, | 3.1, | 4.0 |

21. The protein content in two varieties A and B of wheat is given for samples of size 7 and 9. They are normally distributed with known variances of 9.0 and 11.0, respectively. Find a 90% confidence interval for the mean difference:

| A: | 8.25, 9.23, 8.01, 10.95, 11.39, 9.95, 3.85 |
| B: | 11.47, 9.70, 9.64, 12.55, 10.57, 7.25, 7.85, 9.41, 8.70 |

Confidence Intervals for the Difference of Two Proportions

Let X_1 be the number of successes in n trials with p_1 as the probability of success:

$$\hat{p}_1 = \frac{X_1}{n}$$

Similarly, let X_2 be the number of successes in m trials with p_2 as the probability of success:

$$\hat{p}_2 = \frac{X_2}{m}$$

We assume as before that n and m are large so that $\hat{p}_1$ and $\hat{p}_2$ are normally distributed. Also $\hat{p}_1 - \hat{p}_2$ is normally distributed with mean $p_1 - p_2$ and variance

$$\frac{p_1(1-p_1)}{n} + \frac{p_2(1-p_2)}{m}$$

The estimate of this variance is,

$$D_1 = \frac{\hat{p}_1(1-\hat{p}_1)}{n} + \frac{\hat{p}_2(1-\hat{p}_2)}{m}$$

Another estimate of the variance can be obtained by the pooled estimate of the common proportion p,

$$\hat{p} = \frac{X_1 + X_2}{n + m} = \frac{n\hat{p}_1 + m\hat{p}_2}{n + m}$$

given by $D_2 = \hat{p}(1 - \hat{p})(1/n + 1/m)$.

The confidence interval for $p_1 - p_2$ at the $(1 - \alpha)$ level is:

$$(\hat{p}_1 - \hat{p}_2 - z_{1-\alpha/2}\sqrt{D_2}, \hat{p}_1 - \hat{p}_2 + z_{1-\alpha/2}\sqrt{D_2}) \qquad (6.14)$$

Sometimes D_1 is used instead of D_2. There are situations when D_1 gives better results than D_2 (see Eberhardt and Fligner [1977]).

Note: When the assumption of normality is not valid, confidence intervals for the mean of a population can still be obtained if the sample size is large. An important result in probability theory, known as the Central Limit Theorem, provides the approximate distribution of the average of a random sample from *any* population. Roughly speaking the average of a large sample, suitably standardized, has the standard normal distribution. This fact enables us to utilize normal theory-confidence intervals even when normal assumptions do not hold. Procedures for nonnormal populations in general are discussed under nonparametric inference in Chapter 8.

Example 6.19: The proportion of stocks that went up on a given day on the New York Stock Exchange, based on a sample of 100 stocks, was .6. The proportion of stocks that went up on the American Stock Exchange was .5, based on a sample of 80 stocks. Find a 90% confidence interval for the difference in proportions.

Here
$$\hat{p}_1 = .6 \qquad n = 100$$
$$\hat{p}_2 = .5 \qquad m = 80$$
$$\hat{p} = \frac{100 \times .6 + 80 \times .5}{100 + 80} = .556$$

Hence, the 90% confidence interval for $p_1 - p_2$ from Equation (6.14) is:

$$= \left[.6 - .5 - 1.645 \times \sqrt{.556 \times .444 \times \left(\frac{1}{100} + \frac{1}{80}\right)}, \right.$$

$$\left. .6 - .5 + 1.645 \sqrt{.556 \times .444 \times \left(\frac{1}{100} + \frac{1}{80}\right)} \right]$$

$$= [.1 - .12, .1 + .12] = [-.02, .22]$$

Exercises

22. The proportion of grass seeds germinating is obtained as .8 in a sample of $n = 150$ on variety A. However, for blue grass, variety B has a rate of .90 for germination based on a sample of 200. Find the 95% confidence interval for the difference between proportions of varieties A and B.

23. The following table gives the number cured by drugs A and B for the same disease in a clinical trial:

drug	n	number cured
A	150	17
B	250	20

Find the 98% confidence intervals for the difference between the cure rates of drugs A and B.

24. Two methods, conventional (C) and experimental (E), were used in teaching mathematics at the elementary level. The following table gives the number of students who passed a common standard test. Find the 96% confidence intervals for the difference of proportions of the two methods.

method	n	number passed
C	240	220
E	300	250

Chapter Exercises

1. In an opinion poll, President Reagan's economic policy was approved by 45 out of 150 persons polled, whereas the rest disapproved. What is the estimate of the proportion who disapproved? Find the 95% confidence interval for the proportions of those who approved as well as of those who disapproved of Reagan's policy.

2. Suppose it is necessary to estimate the proportions in Exercise 1 with 2% of the true proportion. How big a sample should be taken if it is known that the proportion of approvals is close to .4? What is the sample size if we do not assume anything about the proportion?

3. A drug to reduce blood pressure was clinically tried on 15 patients. Assuming that the changes are normally distributed, find the 95% confidence interval for the change:

$$-7, \; -1, \; 0, \; 5, \; -2, \; -3, \; -10, \; 5$$
$$15, \; -10, \; -5, \; 10, \; -8, \; -9, \; -11$$

4. Two machines were tested for the number of defective items produced. The data are:

Machine	Number Examined	Defectives
I	25	3
II	75	8

Find the 90% confidence interval for the difference between proportion defectives. Does the interval contain zero? What does it signify if it does?

5. The nicotine content in cigarettes of two brands A and B are given below (in milligrams). Find a 95% confidence interval for the difference between their means? If both samples come from the same population, what is the 90% confidence interval of the variance, assuming normality:

Brand A: 27, 25, 22, 28, 19, 17, 15, 27
Brand B: 19, 23, 18, 17, 15, 13, 10, 15

6. The data on scores in the final examinations of a statistics course at The Ohio State University given at 8:00 A.M. and 12:00 noon taught by the same instructor are given below. Find the 90% confidence interval for the difference between the mean scores:

8: A.M. class	$\bar{X} = 67$	$n = 150$	$s_x^2 = 90$
12:00 noon class	$\bar{Y} = 83$	$m = 215$	$s_y^2 = 75$

Summary

Parametric inference is concerned with *point estimation, interval estimation,* and *hypotheses testing* about the parameters in the probability model. Point estimates are obtained by the *maximum likelihood method* or the *method of moments,* among many others. The maximum likelihood estimate maximizes the likelihood that is obtained by the product of densities evaluated at sample values. The method of moments requires equating *sample moments* to population moments. An estimate is *unbiased* if the mean of the distribution of the estimate is the value to be estimated. An estimate is more *efficient* than the other if its variance is smaller. *Interval estimates* are random intervals that contain the unknown parameter with preassigned probability. This preassigned probability is known as the *confidence coefficient,* and the intervals are known as *confidence intervals.* Interval estimates for the mean and variance of the normal distribution are given; so are the confidence intervals for the *difference of means* of two normal populations. Confidence intervals for the *difference of two proportions* are also given for large samples.

When an estimate of the mean is desired to be within *d* units of the true mean and the population has a known variance, the *sample size* can be determined; it can also be determined in the case of estimating a proportion.

References

Beyer, William H. (Editor). *CRC Handbook of Tables for Probability and Statistics*, Cleveland, Ohio: Chemical Rubber Co., 1966.

Eberhardt, Keith, R., and Fligner, Michael A. A comparison of two tests for equality of proportions, *Am. Statist.*, 1977; 151–155.

Efron, Bradley. *The Jackknife, the Bootstrap and other resampling plans.* Philadelphia: Society of Industrial and Applied Mathematics, 1982.

Hogg, R. V., and Craig, A. T. *Introduction to Mathematical Statistics.* New York: Macmillan, 1970.

chapter seven

Statistical Inference— Hypothesis Testing

Frequently an experimenter wants to test conjectures or claims about the parameters of a model. For example, he may want to decide whether or not a group of persons under study has normal blood pressure, or whether the weights of pigs on a new diet are higher than those on a regular diet. There are problems of practical importance in testing whether the proportion of defective items produced by a machine is less than 5 percent or whether a new instrument that measures eye pressure makes more precise determinations than the conventional one. In scientific terminology, such problems belong to the area of *hypothesis testing*. Various methods of testing hypotheses are discussed here for parameters. Nonparametric procedures for testing hypotheses are discussed in Chapter 8.

The scientific questions proposed by the experimenter can be translated in terms of claims about the parameters of the proposed probability model. The appropriateness and tenability of the hypotheses to be tested should be carefully studied before any further statistical solution of the problem is proposed. The hypothesis tested about the parameter is generally called a *null hypothesis*, denoted here by H, and the *alternative hypothesis*, denoted here by A. The hypothesis is called *simple* if it specifies the probability model completely; otherwise, it is called *composite*. When we test the hypothesis that the mean of a normal distribution is 5 while the variance is already known, we have a simple hypothesis: $H: \mu = 5$. So under the null hypothesis, the distribution is completely known. The alternative hypothesis may be any value of the mean larger than 5 and it is composite:

$$A: \mu > 5$$

The statistician proposes a *decision rule* to test a given hypothesis. Usually the rule is given in terms of a *test statistic* that gives a region of values for which the null hypothesis should be rejected. This region is called a *rejection region* or *critical region* of the test statistic.

The decision maker commits errors in applying the decision rule, since he may reject a true hypothesis or accept a false hypothesis. These errors are called *type I* and *type II errors*, and they play an important role in developing decision rules. Table 7.1 gives these errors:

Table 7.1
Errors in Testing Hypotheses

Decision True Hypothesis	Reject *H*	Accept *H*
H true	type I error	no error
H not true	no error	type II error

The *significance level* of a test is the probability of a type I error. That is, the probability of rejecting the hypothesis when it is true is called the significance level of the test. The probability of a type II error is called the *operating characteristic* of the test. The probability of not committing a type II error is called the *power* of the test. When the alternative is a composite, the power has several values and is known as *power function*. The probabilities of type I and type II errors are designated by α and β.

$$\alpha = P\{\text{type I}\} = P\{\text{Rejecting } H \text{ when } H \text{ is true}\}$$
$$\beta = P\{\text{type II}\} = P\{\text{Accepting } H \text{ when } H \text{ is false}\} \qquad (7.1)$$

A type I error is also referred to as α-error and a type II as β-error.

Suppose we want to test the hypotheses about the mean of the normal distribution

$$H: \mu = 5$$
$$A: \mu = 10$$

and let H be rejected when $\bar{X} > 7$. The probability of type I and II errors are shown in Figure 7.1.

Example 7.1: The ball bearings produced by a machine are inspected and their diameter is measured. It is assumed that they are normally distributed with $\sigma = .003$. The production of the machine is accepted if the sample mean, based on nine measurements of randomly selected ball bearings, is less than 1.002 cm; otherwise, it is rejected. The hypothesis tested here is:

$$H: \mu = 1$$
$$A: \mu = 1.005$$

Decision rule: Reject H when $\bar{X} > 1.002$.

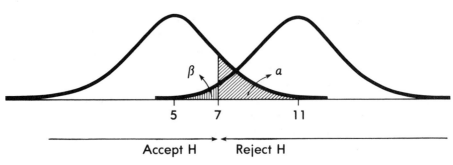

Figure 7.1 Errors of type I and II

Since $\bar{X}$ is normally distributed with mean 1 and standard deviation $\frac{1}{3}(.003)$ under the null hypothesis, we can find α as follows:

$$\alpha = P\{\bar{X} > 1.002 | \mu = 1\}$$

$$= P\left\{\frac{3(\bar{X}-1)}{.003} > \frac{3(1.002-1)}{.003}\right\}$$

$$= P\{Z > 2\} = .0228$$

Similarly, under the alternative hypothesis, $\mu = 1.005$, so that:

$$\beta = P\{\bar{X} < 1.002 | \mu = 1.005\}$$

$$= P\left\{\frac{3(\bar{X}-1)}{.003} < \frac{3(1.002-1.005)}{.003}\right\}$$

$$= P\{Z < -3\} = .0013$$

At $\mu = 1.005$, Power $= 1 - \beta = .9987$.

One of the most important problems in statistical theory is to find the most powerful tests. The theory of testing hypotheses, especially the concepts of *power* and of *most powerful tests*, was developed in 1920s by J. Neyman and E. S. Pearson, and their theory is used extensively in practice.

Usually, for a reasonable test, one would like to have small probabilities for both types of errors. However, given a fixed sample, these probabilities cannot be simultaneously controlled. Neyman and Pearson theory of testing hypotheses makes a compromise. We choose that test procedure among a class of tests having the same significance level that is most powerful. This can be done by using a ratio of likelihoods under the null and the alternative hypotheses.

The Neyman-Pearson Criterion: Given a random sample $X_1, X_2, \ldots, X_n$ from $f(x, \theta)$, let $H: \theta = \theta_0$ versus $A: \theta = \theta_1$. Among all tests of the same significance level, the test for which the critical region is given by the likelihood ratio, $L(\theta_0)/L(\theta_1) < k$, is most powerful.

Example 7.2: A test is required for the parameter of the exponential distribution with probability density function:

$$f(x,\theta) = \begin{cases} \theta e^{-\theta x}, & x > 0, \\ 0, & \text{otherwise} \end{cases}$$

Let $H:\theta = \theta_0$, $A:\theta = \theta_1$ and $\theta_0 > \theta_1$. Now the likelihood ratio is:

$$\frac{L(\theta_0)}{L(\theta_1)} = \frac{\theta_0^n e^{-\theta_0 \Sigma X_i}}{\theta_1^n e^{-\theta_1 \Sigma X_i}}$$

So $L(\theta_0)/L(\theta_1) < k$ means that

$$\left(\frac{\theta_0}{\theta_1}\right)^n e^{-(\theta_0 - \theta_1)\Sigma X_i} < k$$

Taking logarithms of both sides, we find that the criterion for rejecting H is:

$$\Sigma X_i > k_1$$

or

$$\bar{X} > k_2$$

To determine k_2, we need to know the probability distribution of $\bar{X}$ when the null hypothesis is true. For most of the problems in this book, we know the distribution, so in testing hypotheses, we shall only give the *decision rule* in terms of a test statistic and its distribution under the null hypothesis. Most often, the likelihood ratio test reduces to statistics that are intuitively suggested for testing the hypothesis anyway.

Exercises

1. A random sample of 25 observations from a normal population with mean μ and variance 16 is given. Test the following hypothesis: $H:\mu = 10$ vs. $A:\mu = 12$. Find α and β for the decision rules:

   (a) Reject H when $\bar{X} > 10.5$
   (b) Reject H when $\bar{X} > 11$
   (c) Reject H when $\bar{X} > 11.2$
   (d) Reject H when $\bar{X} > 11.5$

2. Find α in Exercise 1 if the variance is estimated from the sample and is found to be $s^2 = 9$.

3. Are the following hypotheses simple or composite? The sample is from a normal population with a variance of 16.

   (a) $H:\mu = 5$, $A:\mu > 5$
   (b) $H:\mu = 10$, $A:\mu < 10$

(c) $H: \mu \leq 10, \ A: \mu > 10$
(d) $H: \mu = 10, \ A: \mu \neq 10$

4. An inspection rule for the diameter of a ball bearing requires that the items produced be rejected if a sample of five gives an average diameter of 3.5 cm or more. The hypothesis being tested is $H: \mu = 3.4, \ A: \mu > 3.4$. Assume that the diameters of ball bearings are normally distributed with a variance of .01. Give five points on the operating characteristic curve. That is, calculate β for $\mu = 3.41, 3.42, 3.44, 3.46$ and 4.0.

5. The life of electron tubes has an exponential distribution with mean $1/\theta$. Test the following hypotheses, using the likelihood ratio test and give a decision rule in general:

$$H: \theta = .001$$
$$A: \theta = .002$$

Testing Hypotheses for the Mean of a Normal Distribution

Many applications use data from normal distributions. We consider tests of hypothesis if a single sample from a normal distribution with mean μ and variance σ^2 is given. We shall consider tests for the mean first and then for the variance. When two populations are compared, the theory of testing hypotheses is similar and will be discussed in a later section. Cases for known σ^2 and unknown σ^2 will be considered separately when testing the mean.

Test for the Mean of a Normal Distribution

Case I: σ^2 known

Suppose we want to test a simple null hypothesis, $H: \mu = \mu_0$, against a simple alternative, $A: \mu = \mu_1, \ \mu_0 < \mu_1$. Let the significance level be α.
Decision rule: Reject H if $\bar{X} > k$.
Determination of k:

$$\alpha = P(\bar{X} > k | \mu = \mu_0)$$

$$= P\left\{ \frac{\sqrt{n}(\bar{X} - \mu_0)}{\sigma} > \frac{\sqrt{n}(k - \mu_0)}{\sigma} \right\}$$

$$= P\left\{ Z > \frac{\sqrt{n}(k - \mu_0)}{\sigma} \right\}$$

$$= P\{Z > z_{1-\alpha}\}$$

Hence, $\sqrt{n}(k - \mu_0)/\sigma = z_\alpha$, giving:

$$k = \mu_0 + z_\alpha \frac{\sigma}{\sqrt{n}} \qquad (7.2)$$

The power of the test $= P\{\bar{X} > k | \mu = \mu_1\}$

$$= P\left\{ \frac{\sqrt{n}(\bar{X} - \mu_1)}{\sigma} > \frac{\sqrt{n}(k - \mu_1)}{\sigma} \right\}$$

$$= P\left\{ Z > \frac{\sqrt{n}(k - \mu_1)}{\sigma} \right\} \qquad (7.3)$$

Note that the above decision rule is also valid for testing $H: \mu = \mu_0$ versus $A: \mu > \mu_0$. The power will have many values in this case, since the alternative is composite.

For the two-sided alternative, let:

$$H: \mu = \mu_0$$
$$A: \mu \neq \mu_0$$

Decision rule: Reject H if $\bar{X} > k_2$ or $\bar{X} < k_1$.

Determination of k_1, k_2:

$$\alpha = P\{\bar{X} < k_1 \text{ or } \bar{X} > k_2 | \mu = \mu_0\}$$

$$= P\left\{ \frac{\sqrt{n}(\bar{X} - \mu_0)}{\sigma} < \frac{\sqrt{n}(k_1 - \mu_0)}{\sigma} \right.$$

$$\text{or} \qquad \left. \frac{\sqrt{n}(\bar{X} - \mu_0)}{\sigma} > \frac{\sqrt{n}(k_2 - \mu_0)}{\sigma} \right\}$$

$$= P\{Z < -z_{\alpha/2} \text{ or } Z > z_{\alpha/2}\}$$

So that:

$$\frac{\sqrt{n}(k_1 - \mu_0)}{\sigma} = -z_{\alpha/2}$$

$$k_1 = \mu_0 - z_{\alpha/2} \frac{\sigma}{\sqrt{n}} \qquad (7.4)$$

Similarly,

$$k_2 = k_0 + z_{\alpha/2} \frac{\sigma}{\sqrt{n}} \qquad (7.5)$$

The power function can be calculated by using k_1 and k_2 from Equations (7.4) and (7.5).

Example 7.3: Suppose a random sample of size 9 is given from a normal distribution with a variance of 1, and we test the hypothesis at $\alpha = .1$:

$$H: \mu = 5$$
$$A: \mu \neq 5$$

Decision rule: Reject H when $\bar{X} < k_1$ or $\bar{X} > k_2$

where $k_1 = 5 - 1.645 \times \dfrac{1}{3} = 4.452$

and $k_2 = 5 + 1.645 \times \dfrac{1}{3} = 5.548$.

That is, we accept H when:

$$4.452 < \bar{X} < 5.548$$

The β error depends on the value of μ for the alternative hypothesis. Since the hypothesis is composite, we denote it by $\beta(\mu)$. Then:

$$\beta(\mu) = P\{4.452 < \bar{X} < 5.548 | \mu\}$$
$$= P\{3(4.452 - \mu) < Z < 3(5.548 - \mu)\}$$
$$= \Phi[3(5.548 - \mu)] - \Phi[3(4.452 - \mu)]$$

A few values are tabulated below for $\beta(\mu)$:

μ	Operating Characteristic $\beta(\mu)$	Power $1 - \beta(\mu)$
4	.0869	.9131
4.5	.5549	.4451
4.8	.8383	.1617
5	.9	.10
5.1	.8837	.1163
6.0	.0869	.9131
6.5	.0021	.9979

Notice that the operating characteristic decreases as μ moves away from the null hypothesis. The power, that is the complement of the operating characteristic, increases when μ moves away from the hypothesized value.

Case II: σ^2 unknown

When the variance is unknown, the test statistic uses s in place of $\sigma \cdot A$ different distribution of the test statistic is required to determine k. We use the distribution of $\sqrt{n}(\bar{X} - \mu)/s$, which now is a t-distribution with $n - 1$ degrees of freedom. For example:

$$H: \mu = \mu_0$$
$$A: \mu = \mu_1, \quad \mu_1 > \mu_0$$

Decision rule: Reject H if $\bar{X} > k$ where

$$k = \mu_0 + t_{n-1,\alpha} \frac{s}{\sqrt{n}} \qquad (7.6)$$

The probability of accepting the hypothesis when it is not true requires the distribution of a noncentral t. Because this is beyond the scope of the book, we do not give the β error in this case. The computations for two-sided alternative are similar.

Example 7.4: Assume that serum cholesterol levels are normally distributed with a mean of 250 for healthy subjects. A new drug is claimed to reduce cholesterol levels, and a random sample of nine subjects who were administered this drug gave $\bar{X} = 245$ and $s^2 = 100$. Test the claim that the drug reduces cholesterol levels at $\alpha = .05$. Test:

$$H: \mu = 250$$
$$A: \mu < 250$$

Decision rule: Reject H if $\bar{X} < k$ where

$$k = 250 - t_{8,.05} \tfrac{10}{3} = 243.8$$

We do not reject H, since $245 > 243.8$.

Test of the Variance of Normal Distribution

The precision of the determinations of a quantity made with the help of an instrument is reflected by the variance. A smaller variance signifies more precision, and we are often interested in testing the hypothesis about the variance of a normal distribution. Test the hypothesis:

$$H: \sigma^2 = \sigma_0^2$$
$$A: \sigma^2 > \sigma_0^2$$

Decision rule: Reject H if $s^2 > k$. Since $(n-1)s^2/\sigma^2$ has χ^2_{n-1} we can determine k as follows:

$$\alpha = P\left\{ \frac{(n-1)s^2}{\sigma^2} > \frac{(n-1)k}{\sigma^2} \middle| \sigma^2 = \sigma_0^2 \right\}$$

$$= P\left\{ \chi^2_{n-1} > \frac{(n-1)k}{\sigma_0^2} \right\}$$

$$= P\{\chi^2_{n-1} > \chi^2_{n-1,1-\alpha}\}$$

So that
$$\frac{(n-1)k}{\sigma_0^2} = \chi_{n-1,1-\alpha}^2,$$

giving
$$k = \frac{\sigma_0^2}{n-1} \chi_{n-1,1-\alpha}^2 \qquad (7.7)$$

The power of the test against any specific alternative, say $\sigma^2 = \sigma_1^2$, can be obtained by using the chi-square distribution:

$$\beta = P\{s^2 < k | \sigma^2 = \sigma_1^2\}$$

$$= P\left\{\chi_{n-1}^2 < \frac{(n-1)k}{\sigma_1^2}\right\}$$

Example 7.5: Scores on a standardized test have a normal distribution with a variance of 100. It is claimed now that the variance of scores has increased. Given a sample of 30 scores with $s^2 = 110$, testing the hypothesis at level .05:

$$H: \sigma^2 = 100$$
$$A: \sigma^2 > 100$$

Decision rule: Reject H if $s^2 > k$:

$$k = \chi_{29,.95}^2 \frac{100}{29} = \frac{42.6 \times 100}{29} = 146.9$$

The hypothesis H is not rejected.

To find β for an alternative, say, $\sigma^2 = 266$, we have:

$$\beta = P\{s^2 < 146.9 | \sigma^2 = 266\}$$

$$= P\left\{\chi_{29}^2 < \frac{146.9 \times 29}{266}\right\}$$

$$= P\{\chi_{29}^2 < 16.0\} = .025$$

Testing Hypotheses About a Proportion

As in finding confidence intervals for the proportion, we assume here that the sample size is large so that the normal approximation can be used for the distribution of the estimate. We test here:

$$H: p = p_0$$
$$A: p = p_1, p_1 > p_0$$

Let $\hat{p}$ be the estimate of p. We have seen that $\hat{p}$ is approximately normally distributed with mean p and variance $p(1-p)/n$. We can use the same test statistic that we used for the normal mean with a known variance.

Decision rule: Reject H if $\hat{p} > k$.

$$k = p_0 + z_\alpha \sqrt{\frac{p_0(1 - p_0)}{n}} \qquad (7.8)$$

Calculations for β are similar.

Example 7.6: In a hospital, 85% beds are usually occupied. Test the claim that the occupancy rate has decreased to 75%. A sample of 150 beds is available and $\alpha = .05$:

$$H: p = .85$$
$$A: p = .75$$

Decision rule: Reject H if $\hat{p} < k$:

$$k = .85 - 1.645 \sqrt{\frac{.85 \times .15}{150}} = .80$$

$$\beta = P\{\hat{p} < .80 | p = .75\}$$

$$= P\left\{ Z < \frac{.80 - .75}{\sqrt{\dfrac{.75 \times .25}{150}}} \right\} = P(Z < 1.41)$$

$$= .9207$$

Confidence Intervals and Testing Hypotheses

There is almost no difference in obtaining $(1 - \alpha)$ level confidence interval for the mean and testing a hypothesis about the mean against two-sided alternatives at significance level α. The confidence interval acts as an acceptance region of the null hypothesis. If the hypothesized value is in the confidence interval, we accept the null hypothesis; otherwise, we reject it. The identification of one-sided alternative hypotheses also could be made if we assume that there is a one-sided confidence interval. Consider the following example.

Example 7.7: Find the 90% confidence interval for the mean of weights of a week-old chicken, assuming a normal distribution with a known variance of 9 (grams)^2. The sample of 16 chickens gave an average weight of 18 grams. The confidence interval is:

$$\left(18 - 1.645 \sqrt{\frac{9}{16}}, 18 + 1.645 \sqrt{\frac{9}{16}} \right)$$

$$= (16.8, 19.2)$$

Supposing a two-sided alternative is tested for the hypothesis

$$H:\mu = 20$$
$$A:\mu \neq 20$$

at $\alpha = .1$, we have the rejection region given by $\bar{X} > k_2$ or $\bar{X} < k_1$, where

$$k_2 = 20 + 1.645 \sqrt{\frac{9}{16}} = 21.2$$

$$k_1 = 20 - 1.645 \sqrt{\frac{9}{16}} = 18.8$$

We reject the hypothesis, since the sample average, 18, is less than 18.8.

Using the confidence interval, we find that the hypothesized value of the mean, $\mu = 20$, is not in the confidence interval. Thus we reject it.

Arguments such as the above can be repeated for all the situations of testing hypotheses and confidence intervals discussed so far. Essentially, there is no difference between the two approaches.

Exercises

6. Suppose p is the probability of having cancer in a given population. A random sample of 1,000 persons is available. Test the hypothesis at $\alpha = .05$, that:

$$H:p = .005 \text{ vs. } A:p = .01$$

7. In Example 7.3, test the hypothesis at level $\alpha = .05$, $H:\mu = 5$ vs. $A:\mu > 5$. Obtain the power function of the test at four values of μ.

8. Serum triglycerides are assumed to have a normal distribution. A sample of 12 subjects gave the values:

$$207, \ 89, \ 125, \ 87, \ 140, \ 81,$$
$$103, \ 151, \ 74, \ 191, \ 88, \ 110$$

(a) Test the hypothesis that the mean is 120 versus the alternative that it is greater than 120 at $\alpha = .05$.

(b) At level .05, test the hypothesis that the variance is 100 versus that it is less than 100.

9. The yield of strawberries is a normal random variable. A farmer wants to test the hypothesis at level .01 that the mean is 200 grams versus that it is smaller. Given a sample of 15 plants with $\bar{X} = 190$, and $s^2 = 64$, what is the decision?

10. The police chief of a midwestern city claims that the crime rate has decreased in his city from that of .008 in previous years. The present crime rate based on a sample of 1,000 is .005. Test his claim at level .05.

11. The price of regular gasoline is normally distributed. A sample of eight gas stations gave the following prices per gallon:

$$\$1.27, \$1.19, \$1.23, \$1.85, \$1.25, \$1.28, \$1.13, \$1.75$$

Find a 95% confidence interval of the mean and use it to test the following hypothesis at $\alpha = .05$:

$$H:\mu = 1.30 \text{ vs. } A:\mu \neq 1.30$$

12. Suppose the price of a stock is distributed normally and we are given seven observations as follows: $7.25, \$8.50, \$9.50, \$7.875, \$8.75, \$6.50, \$7.50$. Test the hypothesis that the mean price of the stock is $8.00 at $\alpha = .05$. What is your alternative? Why?

Testing Hypotheses of the Equality of Two Means

In this section, we assume that we have samples from two different normal populations and we are interested in testing hypotheses whether their means are equal or unequal. We have already considered the confidence intervals for the difference of the means of two normal distributions. The procedures for testing hypotheses are similar and the statistics involved are also the same.

We want to study two normal populations with means μ_1 and μ_2 and variances σ_1^2 and σ_2^2. As before, we have samples of size n and m with sample averages $\bar{X}$ and $\bar{Y}$ and sample variances s_x^2 and s_y^2, respectively. At significance level α, consider the test:

$$H:\mu_1 = \mu_2$$
$$A:\mu_1 > \mu_2$$

We give results for this alternative, but the results are similar if we consider other alternatives, such as $\mu_1 < \mu_2$ or $\mu_1 \neq \mu_2$. Notice that all of these hypotheses are composite. Three cases are considered.

Case I: σ_1^2, σ_2^2 known

Decision rule: Reject H if $\bar{X} - \bar{Y} > k$ (see Equation [6.11]) where

$$k = z_\alpha \sqrt{\frac{\sigma_1^2}{n} + \frac{\sigma_2^2}{m}}$$

Case II: $\sigma_1^2 = \sigma_2^2 = \sigma^2$ unknown

We use the pooled variance as an estimate of σ^2, as in Equation (6.12); hence, our test is given by the following decision rule.

Decision rule: Reject H if $\bar{X} - \bar{Y} > k$ where

$$k = s_p t_{m+n-2,\alpha} \sqrt{\frac{1}{m} + \frac{1}{n}}$$

The adjustment to cut off points z and t can be made if the alternative is two-sided; see the confidence intervals in Equations (6.11) and (6.13).

Case III: σ_1^2, σ_2^2 both unequal and unknown

The test of this hypothesis is involved. The problem is known as the Behrens–Fisher problem, and it has been an object of intensive study in statistics. Here we give an approximate solution under the assumption that m, n are large.

Decision rule: Reject H if $\bar{X} - \bar{Y} > k$ where

$$k = z_\alpha \sqrt{\frac{s_x^2}{n} + \frac{s_y^2}{m}}$$

Example 7.8: Two brands of tires (A and B) are being used by a taxi company. The company wants to buy Brand A tires if they give more wear. Suppose the life of tires is normally distributed with variances of 3 and 5, respectively. Samples of sizes 15 and 25 were obtained. Test the hypothesis at $\alpha = .05$:

$$H: \mu_1 = \mu_2$$
$$A: \mu_1 > \mu_2$$

Decision rule: Reject H if $\bar{X} - \bar{Y} > k$ where

$$k = 1.645 \sqrt{\frac{3}{15} + \frac{5}{25}} = 1.04$$

Hence, if $\bar{X} - \bar{Y} > 1.04$, Brand A is preferable.

Example 7.9: Suppose in Example 7.8 the variances were estimated from the samples as $s_x^2 = 4.286$ and $s_y^2 = 4.5$ and the hypothesis of the equality of the mean is tested, assuming that the unknown variances are equal. The pooled estimate of σ^2 is:

$$s_p^2 = \frac{(14 \times 4.286) + (24 \times 4.5)}{14 + 28}$$
$$= 4$$

Decision rule: Reject H if $\bar{X} - \bar{Y} > k$ where

$$k = 1.687 \times 2 \sqrt{\frac{1}{15} + \frac{1}{25}}$$
$$= 1.102$$

Hence, Brand A is preferable if $\bar{X} - \bar{Y} > 1.102$.

Example 7.10: Two methods of teaching a course in statistics are used. Method *A* uses the traditional lecture method, while in Method *B* the students are exposed to actual experimentation in class. The scores on similar examinations based on samples of sizes 35 and 55 give:

$$\bar{X} = 65 \qquad s_x^2 = 70$$
$$\bar{Y} = 70 \qquad s_y^2 = 110$$

Test at level .05 the equality of the mean scores versus that Method *B* leads to higher scores than Method *A*. Assume that the scores are normally distributed with unknown and unequal variances. We use the approximate test as in Case III.

Decision rule: Reject *H* if $\bar{Y} - \bar{X} > k$ where

$$k = 1.645 \times \sqrt{\frac{70}{35} + \frac{110}{55}}$$

$$= 3.29$$

Since the difference of the sample averages, $70 - 65 = 5$, is greater than 3.29, we reject the null hypothesis of the equality of the mean scores.

Paired Comparisons

When measurements are made on the same subject at different times or the observations are made in pairs, we can reduce the two population problems to that of one population. For example, to measure the effect of a drug on the blood pressure of a subject, we measure blood pressure before and after the administration of the drug. The increase or decrease in blood pressure is the quantity of interest. This method removes the dependence between two observations taken on the same individual. That is, if *X* and *Y* are measured on the same individual and have normal distributions with the same mean, then $X - Y$ is also normally distributed with a mean of zero. We test the hypothesis of the equality of means by testing that the mean of the population of difference is zero. We assume that the variance is unknown. This is given by the *t*-test used for one population earlier in this chapter.

Example 7.11: The pulse rate per minute of ten subjects was measured before and after exercise. Test at level .10 significance if the exercise increases pulse rate.

Before (X)	71	74	75	80	85	72	73	81	79	76
After (Y)	81	82	80	82	75	81	75	78	89	78

Let the mean of $Z = Y - X$ be μ and Var σ_Z^2; the variance is unknown:

$$H: \mu_Z = 0$$
$$A: \mu_Z > 0$$

Decision rule: Reject if $\bar{Z} > k$ where

$$k = t_{9,.90}\sqrt{10}$$

The paired differences are:

$$10, \ 8, \ 5, \ 2, \ -10, \ 9, \ 2, \ -3, \ 10, \ 2$$
$$\bar{Z} = 3.5 \qquad s_z^2 = 40.94$$

So that $k = 1.383 \times \sqrt{\dfrac{40.94}{10}} = 2.8$. Since $\bar{Z}$ is larger, we reject the hypothesis that the difference is zero.

Exercises

13. Using the same data as in Example 7.8, give a test of the following hypotheses at level .05:

    (a) $H: \mu_1 = \mu_2$
       $A: \mu_1 \neq \mu_2$
    (b) $H: \mu_1 = \mu_2$
       $A: \mu_1 < \mu_2$

14. In studying the effect of reserpine on pituitary adenine nucleotide content, the following two groups of observations on ATP (μ moles/grain net weight) were made:

Control Group		Resperpine Group	
2.33	2.78	1.90	2.05
2.52	2.16	1.85	2.10
2.41	1.90	2.00	1.95
2.23	3.30	1.70	1.80

    Assuming normality but an unknown common variance, determine whether reserpine has an effect at $\alpha = .05$.

15. The protein content of two varieties of wheat (A and B) is given below. Assume that it is normally distributed with unequal and unknown variances. Test the hypothesis that the mean protein content of the two varieties is the same at level .05. The sample data given:

Variety A	$\bar{X} = 15.4$	$s_x^2 = 5.2$	$n = 35$
Variety B	$\bar{Y} = 17.8$	$s_y^2 = 8.3$	$m = 45$

16. Compare the reduction in blood glucose levels of subjects on two drugs I and II. Assume normality and test the hypothesis that the mean reductions are equal at level .05. Assume that the variances are unknown but equal:

$$\text{Drug I} \quad \bar{X} = 2.5 \quad s_x^2 = 2.3 \quad n = 10$$
$$\text{Drug II} \quad \bar{Y} = 3.5 \quad s_y^2 = 3.1 \quad m = 15$$

17. Samples from the same cigarette filters were sent to two different laboratories to determine nicotine content. Test the hypothesis that there is no laboratory difference on the average. Use a paired t-test and $\alpha = .05$.

Samples

	1	2	3	4	5	6
Laboratory A	.161	.192	.373	.663	.692	.952
Laboratory B	.157	.193	.366	.660	.688	.965

Data abstracted from: Wagner, J. R., and Thaggard, N. A., Gas-liquid chromatographic determination of nicotine contained on Cambridge filter pads, J. Assoc. Off. Anal. Chem., 1979, 62, 229–36.

18. The death rates in England and Wales as compared to those in Sheffield are given for ten years during 1970–79. Test the hypothesis that the death rates are the same, assuming that they are normally distributed. Use a paired t-test:

	England and Wales (per 10,000)	Sheffield (per 10,000)
1970	31.9	35.3
1971	30.8	35.5
1972	30.8	38.9
1973	30.6	30.0
1974	30.0	24.2
1975	28.8	15.6
1976	25.7	23.5
1977	27.1	33.0
1978	26.0	24.6
1979	27.1	36.0

Source: Carpenter, R. G., Scoring to provide risk-related primary health care: Evaluation and updating during use, J.R. Stat. Soc. Series A, 1983; 1–32.

Testing the Equality of Two Variances

Sometimes populations do not differ in their means but have different variances. The mean blood pressure of patients and normal subjects may be the same but have different variances. The precision of instruments is usually obtained from the variance of the quantity measured. One would like to choose an instrument with higher precision. In these problems, we are concerned with tests of the equality of the variances of two populations. We assume that the populations are normal with a variance of σ_1^2 and σ_2^2 and we test at level α:

$$H: \sigma_1^2 = \sigma_2^2$$
$$A: \sigma_1^2 > \sigma_2^2$$

Suppose that we have random sample of $X_1, X_2, \ldots, X_n$ of size n from a normal population with a variance of σ_1^2 and a random sample $Y_1, Y_2, \ldots, Y_m$ of size m from another normal population with a variance of σ_2^2.

Decision rule: Reject H if $F_{n-1, m-1} > k$ where

$$k = F_{n-1, m-1, 1-\alpha}$$

$$F_{n-1, m-1} = \frac{s_x^2}{s_y^2}$$

The ratio of the sample variances can be intuitively justified as the statistic for testing the equality of variances, since essentially we are testing that the ratio of population variances is one.

Example 7.12: The sizes of nails produced by two machines are measured. The machines were set to make 1″ nails. Test at level .05 that their variances are equal. It is given that:

$$s_x^2 = .03 \qquad n = 16$$
$$s_y^2 = .02 \qquad m = 9$$

Decision rule: Reject H if $F_{15,8} > k$ where $k = 3.22$.

In the above case $\dfrac{.03}{.02} = 1.5$, so we do not reject H.

Example 7.13: Two laboratories (A and B) determined nicotine content in certain samples of cigarettes and the data were given as:

$$\text{Laboratory A} \qquad s_x^2 = .2 \qquad n = 10$$
$$\text{Laboratory B} \qquad s_y^2 = .4 \qquad m = 8$$

Test the hypothesis that their variances are the same at level .10. Assume a one-sided alternative: $\sigma_2^2 > \sigma_1^2$.

Decision rule: Reject H if $F_{7,9} > k$ where $k = 3.29$. Since $F_{7,9} = \dfrac{.4}{.2} = 2$, we accept the null hypothesis.

Note that the alternative was chosen so that the bigger variance was in the numerator.

Exercises

19. The protein content of two varieties of wheat is given below. Test the hypothesis of the equality of variances of the two populations at $\alpha = .10$:

> Variety 1: 9.23, 8.01, 10.95, 11.67, 10.41, 9.51
> Variety 2: 9.01, 10.23, 12.67, 10.92, 11.27, 12.35

Choose your own alternative hypothesis.

20. The scores in an examination given to two different sections of mathematics course gave the following data. Test the hypothesis of the equality of variance of the two populations at $\alpha = .05$:

> Section I: 69, 75, 72, 82, 95, 73, 41, 32
> Section II: 15, 92, 82, 13, 40, 55, 73, 84, 35, 97, 25, 27

21. Two watches lose or gain a variable amount of time each day. Test the hypothesis of the equality of variance at level .05:

> Watch 1 (in seconds): -10 -15 -15 -20 -13 -17
> Watch 2 (in seconds):   25   35   20   10   30   15

Testing the Hypotheses of the Equality of Two Proportions

Confidence intervals for the difference of two proportions were discussed in Chapter 6. We use the same argument in testing the hypothesis:

$$H: p_1 = p_2$$
$$A: p_1 \neq p_2$$

We assume that n and m are large. The α level test is given by:

Decision rule: Reject H if $\hat{p}_1 - \hat{p}_2 < k_1$ or $\hat{p}_1 - \hat{p}_2 > k_2$ where

$$k_1 = z_{\alpha/2}\sqrt{D_2} \quad \text{and} \quad k_2 = z_{1-\alpha/2}\sqrt{D_2}$$

Note that D_1 also can be used in place of D_2.

Example 7.14: Consider Example 6.19 and test the hypothesis that p_1, the proportion of stocks that went up on the New York Stock Exchange, is equal to the proportion p_2, the proportion of stocks that went up on the American Stock Exchange. We know that:

$$n = 100, \quad \hat{p}_1 = .6$$
$$m = 80, \quad \hat{p}_2 = .5$$
$$\text{and} \quad \hat{p} = .556. \quad \text{Let} \quad \alpha = .1$$

Hence, $k_1 = -.12$ and $k_2 = .12$. Since $\hat{p}_1 - \hat{p}_2 = .1$, we do not reject the hypothesis that the proportions are equal.

Note that the confidence interval gives the same result, since, under equality, $p_1 - p_2 = 0$. That is, 0 is not included in the acceptance region given by the confidence interval (.03, .17).

Example 7.15: The number of males in favor of legalizing abortion in an opinion poll based on 400 interviews resulted in 125 favoring it, while the number of females favoring it was 150 out of 300 interviewed. Test at level .05 the hypothesis of the equality of the proportions favoring the proposal:

$$\hat{p}_1 = \tfrac{125}{400} = .3125$$

$$\hat{p}_2 = \tfrac{150}{300} = .5$$

$$\hat{p} = \frac{125 + 150}{400 + 300} = \frac{275}{700} = .393$$

$$\sqrt{D_2} = \sqrt{.313 \times .607 \times \left(\frac{1}{400} + \frac{1}{300}\right)} = .037$$

Decision rule: Reject H if $\hat{p}_1 - \hat{p}_2 < -.06$ or $\hat{p}_1 - \hat{p}_2 > .06$. Since $p_1 - p_2 = -.1875$, we reject H. That is, at level .05 male and female proportions differ on matters of abortion.

The test of the equality of two proportions can also be performed by using a chi-square test resulting from the consideration of a 2×2 contingency table. The data in Example 7.15 can be written as:

Sex \ Opinion	Favor	Do not favor	Total
Males	125	275	400
Females	150	150	300
Total	275	425	700

Such tables are also called *contigency tables*. The test of independence in a contigency table is equivalent to the test of the equality of proportions of two populations.

Consider a 2×2 table that gives counts for the presence or absence of two attributes A and B. Let the frequencies be denoted by a, b, c, and d as given in the 2×2, or fourfold, table.

A \ B	Present	Absent	Total
Present	a	b	$a + b$
Absent	c	d	$c + d$
	$a + c$	$b + d$	$a + b + c + d = N$

Let the total be N. Then the statistic

$$\chi_1^2 = \frac{N\left\{|ad-bc|-\dfrac{N}{2}\right\}^2}{(a+b)(c+d)(a+c)(b+d)} \qquad (7.9)$$

has a chi-square distribution with one degree of freedom and is used to test the hypothesis of independence between attributes A and B:

$$H: A \text{ and } B \text{ are independent}$$

$$A: A \text{ and } B \text{ are not independent}$$

Decision rule: Reject H if

$$\chi_1^2 > k$$

where $k = \chi_{1,1-\alpha}^2$ giving a α-level test. Note that since the square of a standard normal random variable has a chi-square distribution with one degree of freedom, this test is similar to the one proposed above.

Example 7.16: Consider Example 7.15.

$$\chi_1^2 = \frac{700[|125 \times 150 - 150 \times 275| - 350]^2}{275 \times 425 \times 300 \times 400}$$

$$= 24.49$$

Since $\chi_{1,.95}^2 = 3.84$, we reject the hypothesis of independence at level .05. This agrees with the solution of Example 7.15.

Example 7.17: The data on the survival of patients with a certain disease that is being treated with two treatments A and B are given in the following table. Test whether the proportions of survivors are the same for A and B.

Treatments	Outcome		Total
	Survival	Death	
A	85	15	100
B	95	5	100
Total	180	20	200

We have:

$$\chi_1^2 = \frac{200\{|85 \times 5 - 15 \times 95| - 100\}^2}{180 \times 20 \times 100 \times 100}$$

$$= 4.5$$

However, since $\chi_{1,.95}^2 = 3.86$, we reject the hypothesis of the equality of proportion or independence of treatment and outcome at level .05 of significance.

Tests of hypothesis of the equality of several proportions and tests of independence in higher order contingency tables will be discussed later. The chi-square test is only valid when N is large. For small sample sizes, the test is based on the hypergeometric distribution and is called *Fisher's exact test.*

Exercises

22. Data are obtained on infants fed by two methods, and the incidence of the malocclusion of their teeth is tested. Test the hypothesis of the equality of the proportions of normal teeth under the two methods at level .05:

	Normal teeth	Malocclusion
Breast-fed	39	141
Bottle fed	12	198

23. Students were surveyed for preference of Course A against Course B in statistics. A student was classified as a biology major or a sociology major. Test the hypothesis of the equality of proportions of preference to Course A by each major at level .05:

	Biology major	Sociology major
Prefer A	30	20
Prefer B	50	30

24. A drug is tested to ascertain its effectiveness in reducing vomiting after surgery. Given the following data on 300 patients, test at level .05 if the drug is effective. That is, test the hypothesis of the equality of proportions of vomiting episodes under the drug and the placebo:

	Vomit	Do not vomit
Drug	130	50
Placebo	110	10

25. In a public-opinion survey, answers to two questions A and B are obtained as follows. Test the hypothesis of independence at level .05:

		Question A Yes	No
	Yes	80	20
Question B	No	40	60

Chapter Exercises

1. The unemployment rate obtained from a random sample of 900 persons in a county is 10.5%. In the previous year the rate was 9.7%. Test at level .01 if there has been a significant increase in the unemployment rate in this county.

2. The incomes of a sample of 40 families gave the sample average of $23,500 and a variance of $(950)^2$. (a) Test the hypothesis that the mean income of the population is $24,000. Choose your own alternative and use level .05. (b) Test if the variance of the incomes is 900^2 against the alternative that it is greater than 900^2 at level .05.

3. To see the effect of insulation in an attic in saving energy, a sample of ten households was taken. The amount of energy used before and after insulation is given below. Assume the significance level to be .01. Is there an effect?

Sample	1	2	3	4	5
Before	541	567	1,011	238	356
After	582	601	910	210	307

Sample	6	7	8	9	10
Before	786	513	2,122	178	183
After	730	427	1,809	192	151

Summary

An important area of statistical inference is the *test of hypotheses*. When the *null hypothesis* is completely specified, it is called *simple*; otherwise, it is called *composite*. The *decision rule* to reject or accept the null hypothesis is given in terms of a *test statistic*. The probability of rejecting the hypothesis when it is true is the *significance level* of the test, or *α-error*. The probability of accepting the null hypothesis when the alternative is true is the *operating characteristic* of the test, or *β-error*. The *power* of the test is the probability of rejecting the null hypothesis when it is not true and is $1 - \beta$. Most powerful tests are obtained by the *likelihood ratio tests* developed by Neyman and Pearson. The theory is applied to problems of testing means, variances, and *equality of means* of two normal populations. The equality of proportions of *two populations* is tested when sample sizes are large.

chapter eight

Nonparametric Methods

Most of the procedures discussed so far have been developed for models that are partially known to be normal or exponential but with unknown parameters. We have discussed, for example, estimation and tests of hypotheses for the mean and variance of normal distribution, as well as confidence intervals and tests of hypotheses about the difference of means of two normal populations. There are many situations where the precise nature of the probability law is not known, and thus no parametric inference procedure can be used. Procedures that do not use any specified form of distribution are called *nonparametric*.

Nonparametric procedures are simple to perform and can be conveniently used for preliminary analysis of data. We give a few nonparametric techniques that are commonly used in practice. Nonparametric methods must be used when the data have an ordinal scale or are available only in the form of ranks such as the color of a human tissue, the severity of pain, the social status of an individual, the degree of liking of a painting, and so on.

Nonparametric methods are increasingly being applied in the social and behavioral sciences, education, and medicine. There are several introductory books on applied nonparametric methods: for example, Conover (1980), Hollander and Wolfe (1973), and Lehmann (1975).

Procedures for the Median

One of the simplest nonparametric methods is the estimation of the population median and other quantiles. The *median* of a random variable X is defined to be η if:

$$P(X \leq \eta) = P(X \leq \eta) = \tfrac{1}{2} \tag{8.1}$$

The median is unique if X is a continuous random variable.

Suppose X_1, X_2,... X_n is a random sample from a population with a probability density function of $f(x)$. The *sample median,* which has been defined to be the middle value if the number of observations is odd and the average of the middle two values if the number of observations is even, can be taken to be an *estimate* of η.

The Confidence Interval for η

Let $Y_1 \le Y_2 \le \ldots \le Y_n$ be the ordered values of the sample. That is:

$$Y_1 = \text{smallest observation}$$
$$Y_2 = \text{second smallest observation} \ldots$$
$$Y_n = \text{largest observation}$$

Y_1, Y_2,..., Y_n are called the *order statistics* of the sample. We assume that $f(x)$ is continuous. We can find confidence intervals for the median in terms of order statistics. One such interval is (Y_1, Y_n). Let:

$$P\{Y_1 < \eta < Y_n\} = 1 - \alpha$$

Then
$$1 - \alpha = 1 - P(Y_n \le \eta) - P(Y_1 \ge \eta).$$

Since Y_1 is the smallest observation, the event that $Y_1 > \eta$ is the same that every X_1, X_2,..., $X_n > \eta$. Since X_1, X_2,..., X_n are independent,

$$P(Y_1 \ge \eta) = P(X_1 \ge \eta)P(X_2 \ge \eta) \ldots P(X_n \ge \eta)$$

$$= \left(\frac{1}{2}\right)^n \text{ from (8.1)}$$

Similarly,
$$P(Y_n \le \eta) = \left(\frac{1}{2}\right)^n$$

Hence,
$$1 - \alpha = 1 - \left(\frac{1}{2}\right)^n - \left(\frac{1}{2}\right)^n$$

$$= 1 - \left(\frac{1}{2}\right)^{n-1} \tag{8.2}$$

For various values of n, Equation (8.2) can be tabulated as follows:

n	2	3	4	5	6	7	8
$1-\alpha$	.5	.75	.875	.938	.969	.984	.992

As we note from the table, we cannot obtain confidence intervals with a prespecified arbitrary level of confidence. Confidence intervals for the median involving other order statistics can be similarly obtained. Table XI in the

Appendix gives the confidence coefficient for the interval (Y_d, Y_{n-d+1}). The equation

$$P(Y_d < \eta < Y_{n-d+1}) = 1 - \alpha$$

is tabulated giving $(1 - \alpha)$-level confidence interval to be (Y_d, Y_{n-d+1}).

Example 8.1: Six measurements on cholesterol level are given:

$$175, \ 324, \ 205, \ 226, \ 185, \ 272$$

We find the order statistics to be:

$$Y_1 = 175$$
$$Y_2 = 185$$
$$Y_3 = 205$$
$$Y_4 = 226$$
$$Y_5 = 272$$
$$Y_6 = 324$$

The sample median is $\frac{1}{2}(205 + 226) = 215.5$.

From the table for equation (8.2) we have $P(Y_1 < \eta < Y_6) = .969$. So the confidence coefficient of the interval (175, 324) is .969. Similarly, Table XI in the Appendix gives (185, 272) to be the confidence interval at level .781.

Example 8.2: The incomes of ten households are given (in thousands of dollars):

$$15, \ 75, \ 25, \ 50, \ 70, \ 120, \ 12, \ 17, \ 28, \ 80$$

From Table XI in the Appendix, we find:

$$P(Y_2 < \eta < Y_9) = .979$$

The ordered sample is:

$$12 < 15 < 17 < 25 < 28 < 50 < 70 < 75 < 80 < 120$$

Therefore, the .979 level confidence interval for the median is (15, 80), and the .891 level confidence interval for the median is (17, 75).

Tests of Hypotheses About the Median

Suppose we test the following hypothesis with a significant level of α:

$$H : \eta = \eta_0$$
$$A : \eta \neq \eta_0$$

The order statistics of the sample are used for testing the hypothesis as in confidence intervals.

Decision rule: Reject H when η_0 is not in the $(1 - \alpha)$ level confidence interval That is, reject H when

$$Y_d > \eta_0 \text{ or } Y_{n-d+1} < \eta_0$$

with appropriate d chosen from Table XI for appropriate α. As we mentioned in the case of confidence intervals, it is not possible to obtain a test with an exact prespecified significance level.

Example 8.3: For data in Example 8.2, test the hypothesis:

$$H : \eta = 25$$
$$A : \eta \neq 25$$

Decision rule: Reject H at level .021 if:

$$Y_2 > 25 \quad \text{or} \quad Y_9 < 25$$

Since $Y_2 = 17$ and $Y_9 = 80$, we do not reject the hypothesis H.

For one-sided hypotheses, see Table XI in the Appendix for the appropriate significance levels.

The Sign Test

Another way to test a hypothesis about the median of a population is by using the sign test. Let $H : \eta = \eta_0$ and $A : \eta \neq \eta_0$. Also let:

$S_+ = $ number of observations greater than η_0
$S_- = $ number of observations less than η_0

The test statistic for the two-sided alternative is $S = $ smaller of S_+, S_-.

Decision rule: Reject H if $S < d$, where d is determined from Table XI. For one-sided alternatives, use statistic S_+ for the alternative $\eta > \eta_0$ and use S_- for the alternative $\eta < \eta_0$.

For large n, the statistic S has normal distribution with mean $n/2$ and variance $n/4$ when H is true.

Example 8.4: Consider the data in Example 8.1. Let the hypothesis be:

$$H : \eta = 250$$
$$A : \eta \neq 250$$

Substracting 250 from each observation and counting the number of positive and negative signs, we find:

$$S_+ = 2$$
$$S_- = 4$$

Therefore, $S = 2$.

From Table XI, we find $d = 1$, for $\alpha = .031$. That is, we reject H at level .031 if $S < 1$.

In this case, we accept the hypothesis.

Note that the confidence interval for the median is (175, 324) with level
.969. Since 250 is in this interval, we also accept the hypothesis that $\eta = 250$ at
level $1 - .969 = .031$.

The sign test does not use the magnitude of the observations. It is based only
on the count of observations smaller or larger than the hypothesized median.
An improvement over the sign test is the Wilcoxon signed rank test, which is
given in the next section.

Exercises

1. A random sample of seven cholesterol levels gave:

$$180, \ 171, \ 225, \ 265, \ 300, \ 190, \ 270$$

   (a) Find the confidence level for the median using:

   (a) (Y_2, Y_6)
   (b) (Y_3, Y_5)

   (b) Test the hypothesis that the median is 250 against the alternatives:

   (a) $\eta \neq 250$
   (b) $\eta < 250$
   (c) $\eta > 250$

   Give the appropriate level.
2. The income of ten randomly chosen office workers in the San Francisco
   Bay Area is (in thousand dollars):

$$17, \ 27, \ 32, \ 18, \ 21, \ 23, \ 28, \ 19, \ 28, \ 30$$

   (a) Find the confidence level using the intervals:

   (i) (Y_3, Y_7)
   (ii) (Y_4, Y_6)

   (b) Test the hypothesis:

$$H: \eta = 25$$
$$A: \eta \neq 25$$

   Use the sign test and give the significance level.
   (c) Test the hypothesis:

$$H: \eta = 25$$
$$A: \eta \neq 25$$

   Use the confidence interval of the median and give the significance level.
3. Suppose an experimental drug for reducing blood pressure was used on 15

subjects. The blood pressure of ten subjects was reduced, whereas five showed an increase. Test the hypothesis that the drug is effective.

4. Ten judges were asked to rate the two finalist candidates in a beauty contest. Four judges rated candidate A better than B, and the others rated B better than A. Does this show a significant difference in preferences of one candidate over the other?

5. In market research for a new detergent, eighty five housewives were given free samples and asked if they preferred the new detergent to the one they were using. Of these fifty five preferred the new detergent, whereas the others still liked the one they were using. Using a large sample distribution of the sign test statistic, test the hypothesis that there is no preference for one detergent over the other. Give the significance level.

The Wilcoxon Signed Rank Test

The sign test does not use the magnitude of the difference when testing a hypothesis about the median. The modification by Wilcoxon utilizes the ranks of the differences. We test:

$$H: \eta = \eta_0$$
$$A: \eta \neq \eta_0$$

The quantities $X_1 - \eta_0, X_2 - \eta_0, \ldots, X_n - \eta_0$ are ranked in their absolute value, and the following are calculated:

$$T_+ = \text{sum of ranks from positive numbers}$$
$$T_- = \text{sum of ranks from negative numbers}$$
$$T = \text{smaller of } T_+ \text{ and } T_-$$

Decision rule: Reject H if $T < d$. Critical values of d are given in Table XII. For one-sided alternatives, we use T_+ or T_- depending on the alternative.

For large n, T is known to have a normal distribution with a mean of $n(n + 1)/4$ and a variance of $n(n + 1)(2n + 1)/24$.

Note that the tests described above assume that the sample comes from a continuous distribution. However, the data still may have values that are equal, and hence there may be ties when we rank them. The convention in such cases is to use the average of the ranks for both observations. That is, let the ordered values be:

$$2, 7, 9, 9, 10, 11$$

Then the ranks will be

$$1, 2, 3.5, 3.5, 5, 6$$

where the tied observations have been assigned the same ranks, being the average of 3 and 4.

Example 8.5: For the Example 8.1, we test the same hypothesis but apply the Wilcoxon signed rank test. The differences from the median are:

$$-40.5, \ 108.5, \ -10.5, \ 10.5, \ -30.5, \ 56.5$$

Hence, we rank first:

$$40.5, \ 108.5, \ 10.5, \ 10.5, \ 30.5, \ 56.5$$

The ranks are:

$$4, \ 6, \ 1.5, \ 1.5, \ 3, \ 5$$

So:

$$T_+ = 6 + 1.5 + 5 = 12.5$$
$$T_- = 4 + 1.5 + 3 = 8.5$$
$$T = 8.5$$

For level .31, we have $d = 1$ (from Table XII), and thus we do not reject H.

Example 8.6: A store owner wants to test the hypothesis that the median number of items purchased by a customer is 15. The following data have been observed for seven customers on items purchased:

$$17, \ 10, \ 21, \ 29, \ 3, \ 8, \ 18$$

Test his hypothesis at an appropriate level. Here the differences are:

$$2, \ -5, \ 6, \ 14, \ -12, \ -7, \ 3$$

The ranks of the absolute differences are:

$$1, \ 3, \ 4, \ 7, \ 6, \ 5, \ 2$$

We have now:

$$T_+ = 1 + 4 + 7 + 2 = 14$$
$$T_- = 3 + 6 + 5 = 14$$
$$T = 14$$

From Table XII, we see that for $n = 7, d = 5$ at level .109; thus we reject the hypothesis.

Exercises

6. The scores in a test are given as:

$$67, \ 85, \ 32, \ 78, \ 95, \ 17, \ 83, \ 69, \ 43, \ 58$$

Test the hypothesis that the median is 60 using the Wilcoxon signed rank test at the appropriate level.
7. Test the hypothesis in Exercise 1 using the Wilcoxon signed rank test.
8. Test the hypothesis in Exercise 2 using the Wilcoxon signed rank test.
9. Find the confidence interval for the median using the Wilcoxon signed rank statistic for the data:

$$17, \ 27, \ 10, \ 15, \ 20, \ 18, \ 19$$

Estimates of the Cumulative Distribution Function

Let $X_1, X_2, \ldots, X_n$ be a random sample from a population with a cumulative distribution function of $F(x)$. To estimate $F(x)$ at each point x, we use the *empirical distribution function* or *sample distribution function* $F_n(x)$ defined below.

$$F_n(x) = \frac{1}{n}\{\text{Number of} \quad X_i \leq x\} \tag{8.3}$$

The sample distribution function gives the proportion of observations at each value of x. At each observation X_i, the function takes a jump of size $1/n$, resulting in a step function.

Example 8.7: Let a random sample of heights of marigold plants be:

$$2, \ 5, \ 25, \ 37, \ 21, \ 20, \ 10, \ 15, \ 7, \ 17$$

What is the sample distribution function of X? We first order the observations as:

$$3, \ 5, \ 7, \ 10, \ 15, \ 17, \ 20, \ 21, \ 25, \ 37$$

There are jumps of size $\frac{1}{10}$ at each of these points; the rest of the function is flat. The graph is given in Figure 8.1.

The Kolmogorov Statistic

To obtain confidence intervals for the distribution function, we need the Kolmogorov statistic, which gives the maximum deviation of the distribution function from the sample distribution function. That is:

$$D_n = \max_x |F_n(x) - F(x)| \tag{8.4}$$

D_n is called the *Kolmogorov statistic*, and it is *distribution free*; that is, its distribution does not depend on the distribution $F(x)$ when $F(x)$ is continuous. D_n is used in several nonparametric tests, such as tests of goodness of fit. For a given confidence level $1 - \alpha$, we obtain a value for D_n from the Table

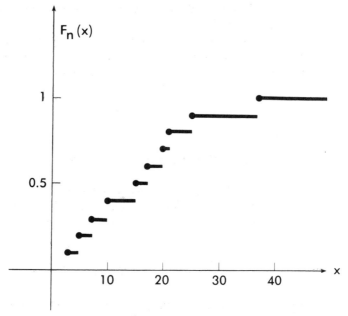

Figure 8.1. Sample Distribution Function.

XIII in the Appendix and then find the lower and upper estimates of the distribution function at x. This procedure provides the band in which the unknown distribution is situated. The following example illustrates the procedure.

Example 8.8: The times of death of ten mice in a toxicological experiment are (in days):

$$10, \ 13, \ 14, \ 20, \ 16, \ 27, \ 28, \ 25, \ 31, \ 29$$

The sample distribution function $F_n(x)$ is the step function with jumps of size $1/10$ at each observation. For .90-level confidence interval, we find from Table XIII in the Appendix;

$$D_{10} = .368$$

In Figure 8.2 the sample distribution function is denoted by the solid line and the upper and lower limits by dotted lines.

Exercises

10. A train arrives at a station at random between 8:00 A.M. and 8:15 A.M. A sample of six times observed is:

$$8:01, \ 8:07, \ 8:03, \ 8:04, \ 8:14, \ 8:09$$

Graph the sample distribution function and give a 90% level confidence interval of the distribution function.

11. Customers arrive at a store counter at random at intervals of (in minutes):

$$2, 7, 5, 3, 10, 11, 12, 17$$

What is the sampling distribution function of interarrival times? Give a 90% confidence interval for $F(x)$.

12. The speed of passing cars was observed as:

$$48.7, 50.5, 60.5, 57.6, 54.3, 62.7, 53.8, 63.4$$

Give the sample distribution function of the speed limits and a 90% confidence interval for the distribution function.

Tests of Goodness of Fit

The hypothesis that a proposed model is valid in a given set up is important in may situations. Assumptions such as whether a coin is fair, a die is unbiased, or a distribution is normal are needed before they can be statistically verified. The tests of goodness of fit are concerned with the basic question of whether the sample comes from a normal distribution or a Poisson distribution, or whether the progeny corresponds to given genetic frequencies. For example, the hypothesis that a die is fair can be reformulated in terms of probabilities

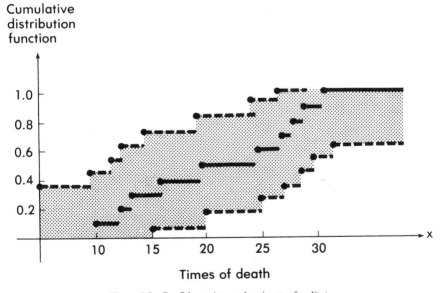

Figure 8.2. Confidence interval estimates for $F(x)$.

$p_1, p_2, \ldots, p_6$ of the occurrence of the dots $1, 2, \ldots, 6$ on the die. We test:

$$H: p_1 = p_2 = p_3 = p_4 = p_5 = p_6 = \tfrac{1}{6}$$
$$A: \text{not equal} \tag{8.5}$$

One such test statistic is the chi-square.

The Chi-Square Test Statistic

Let $O_1, O_2, \ldots, O_c$ be the number of *observed* frequencies in classes $1, 2, \ldots, c$. Let $E_1, E_2, \ldots, E_c$ be the number of frequencies *expected* under the model in null hypothesis H. Then the *chi-square* test statistic is defined as:

$$\chi^2_{c-1} = \sum_{i=1}^{c} \frac{(O_i - E_i)^2}{E_i} \tag{8.6}$$

The distribution of χ^2_{c-1} is chi-square with $c - 1$ degrees of freedom.

For Hypothesis (8.5), suppose we have a sample of n tosses of the die with observed frequencies $X_1, X_2, \ldots, X_6$ so that $X_1 + X_2 + \cdots + X_6 = n$. We also have $E_i = np_i, i = 1, 2, \ldots, 6$.

Decision rule: Reject H if $\chi^2_{c-1} > k$ where k is determined from a chi-square distribution with 5 degrees of freedom with a preassigned α.

In this case, we can easily calculate:

$$\chi^2_5 = \sum_{i=1}^{6} \frac{(X_i - np_i)^2}{np_i} = \sum_{i=1}^{6} \frac{(X_i - n/6)^2}{n/6} \tag{8.7}$$

Example 8.9: A die is thrown 300 times and the following frequencies are observed. Test the hypothesis that the die is fair at level .05:

$$X_1 = 57$$
$$X_2 = 43$$
$$X_3 = 59$$
$$X_4 = 55$$
$$X_5 = 63$$
$$X_6 = 23$$

The expected frequencies are all equal to $300 \times \tfrac{1}{6} = 50$ under the null hypothesis:

$$\chi^2_5 = \frac{(57 - 50)^2}{50} + \frac{(43 - 50)^2}{50} + \frac{(59 - 50)^2}{50}$$

$$+ \frac{(55 - 50)^2}{50} + \frac{(63 - 50)^2}{50} + \frac{(23 - 50)^2}{50}$$

$$= 22.04$$

Since $\chi^2_{5,.95} = 11.07$, we reject the hypothesis at level .05.

Example 8.10: In a breeding system, the progeny has the genetic frequencies in ratio $1:3:3:1$. A sample of 160 provided the observed frequencies as 15, 65, 73, and 7. Do the observations conform to the model at level .05? Here we have:

O_i	15 65 73 7	160
E_i	20 60 60 20	160

$$\chi_3^2 = \frac{(15-20)^2}{20} + \frac{(65-60)^2}{60} + \frac{(73-60)^2}{60} + \frac{(7-20)^2}{20}$$

$$= 12.93$$

Since $\chi_{3,.95}^2 = 7.82$, we reject the hypothesis.

Example 8.11: The data in the following table give the number of auto accidents observed on 100 days at a city intersection. Test the hypothesis that the accidents follow a Poisson distribution with parameter $\lambda = 2$ per day at level .05.

Number of accidents	0	1	2	3	4	≥ 5
Frequency	16	24	22	21	11	6

Assuming a Poisson distribution with $\lambda = 2$, we have the probabilities given by:

X	0	1	2	3	4	≥ 5
$p_X(x)$	.1353	.2707	.2707	.1804	.0902	.0527
E_i	13.5	27.1	27.1	18.0	9.0	5.3

The test statistic is:

$$\chi_5^2 = \frac{(16-13.5)^2}{13.5} + \frac{(24-27.1)^2}{27.1} + \frac{(22-27.1)^2}{27.1}$$

$$+ \frac{(21-18)^2}{18} + \frac{(11-9)^2}{9} + \frac{(6-5.3)^2}{5.3}$$

$$= 2.81$$

From Table IV in the Appendix we find that $\chi_{5,.95}^2 = 11.07$, and hence we accept the hypothesis of a Poisson distribution with a mean of 2.

If the parameter of the Poisson distribution is not given, we can estimate it from the data and use it in computing Poisson probabilities. However, the resulting chi-square statistic will have one less degree of freedom. Similarly, if two parameters are calculated using the data and these estimated values are used in the chi-square statistic, the degrees of freedom should be reduced by 2.

Example 8.12: The following frequency distribution gives the data for the measurement errors (in cm) of a rod given to a class of 75 students to measure. Test the hypothesis that measurement errors are normal with a mean of 0 and a variance of 1. That is:

H: Sample comes from standard normal distribution
A: Sample does not come from standard normal distribution

Class Intervals	Observed Frequency
− 1.5, and below	2
− 1.5, − 1	15
− 1, .5	15
− 0.5, .5	30
0.5, 1.0	20
1.0, 1.5	8
1.5, and above	10
Total	100

From tables of the normal, the probabilities, and hence the corresponding expected values, are found as follows:

Class Intervals	Probabilities	Expected Frequency
− 1.5, and below	.0668	6.7
− 1.5, − 1	.0919	9.2
− 1,. − .5	.1498	15.0
− 0.5, .5	.3830	38.3
0.5, 1.0	.1498	15.0
1.0, 1.5	.0919	9.2
1.5, and above	.0668	6.7

The chi-square statistic is:

$$\chi_6^2 = \frac{(2 - 6.7)^2}{6.7} + \frac{(15 - 9.2)^2}{9.2} + \frac{(15 - 15)^2}{15} + \frac{(30 - 38.3)^2}{38.3}$$

$$+ \frac{(20 - 15)^2}{15} + \frac{(8 - 9.2)^2}{9.2} + \frac{(10 - 6.7)^2}{6.7}$$

$$= 3.30 + 3.66 + 0 + 1.8 + 1.67 + .16 + 1.63$$

$$= 12.22$$

Now $\chi^2_{6,.90} = 10.6$, and we reject the hypothesis of normality at a 10% level of significance.

The *Kolmogorov Statistic*: An alternative method of testing the goodness of fit is by D_n. Suppose we test the hypothesis that the data come from a known distribution, $F_0(x)$, against the alternative that it does not. That is

$$H : F(x) = F_0(x)$$
$$A : F(x) \neq F_0(x)$$

Decision rule: Reject H if $D_n > k$, where k is determined from the tabulated distribution of D_n discussed earlier.

Example 8.13: The times of arrival of a train between 8:00 and 8:15 A.M. are noted by a commuter as 8:01, 8:03, 8:07, 8:04, 8:11, and 8:14. Test the hypothesis that the train arrives at random between these times at level .05. That is, test the hypothesis that the distribution of times of arrival is uniform over (0, 15). Here:

$$F_0(x) = \begin{cases} 0 & x < 0 \\ \dfrac{x}{15}, & 0 \le x < 15 \\ 1 & x \ge 15 \end{cases}$$

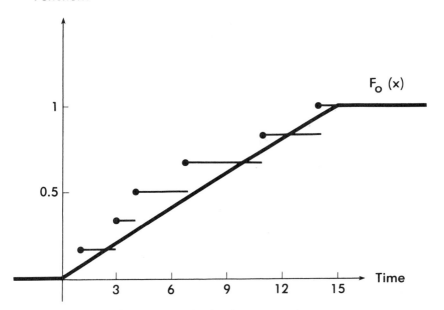

Cumulative Distributuion Functions

Figure 8.3. Sample distribution function and $F_0(x)$ for Example 8.11.

The sampling distribution function obtained from the data has jumps of size $\frac{1}{6}$ at 1, 3, 4, 7, 11, and 14 minutes:

$$D_6 = \max |F_n(x) - F_0(x)|$$

At the points of jumps, we have the values of D_n given by Figure 8.3. $F_n(x+)$ gives the value of $F_n(x)$ after the jump and $F_n(x-)$ before the jump:

x	$F_n(x+) - F_0(x)$	$F_0(x) - F_n(x-)$
1	$\frac{1}{6} - \frac{1}{15} = \frac{3}{30}$	$\frac{1}{15} - 0 = \frac{2}{30}$
3	$\frac{2}{6} - \frac{3}{15} = \frac{4}{30}$	$\frac{3}{15} - \frac{1}{6} = \frac{1}{30}$
4	$\frac{3}{6} - \frac{4}{15} = \frac{7}{30}$	$\frac{4}{15} - \frac{2}{6} = -\frac{2}{30}$
7	$\frac{4}{6} - \frac{7}{15} = \frac{6}{50}$	$\frac{7}{15} - \frac{3}{6} = -\frac{1}{30}$
11	$\frac{5}{6} - \frac{11}{15} = \frac{3}{30}$	$\frac{11}{15} - \frac{4}{6} = \frac{2}{30}$
14	$1 - \frac{14}{15} = \frac{2}{30}$	$\frac{14}{15} - \frac{5}{6} = \frac{3}{30}$

Hence $D_6 = \frac{7}{30}$. Since $D_{6,.90} = .47$, we accept the null hypothesis.

Exercises

13. The number of inert particles is observed in 30 samples of a liquid and found to have the following frequencies:

Number of inert particles	0	1	2	≥ 3
Frequencies	20	5	3	2

Test the hypothesis that they have a Poisson distribution with $\lambda = 1$ at level .10 of significance.

14. The scores in an examination have a normal distribution with a mean of 65 and a standard deviation of 10. Test the hypothesis that the following data satisfy the above assumption at level .05:

Scores	Frequency
0–40	5
40–50	20
50–60	35
60–70	28
70–80	10
80–100	2

15. The guests at a party arrive at random within intervals of 2, 3, 5, 8, 4, and 15 minutes. Test the hypothesis that interarrival times are exponentially distributed with mean of 2. Note that the parameter of the exponential is $\theta = \frac{1}{2}$.

Two-Sample Nonparametric Procedures

Procedures for comparing two populations under parametric assumptions are generally given in terms of comparing their means or variances or both. A nonparametric procedure may test the equality of their medians. In a paired-sample case, we can use one sample procedures as in the parametric case. For testing the hypothesis that a drug lowers cholesterol levels, for example, we can test the hypothesis that the median of the difference of cholestrol levels before and after the administration of the drug is zero against the alternative that it is positive, so tests under the heading "Procedures for the Median" at the beginning of the chapter can be used. Below are a few tests for comparing two populations.

Median Test

Let the medians of two populations with distributions $F(x)$ and $G(x)$ be η_1 and η_2 and test:

$$H:\eta_1 = \eta_2$$
$$A:\eta_1 \neq \eta_2$$

The test statistic in this case is obtained from a 2×2 table by counting the number of observation in each sample above and below the common median of the combined samples. Let there be n observations from $F(x)$ and m observations from $G(x)$. We have the following counts:

Population	Greater than the common median	Smaller than the common median
$F(x)$	a	$n - a$
$G(x)$	b	$m - b$

Let χ_1^2 be the usual chi-square statistic as obtained for the 2×2 contingency table in Chapter 7.

Decision rule: Reject H if $\chi_1^2 > k$.

Example 8.14: The cholesterol levels of seven normal and eight sick human subjects are given:

Normal	320, 280, 270, 265, 310, 190, 180
Sick	350, 330, 275, 280, 260, 310, 190, 230

The common median is $M = 275$. We now have the table:

	≥ 275	< 275
Normal	3	4
Sick	5	3

$$\chi_1^2 = \frac{15\{|3 \times 3 - 4 \times 5| - \frac{15}{2}\}^2}{7 \times 8 \times 8 \times 7} = .059$$

Since χ_1^2 is not significant a level .05, we accept the hypothesis of the equality of the medians.

The Wilcoxon-Mann-Whitney Test

To test the hypothesis $H: F(x) = G(x)$, versus $A: F(x) \neq G(x)$ we now give a statistic based on ranks. Let a random sample of size m be given from $F(x)$ and of size n from $G(y)$. Let the Xs and Ys be jointly ordered and let:

$U_x = $ Number of times X-observations are larger than Y-observations

$U_y = $ Number of times Y-observations are larger than X-observations

Then the *Wilcoxon-Mann-Whitney* statistic is:

$$U = \min(U_x, U_y) \tag{8.8}$$

The distribution of U for small values of m and n is tabulated in Table XV in the Appendix. For $m, n \geq 8$, the normal approximation is good. The result:

For large m and n, the distribution of U is normal with mean $\dfrac{mn}{2}$ and variance $\dfrac{mn(m + n + 1)}{12}$

Example 8.15: Suppose the clotting time (in minutes) of plasma is given under two different methods of treatment of subjects chosen at random:

Method I: 8.7, 10.5, 9.3, 12.3, 10.7

Method II: 9.5, 11.4, 9.0, 13.1, 11.5, 12.7, 8.9

Both samples are ordered:

8.7, 8.9, 9.0, 9.3, 9.5, 10.5, 10.7, 11.4, 11.5, 12.3, 12.7, 13.1

If we denote Method I by X and Method II by Y, we have the observations in the two samples:

$$X \ Y \ Y \ X \ Y \ X \ X \ Y \ Y \ X \ Y \ Y$$

Hence,
$$U_x = 2 + 3 + 3 + 5 = 13$$
$$U_y = 1 + 1 + 2 + 4 + 4 + 5 + 5 = 22$$

so that
$$U = 13$$

From Table XV in the Appendix, we find that for $m = 5, n = 7$, we reject the hypothesis of the equality of two treatments if $U > 6$ at 0.048 level of significance. Since $U = 13$, we reject the hypothesis and claim that the two treatments are different.

The U statistic can also be obtained in terms of the ranks of the X-observations or Y-observations in the combined sample. It can be shown that

$$U_x = R_x - \frac{m(m+1)}{2} \tag{8.9}$$

where R_x is the sum of ranks assigned to X-observations in the combined sample. Similarly, we can define:

$$U_y = R_y - \frac{n(n+1)}{2} \tag{8.10}$$

For instance, in the above example:

$$R_x = 1 + 4 + 6 + 7 + 10 = 28$$

Hence,
$$U_x = 28 - \frac{5 \times 6}{2} = 13$$

and

$$R_y = 2 + 3 + 5 + 8 + 9 + 11 + 12 = 50$$

So that

$$U_y = 50 - \frac{7 \times 8}{2} = 22$$

and therefore
$$U = 13$$

The Kolmogorov-Smirnov Test

Let the sample distribution function based on the sample of Xs be $F_m(x)$ and the sample of Ys be $G_n(x)$. The hypothesis to be tested is:

$$H : F(x) = G(x)$$
$$A : F(x) \neq G(x)$$

The Kolmogorov-Smirnov statistic is defined by:

$$D_{m,n} = \max |F_m(x) - G_n(x)| \qquad (8.11)$$

Decision rule: Reject H when $D_{m,n} > k$, k is determined by Table XIV in the Appendix for m, $n \leq 40$. For higher values of m and n, Kraft and Van Eeden (1968) have tabulated the distribution of $D_{m,n}$.

Example 8.16: Suppose we have the same data as in Example 8.15. Then $F_5(x)$ gives a step function with jumps of $1/5$ at the x-sample values, and $G_7(x)$ gives a step function with jumps of $1/7$ at y-values. The distributions are graphed in Figure 8.4. Again, as we saw in one sample case, the statistic $D_{m,n}$ can be obtained by finding the largest difference at the jump points. We find that:

$$D_{5,7} = \tfrac{4}{5} - \tfrac{3}{7} = \tfrac{13}{35}$$

From the tables, the 95th percentile of $D_{5,7}$ is $5/7$, and hence we accept the null hyothesis.

Exercises

16. Given the following data on normal persons and hospital patients for their

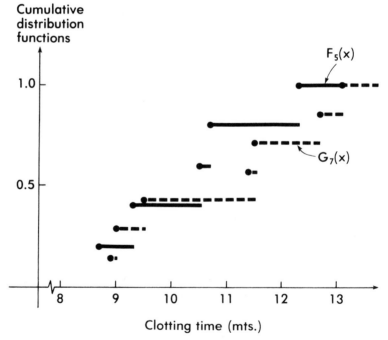

Figure 8.4. Sampling distribution functions for Example 8.16.

cholesterol levels, test the hypothesis of the equality of the populations by using:

(1) The median test
(2) The Wilcoxon-Mann-Whitney test

Normals	250, 217, 321, 280, 370, 325
Hospital patients	320, 287, 371, 210, 327, 285, 290, 355, 260

17. The nicotine content of cigarette smoke of two different brands is given:

Brand I   25, 32, 35, 22, 18, 19
Brand II   31, 37, 21, 18, 19, 27, 28

Test the hypothesis that the two brands have the same amount of nicotine. Use three different alternative hypotheses.

18. The number of admissions for respiratory diseases during high smog-alert days and normal days is given among various hospitals in Los Angeles. Test the claim that smog increases respiratory sickness:

Normal	300, 321, 375, 285, 591, 450, 380
High smog	290, 571, 600, 273, 370

19. A score on a physical fitness test between 0 and 50 signifies the level of fitness, and the higher the better. A faculty fitness program at Ohio State University claims to improve the fitness of those in the program. Two independent samples of those in the program and other faculty were made. Test the hypothesis that the scores of those in the program are higher using the Wilcoxon-Mann-Whitney test at level .10.

Faculty fitness program	17.3, 28.5, 40.3, 37.8, 40.5
Others	19.3, 16.5, 25.8, 32.3, 27.5

20. Two different diets were tested on animals for four weeks and their percent change in weights were noted. Test the hypothesis that both diets are the same at level .05 using the median test:

Diet A	17, 25, 15, 45, 35, 18, 19
Diet B	11, 19, 17, 85, 39, 45, 82

21. Use the Kolmogorov-Smirnov test to solve Exercise 19.
22. In comparing two methods of teaching, two classes consisting of 25 and 35 students were given instruction by the two methods; the Wilcoxon-Mann-Whitney statistic was calculated to be 27. Using normal approximation, test the hypothesis that the two methods are the same at level .05.

Contingency Tables

A brief introduction to 2×2 contigency tables was given in Chapter 7 to test the hypothesis of the equality of two proportions. The extension to $2 \times c$ tables is straightforward when we want to test the hypothesis of the equality of c proportions. We want to test the hypothesis

$$H: p_1 = p_2 = \ldots p_c$$
$$A: \text{at least one inequality}$$

where there are c populations. The counts are given in the table:

Population	1	2	3	...	c	Total
Success	O_{11}	O_{12}	O_{13}	...	O_{1c}	O_1
Failure	O_{21}	O_{22}	O_{23}	...	O_{2c}	O_2
Total	n_1	n_2	n_3	...	n_c	N

The statistic is obtained by the usual chi-square statistic, which compares the observed and expected frequencies as given by (8.6). The sum is to be obtained over all $2c$ cells. The degrees of freedom for this chi-square is $c - 1$. The expected value E_{11} for the first cell in the first row is obtained by:

$$E_{11} = \frac{O_1 n_1}{N}$$

Similarly,

$$E_{12} = \frac{O_1 n_2}{N} \quad \text{and so on}$$

For the expected values in the second row:

$$E_{21} = \frac{O_2 n_1}{N}$$

$$E_{22} = \frac{O_2 n_2}{N} \quad \text{and so on}$$

We then obtain the sum of (observed-expected)^2/expected, over all the classes. Call this χ^2_{c-1}.

Decision rule: Reject the hypothesis if:

$$\chi^2_{c-1} > k$$

Example 8.17: A sample of students in a high school is chosen randomly from an inner-city and a suburban neighborhood. they are classified according to grades. Test the hypothesis that the proportion of inner-city students is the same in each grade at level .05. The data are:

Grades	7th	8th	9th	10th	11th	12th	
Inner city	10	10	12	5	15	8	60
Suburban	20	25	15	25	15	20	120
	30	35	27	30	30	28	180

The table for the expected frequencies under the null hypothesis of the equality of proportions is:

	7	8	9	10	11	12
Inner city	10	11.7	9	10	10	9.3
Suburban	20	23.3	18	20	20	18.7

$$\chi_5^2 = \frac{(10-10)^2}{10} + \frac{(10-11.7)^2}{11.7} + \frac{(12-9)^2}{9} + \cdots + \frac{(20-18.7)^2}{18.7}$$

$$= 9.64$$

Since $\chi_{5,.95}^2 = 11.07$, we accept the hypothesis that the populations have equal proportions at level .05.

When there are more than two rows, we have a general form of the $r \times c$ contingency table. This has r rows and c columns. The observed data are given in the following table, with O_{ij} as the observed counts in the cell of ith row and jth column:

Rows \ Columns	1	2	3	...	c	Total
1	O_{11}	O_{12}	O_{13}	...	O_{1c}	R_1
2	O_{21}	O_{22}	O_{23}	...	O_{2c}	R_2
3	O_{31}	O_{32}	O_{33}	...	O_{3c}	R_3
$\vdots$	$\vdots$	$\vdots$	$\vdots$	$\vdots\vdots\vdots$	$\vdots$	$\vdots$
r	O_{r1}	O_{r2}	O_{r3}	...	O_{rc}	R_r
	C_1	C_2	C_3	...	C_c	N

As we discussed in the previous case, the usual chi-square statistic in the general case has $(r-1)(c-1)$ degrees of freedom. The expected frequencies for (i,j)th cell is obtained is:

$$E_{ij} = \frac{R_i C_j}{N}$$

Then the statistic is:

$$\chi^2_{(r-1)(c-1)} = \sum_{i=1}^{r} \sum_{j=1}^{c} \frac{(O_{ij} - E_{ij})^2}{E_{ij}}$$

The hypothesis that two attributes having r and c different classifications are independent can be tested by $\chi^2_{(r-1)(c-1)}$.

Example 8.18: A random sample of registered voters in the United States was taken to find the extent of approval of President Reagan's policies on South America. The data:

Opinion \ Age	Young Adults Aged 18–35	Adults Aged 36–60	Seniors Aged 60+	Total
Disapprove	105	55	45	205
Slightly approve	75	50	85	210
Strongly approve	10	55	70	135
Total	190	160	200	550

Test the hypothesis that the opinion and the age of the voter are independent at .05 level of significance.

We find the expected frequencies under independence in the same way as before, and they are given by:

$$E_{11} = \frac{205 \times 190}{550} = 70.8 \qquad E_{21} = \frac{210 \times 190}{550} = 72.5$$

$$E_{12} = \frac{205 \times 160}{550} = 59.6 \qquad E_{22} = \frac{210 \times 160}{550} = 61.1$$

The remaining frequencies are obtained by subtraction from the row and column totals. Then we have the table of expected frequencies:

Expected Frequencies E_{ij}

Opinion \ Age	Young Adults	Adults	Seniors	Total
Disapprove	70.8	59.6	74.6	205
Slightly approve	72.5	61.1	76.4	210
Strongly approve	46.7	39.3	49.0	135
Total	190	160	200	550

The chi-square can then be calculated as follows:

$$\frac{(105 - 70.8)^2}{70.8} + \frac{(55 - 59.6)^2}{59.6} + \frac{(45 - 74.6)^2}{74.6} + \frac{(75 - 72.5)^2}{72.5}$$

$$+ \frac{(50 - 61.1)^2}{61.1} + \frac{(85 - 76.4)^2}{76.4} + \frac{(10 - 46.7)^2}{46.7}$$

$$+ \frac{(55 - 39.3)^2}{39.3} + \frac{(70 - 49.0)^2}{49.0}$$

$$= 75.8$$

From tables, $\chi^2_{4,.95} = 9.49$ and hence we reject the hypothesis of independence at level .05. That is, age affects opinion on this matter.

Since the square root of a chi-square with one degree of freedom is a standard normal random variable, one can have an approximate idea of the significance of any chi-square by using the square root of chi-square divided by its degrees of freedom. If we have χ^2_r, then we should consider the statistic $\sqrt{\chi^2_r/r}$, which can be regarded as a standard normal random variable.

The Phi Coefficient

In 2×2 contingency tables, a measure of association has been defined by the phi-coefficient. Suppose the total number of items is N. Then we define the coefficient as,

$$\phi = \sqrt{\frac{\chi^2_1}{N}} \tag{8.13}$$

For an $r \times c$ table, the ϕ-coefficient is:

$$\sqrt{\chi^2_{(r-1)(c-1)}/[(r-1)(c-1)N]}$$

Example 8.18: For Example 8.17, we have:

$$\phi = \sqrt{\frac{75.8}{4 \times 550}} = .19$$

Exercises

23. Three sections of a mathematics course are being taught by professors X, Y, and Z. Suppose the grade distribution in the courses is the following.

Test the hypothesis that the professors' grading policies are the same at level .05:

Professors \ Grades	A	B	C	D	E
X	10	35	45	20	15
Y	15	30	55	15	10
Z	5	25	23	10	5

24. A random sample of individuals in various age groups was tested for the presence of a certain type of bacteria in their stool. The following data were obtained. Test the hypothesis that the proportions are the same in each age group at .05 level of significance:

Age in years	0–10	11–20	21–30	31–40
Presence	10	18	12	6
Absence	60	25	10	2

25. The degree of the positivity of sputa from patients with pulmonary tuberculosis treated with PAS, streptomycin, or a combination of both drugs is given in the following table. Test the independence of treatment and the type of sputum at level .01.

Treatment \ Sputum	Positive Smear	Negative Smear Positive Culture	Negative Smear Negative Culture
PAS	56	30	13
Streptomycin	46	18	20
PAS and streptomycin	37	18	35

Source: Armitage, P. *Statistical Method in Medical Research*, (New York: John Wiley & Sons, 1971) p. 212.

26. Students at a university were polled to find their opinion about legalizing marijuana in the United States. The following table gives the counts for 600 students. Test whether there is independence between the sex of students and their opinion at level .05. Calculate the ϕ coefficient.

Opinion \ Sex	Male	Female	Total
Legalize	250	100	350
Do not legalize	50	200	250
	300	300	600

Chapter Exercises

1. Ten houses in the inner city were being auctioned off by the Housing and Urban Development Department of the U.S. Government. The number of bids received is:

$$10, 7, 3, 8, 2, 4, 9, 11, 1, 12$$

(a) Find the median of the sample
(b) Find the confidence interval for the population median
(c) Test the hypothesis that the population median is 5 at the appropriate level of significance
(d) Use the sign test for testing the hypothesis that the median is 6 against the one-sided alternative that it is greater than 6

2. Use a signed-rank test to test the hypothesis that the median in Exercise 1 is 6.

3. Given a random sample of the prices of gasoline in Ohio, find the sample distribution function of prices. Find a 90% confidence interval estimate of the population distribution function using Kolmogorov's statistic:

$$\$1.25, \ 1.42, \ 1.27, \ 1.53, \ 1.33, \ 1.17, \ 1.23, \ 1.55, \ 1.29, \ 1.25$$

4. A sample of gasoline prices in California is given:

$$\$1.52, \ 1.32, \ 1.62, \ 1.34, \ 1.64, \ 1.18, \ 1.82, \ 1.43$$

(a) Test the hypothesis that Ohio prices of gasoline as givn in Exercise 3 and those of California have the same median
(b) Test the hypothesis that the two states have the same distribution function at level .10

5. To test if driver education in high school has any effect on the accidents of teenage drivers, the following data were collected in the state of Arkansas. Test the hypothesis that the education has no effect on accident rates:

Driver Education \ Accidents in 5 years	0	1	2	≥ 3	Total
Yes	15	5	3	4	27
No	25	7	2	0	34
Total	40	12	5	4	61

Summary

Nonparametric procedures are commonly used in applications where assumptions of normality or other distributions cannot be made. Procedures for estimating and finding confidence intervals for the *median* are given. Tables are provided that can be used to test hypotheses about the median. The *Kolmogorov statistic* is used to test hypotheses of *goodness of fit* of the *sampling distribution function* to a model. Two-sample comparisons are made by various tests, such as the *median test*, the *Wilcoxon–Mann–Whitney* test, and the *Kolmogorov–Smirnov test*. The tables provide the critical values of the test. The *Wilcoxon-signed-rank test* for testing about the median is some improvement on the *sign test*, which also can be used to test the median.

Contingency tables lead to the comparison of observed and expected frequencies under various hypotheses of *independence* of attributes or comparison of *proportions*. The statistic has a chi-square distribution with $(r-1)(c-1)$ degrees of freedom for an $r \times c$ contingency table.

References

Conover, W. J. *Practical Nonparametric Statistics*, 2nd ed. New York: John Wiley & Sons, 1980.

Hollander, Myles, and Wolfe, Douglas. *Nonparametric Statistical Methods*. New York: John Wiley & Sons, 1973.

Kraft, Charles H., and Van Eeden, Constance. *A Nonparametric Introduction to Statistics*. New York: Macmillan, 1968.

Lehmann, E. L. *Nonparametrics: Statistical Methods Based on Ranks*. San Franscisco: Holden-Day, 1975.

Mosteller, Fredrick, and Rourke, Robert E. K. *Sturdy Statistics: Nonparametrics and Order Statistics*. Reading, Mass.: Addison-Wesley, 1973.

Noether, Gottfried, *Introduction to Statistics: A Fresh Approach*. Boston: Houghton Mifflin, 1971.

Owen, D. B. *Handbook of Statistical Tables*. Reading, Mass.: Addison-Wesley, 1962.

Siegel, Sidney. *Nonparametric Statistics for the Behavioral Sciences*. New York: McGraw-Hill, 1956.

chapter nine

Regression Analysis

So far we have been concerned with one variable. However, one of the most important problems in scientific research is to find an association between two or more variables. New scientific theories and many discoveries are made as a result of the discovery of such associations. Decisions in business and industry also require the study of several variables simultaneously. The established relationship between two variables can be used for predicting one variable in terms of the other. In statistics, the study of such associations is made through the theory of regression and correlation.

An English scientist, Sir Francis Galton (1822–1911), introduced the term *regression* while studying the resemblance between parents and children. The famous statistician Karl Pearson studied the relationship of heights of fathers to the heights of their sons. Associations between cigarette smoking and lung cancer, between drinking and liver cirrhosis, and between air pollution and health are important recent examples where the theory of regression and correlation has been successfully employed.

In a relationship between two variables, one is *independent* and the other *dependent*, where the dependent variable is given in terms of some function of the independent variable. When the variables are not random, such relationships are *deterministic*. Relationships given by the classical laws of motion, for example, are deterministic. However, when the relationship involves random variables, we call them *stochastic*. We will study such models in this chapter.

Joint probability distributions of two random variables were introduced in an earlier chapter. There are situations when both of the variables cannot be observed jointly. However, when one is given, say $X = x$, we can obtain the probability distribution of Y, called the conditional probability distribution of Y given $X = x$. The mean of this conditional distribution, $E(Y|x)$, is called the *regression function*. Similarly, the mean of the conditional distribution of X given $Y = y$ is also a regression function.

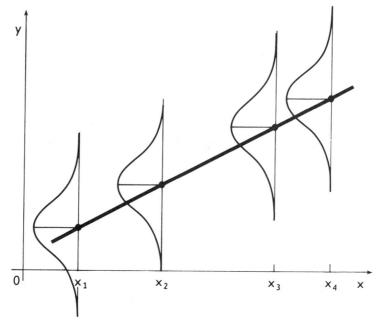

Figure 9.1. Regression function.

The regression line (or curve) is obtained by plotting the means $E(Y|x)$ against x, as in Figure 9.1. We assume that the regression function is linear in x. That is,

$$E(Y|x) = \beta_0 + \beta_1 x \tag{9.1}$$

where β_0 is the *intercept* and β_1 the *slope* of the regression line.

The basic assumptions for the linear regression model are:

(i) the regression $E(Y|x)$ is linear in x
(ii) the variance $V(Y|x)$ is independent of x

When we assume that the random variables X and Y have joint normal distributions, then it turns out that the regression function is linear and the variance of Y given $X = x$ is independent of x. In usual applications, even though this assumption of joint normality does not hold, it is common practice to assume linearity of regression. We shall use the linear regression model to obtain estimates of the parameters β_0 and β_1. The method of least squares, discussed in the next section, is used for estimation.

Since the expected value of the random variable differs from its mean by a random quantity, the model for linear regression is usually assumed to be

$$Y = \beta_0 + \beta_1 x + e \tag{9.2}$$

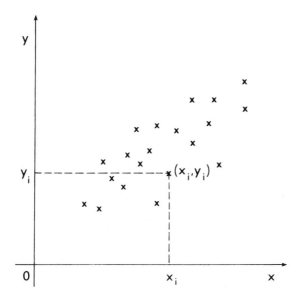

Figure 9.2. Scatter-plot diagram.

where e denotes the random error. We have assumed above that $E(e) = 0$ and $\text{Var}(e) = \sigma^2$. We do not assume any distributional form for the estimation of β_0 and β_1 and σ^2. The method of least squares provides these estimates.

The data for fitting the model in Equation (9.2) arises in the forms of pairs of (X, Y). Let (X_1, Y_1), $(X_2, Y_2), \ldots, (X_n, Y_n)$ be independent observations available to fit the model. Before any analysis is done, it is customary to plot the n data points in two dimensions on (x, y)-plots, or *scatter plots*. An example is given in Figure 9.2.

Exercises

1. A random sample of 15 dogs provides the following values of X, blood hemoglobin (as a percentage of normal), and Y, red blood cells (in millions per cubic millimeter). Make a scatter plot of the data.

X	92	95	103	84	108	94	101	
Y	6.8	7.7	7.5	7.1	7.6	6.7	6.5	
X	94	90	101	63	97	110	102	81
Y	6.9	6.5	7.3	5.9	6.7	7.4	7.6	5.9

2. The yield of wheat per acre (Y) and the amount of fertilizer applied (X) are

given for the ten plots in an experiment. Give a scatter plot of the data:

X	30	35	40	50	60	70	75	80	85	90
Y	15	25	22	35	40	55	49	52	55	60

Least Squares

The method of least squares is applied in estimation when assumptions about the form of probability distributions are not made. The method requires that we minimize the sum of squares of the deviations of observations from their expected value. That is, if $Y_1, \ldots, Y_n$ are independently distributed random variables with means $E(Y_i)$, then we minimize:

$$\sum_{i=1}^{n} [Y_i - E(Y_i)]^2$$

For the regression model

$$Y_i = \beta_0 + \beta_1 X_i + e_i$$

the errors $e_i = Y_i - \beta_0 - \beta_1 X_i$ are squared, since $E(Y_i) = \beta_0 + \beta_1 X_i$. β_0, β_1 are obtained so as to minimize this sum of squares. That is, the least squares estimates of β_0 and β_1 are obtained by minimizing

$$\sum_{i=1}^{n} e_i^2$$

or by minimizing:

$$\sum_{i=1}^{n} (Y_i - \beta_0 - \beta_1 X_i)^2 \qquad (9.3)$$

The minimum is obtained by solving the following equations, which are called *normal equations*.

$$\sum Y_i - n\beta_0 - \beta_1 \sum X_i = 0$$
$$\sum Y_i X_i - \beta_0 \sum X_i - \beta_1 \sum X_i^2 = 0 \qquad (9.4)$$

The estimates are given by

$$\hat{\beta}_0 = \bar{Y} - \hat{\beta}_1 \bar{X} \qquad (9.5)$$

$$\hat{\beta}_1 = \frac{n \sum X_i Y_i - \sum X_i \sum Y_i}{n \sum X_i^2 - (\sum X_i)^2} \qquad (9.6)$$

$$= \frac{\sum (X_i - \bar{X})(Y_i - \bar{Y})}{\sum (X_i - \bar{X})^2},$$

where $\bar{X}$ and $\bar{Y}$ are the sample means of X_is and Y_is.

Example 9.1: The heights and weights of eight infants studied by a pediatrician are given:

Infant	1	2	3	4	5	6	7	8
(Y) Height (inches)	25	23	16	15	10	23	32	19
(X) Weight (pounds)	15	17	8	6	7	11	19	13

The scatter plot of the data is given in Figure 9.3. The assumed linear regression model can be fitted by least squares. The vertical lines from the observations to the assumed line are the errors that are squared and summed. The least squares estimates are obtained using the following:

$$\sum X_i = 96 \qquad \bar{X} = 12$$

$$\sum Y_i = 163 \qquad \bar{Y} = 20.375$$

$$\sum X_i^2 = 1314 \qquad \sum (X_i - \bar{X})^2 = \sum X_i^2 - \frac{(\sum X_i)^2}{n} = 162$$

$$\sum X_i Y_i = 2162 \qquad \sum (X_i - \bar{X})(Y_i - \bar{Y}) = \sum X_i Y_i - \sum X_i \sum Y_i/n = 206$$

$$\sum Y_i^2 = 3649 \qquad \sum (Y_i - \bar{Y})^2 = \sum Y_i^2 - \frac{(\sum Y_i)^2}{n} = 327.875$$

$$\hat{\beta}_1 = 1.27, \qquad \hat{\beta}_0 = 5.14$$

The fitted regression line is:

$$Y = 5.14 + 1.27X$$

Notice that the line passes through $(\bar{X}, \bar{Y})$, that is (12, 20.375).

The difference between the observed value, Y_i, and the fitted value is called the *residual*, $\hat{Y}_i$.

The plot of residuals gives a visual test of the assumptions. If the assumptions are true, the residuals should be randomly distributed about the *x*-axis; otherwise, a pattern will be seen. For example, Figure 9.4 shows that the residuals are random and the assumptions are all correct. In Figure 9.5 the residuals show that the variance does not remain the same with *x*, and something should be done to modify the model. Similarly, the residuals in Figure 9.6 show that they are not random and perhaps there is a quadratic term containing x^2 in the regression model; hence, the linear regression is not adequate.

The residual plot is an important graphical aid in proceeding further with the regression model. The residuals sum to zero and their sum of squares is minimum.

Example 9.2: The residuals for Example 9.1 are obtained below and plotted

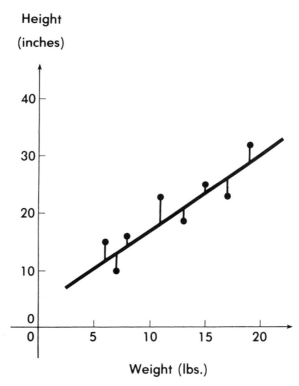

Figure 9.3. Regression of height on weight in Example 9.1.

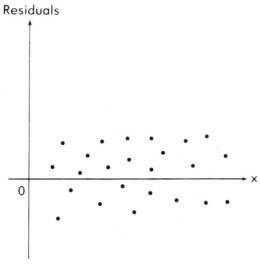

Figure 9.4. Residual plot.

Residuals

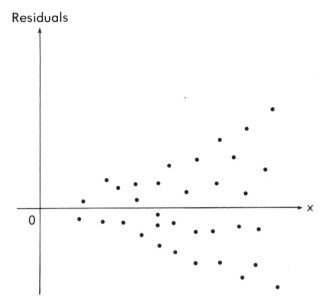

Figure 9.5. Residual plot.

Residuals

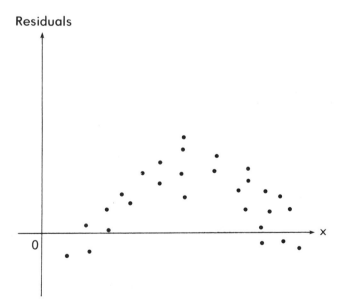

Figure 9.6. Residual plot.

Residuals

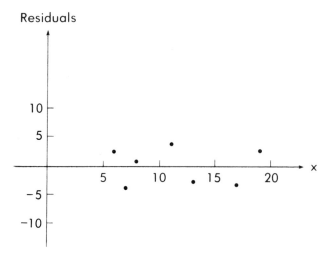

Figure 9.7. Residual plot for Example 9.2.

in Figure 9.7:

X	Observed Y	Fitted Y $Y = 5.14 + 1.27X$	Residual $\hat{Y}$
6	15	12.76	2.24
7	10	14.03	− 4.03
8	16	15.3	.7
11	23	19.11	3.89
13	19	21.65	− 2.65
15	25	24.19	0.81
17	23	26.73	− 3.73
19	32	29.27	2.73

The plot shows a preponderance of the positive residuals (5 out of 8). Overall, the residuals seem random.

Example 9.3: The systolic blood pressure, (Y), of seven white males is given with their age, (X). Fit the linear regression line and find the residuals:

X	20	30	40	50	60	70	80
Y	122	125	128	133	140	147	154

$$\sum X_i = 350 \qquad \bar{X} = 50$$
$$\sum Y_i = 949 \qquad \bar{Y} = 135.6$$
$$\sum X_i^2 = 20{,}300 \qquad \sum(X_i - \bar{X})^2 = 2800$$
$$\sum Y_i^2 = 129{,}507 \qquad \sum(Y_i - \bar{Y}) = 849.71$$

$$\sum X_i Y_i = 48,970 \qquad \sum (X_i - \bar{X})(Y_i - \bar{Y}) = 1520$$
$$\hat{\beta}_1 = .543 \qquad \hat{\beta}_0 = 108.6$$

The regression equation is:

$$Y = 108.6 + .543\,X$$

The residuals are given by:

X	20	30	40	50	60	70	80
Y	122	125	128	133	140	147	154
$\hat{Y}$	119.5	124.9	130.3	135.8	141.2	146.6	152.0
Residuals	2.5	.1	−2.3	−2.8	−1.2	.4	2.0

Estimate of the Variance σ^2

The unbiased estimate of the variance σ^2 under the regression model in Equation (9.2) is obtained as:

$$\hat{\sigma}^2 = \frac{1}{n-2} \sum (Y_i - \hat{\beta}_0 - \hat{\beta}_1 X_i)^2 \tag{9.7}$$

Several other identical expressions are

$$\hat{\sigma}^2 = \frac{1}{n-2} [\sum (Y_i - \bar{Y})^2 - \hat{\beta}_1^2 \sum (X_i - \bar{X})^2] \tag{9.8}$$

$$\hat{\sigma}^2 = \frac{1}{n-2} [\sum Y_i^2 - \hat{\beta}_0 \sum Y_i - \hat{\beta}_1 \sum X_i Y_i] \tag{9.9}$$

when $\quad s_x^2 = \dfrac{1}{n-1} \sum (X_i - \bar{X})^2 \quad$ and $\quad s_y^2 = \dfrac{1}{n-1} \sum (Y_i - \bar{Y})^2.$

Then: $$\hat{\sigma}^2 = \frac{n-1}{n-2} [s_y^2 - \hat{\beta}_1^2 s_x^2] \tag{9.10}$$

Example 9.4: For the estimate of σ^2 in Example 9.3, use Equation (9.8):

$$\hat{\sigma}^2 = \tfrac{1}{5}[849.71 - (.543)^2(2800)]$$
$$= \tfrac{1}{5}[849.71 - 825.58]$$
$$= 4.83$$

For the estimate of σ^2 in Example 9.1, use Equation (9.9):

$$\hat{\sigma}^2 = \tfrac{1}{6}[3649 - (5.14)(163) - (1.27)(2162)]$$
$$= 10.91$$

A regression line fitted by the method of least squares sometimes is not the best line fitted to the data. Visually, when we see the best line, we try to minimize the perpendicular distances from the observed points to the line rather than the vertical distances. This method of finding a regression line that minimizes the sum of perpendiculars from the observed points is called the method of *orthogonal least squares*. Mathematical derivation of the estimates becomes involved in such cases.

Exercises

3. In testing the life (Y) of tires manufactured by a company, a laboratory subjected them to several speeds (X). Fit a regression line to the data and give the residuals:

X	Y
(miles per hour)	(in thousands of miles)
20	27.5
22	25.6
25	28.3
30	29.6
40	24.5
50	22.5
55	24.5
60	23.5
70	21.6

4. The marriage rates (per 1,000 of population) of the United States are given for the period 1900–1960. Assume that the rates are linearly dependent on time, and fit a linear regression to the following data:

Year X	Rate Y
1900	9.3
1905	10.0
1910	10.3
1915	10.0
1920	12.0
1925	10.3
1930	9.2
1935	10.4
1940	12.1
1945	12.2
1950	11.1
1955	9.3
1960	8.5

Source: National Center of Health Statistics, U.S. Department of Health and Human Services

Find the residuals and plot them to see if the assumptions are correct. The years can be coded from 0 for 1900, 1 for 1905, and so on.

5. Crime statistics in a country for the past eight years in terms of the total number of crimes per year (Y) and the number of guns (X) are given below. Fit a linear regression of Y on X and give the residuals:

X	Y
7	18
9	19
8	15
10	17
15	14
19	13
12	16
9	15

6. The preejection period (Y) obtained from recording the electrocardiograms of human subjects are known to be related linearly to heart rate (X). Fit the linear regression line and determine the residuals. Then comment on the assumptions:

X	65	95	95	90	81	86	90	80	85	98
Y	141	150	112	129	126	163	131	135	167	136

7. The price of a stock (Y) is given at the end of every week (t). For the following data, determine the linear relationship of Y in terms of t. Plot the residuals and comment on the assumptions:

t	1	2	3	4	5	6	7	8
Y	25.125	23.5	24.375	27.0	28.875	30.75	29.0	28.625

8. In an experiment, the amount of heavy metals in dry grass was measured along a coastal area. The following data are given in terms of the log content of lumpets (Y) and the log dry weight (X) of seven specimens. Fit a linear regression of Y on X and plot the residuals:

X	Y
$-.7$	-1.7
$-.5$	-1.3
$.1$	-1.1
1.2	$-.5$
1.3	$-.2$
1.9	$-.1$
2.1	$-.3$

Source: Burton, K. W.; Morgan, E.; and Williams, A. T., Trace heavy metals in *Fucus* and *P. Vulgata* along the Glamorgan Heritage Coast, South Wales, *Water, Air and Soil Population 19*, 1983, 377–388.

9. In an experiment with $n = 15$, the following data were obtained. A linear relation is assumed:

$$\bar{X} = 5 \qquad\qquad \bar{Y} = 10 \qquad\qquad \sum(X_i - \bar{X})^2 = 100$$

$$\sum(Y_i - \bar{Y})^2 = 150 \qquad\qquad \sum(X_i - \bar{X})(Y_i - \bar{Y}) = -10$$

   (a) Find the regression equation
   (b) Find the estimate of σ^2

10. For fitting a regression model, we are given the following data:

$$n = 25,\ \bar{X} = 10,\ \bar{Y} = 2.3,$$
$$\sum(X_i - \bar{X})^2 = 175,\ \sum(Y_i - \bar{Y})^2 = 15.25$$
$$\sum(X_i - \bar{X})(\bar{Y}_i - \bar{Y}) = 30.5$$

   (a) Fit the regression of X on Y
   (b) Find the estimate of σ^2

The Inference for Regression Parameters

For the estimation of regression parameters by least squares, no assumption about the distribution of Y was made. The only assumption made in the model given by equation (9.2) was that the errors are independent and have an expectation of zero. To find the confidence intervals for β_0 and β_1 or to test the hypotheses about them, it is necessary to make more assumptions on the error e. For this purpose, assume that the errors are normally distributed with a mean of zero and a variance of σ^2. Without this assumption, we cannot test the hypotheses or give confidence-interval estimates for β_0 and β_1.

Now we have the model of linear regression for $i = 1, 2, \ldots, n$

$$Y_i = \beta_0 + \beta_1 X_i + e_i,$$

where the e_is are independently distributed as normal with a mean of zero and a variance of σ^2. Then the distributions of $\hat{\beta}_0$ and $\hat{\beta}_1$ are also normal. The results:

$\hat{\beta}_0$ is normally distributed with mean β_0 and variance $\dfrac{\sigma^2 \sum X_i^2}{n \sum (X_i - \bar{X})^2}$

$\hat{\beta}_1$ is normally distributed with mean β_1 and variance $\dfrac{\sigma^2}{\sum (X_i - \bar{X})^2}$

This assumption also provides the distribution of $\hat{\sigma}^2$:

$\dfrac{(n-2)\hat{\sigma}^2}{\sigma^2}$ has the chi-square distribution with $n-2$ degrees of freedom

Using the methods in Chapter 7 for developing confidence intervals and testing hypotheses, we can obtain confidence intervals and tests of hypotheses about β_0 β_1 an σ^2.

The Confidence Interval for β_0, β_1 and σ^2

The $(1 - \alpha)$-level confidence interval for β_0 is:

$$\left(\hat{\beta}_0 - \hat{\sigma} t_{n-2, 1-\alpha/2} \sqrt{\frac{\sum X_i^2}{n \sum (X_i - \bar{X})^2}}, \hat{\beta}_0 + \hat{\sigma} t_{n-2, 1-\alpha/2} \sqrt{\frac{\sum X_i^2}{n \sum (X_i - \bar{X})^2}} \right) \quad (9.11)$$

The $(1 - \alpha)$-level confidence intervals for β_1 are:

$$\left(\hat{\beta}_1 - t_{n-2, 1-\alpha/2} \frac{\hat{\sigma}}{\sqrt{\sum (X_i - \bar{X})^2}}, \hat{\beta}_1 + t_{n-2, 1-\alpha/2} \frac{\hat{\sigma}}{\sqrt{\sum (X_i - \bar{X})^2}} \right) \quad (9.12)$$

The $(1 - \alpha)$-level confidence intervals for σ^2 are:

$$\left(\frac{(n-2)\hat{\sigma}^2}{\chi_{n-2, 1-\alpha/2}^2}, \frac{(n-2)\hat{\sigma}^2}{\chi_{n-2, \alpha/2}^2} \right) \quad (9.13)$$

Example 9.5: Consider the data in Example 9.3. Find the 95% confidence interval estimates for β_0, β_1, and σ^2. Since $t_{5,.975} = 2.571$, the confidential interval for β_0 is:

$$\left(108.6 - (2.571) \sqrt{\frac{20,300 \times 4.83}{7 \times 2800}}, 108.6 + (2.571) \sqrt{\frac{20,300 \times 4.83}{7 \times 2800}} \right)$$

$$= (108.6 - 5.75, 108.6 + 5.75) = (102.85, 114.35)$$

The confidence interval for β_1 is:

$$\left(.543 - (2.571) \sqrt{\frac{4.83}{2800}}, .543 + (2.571) \sqrt{\frac{4.83}{2800}} \right)$$

$$= (.543 - .107, .543 + .107) = (.436, .650)$$

The confidence interval for σ^2 is:

$$\left(\frac{5(4.83)}{12.8}, \frac{5(4.83)}{.831}\right) = (1.89, 29.06)$$

Testing the Hypotheses About β_0, β_1, and σ^2

An important application of linear regression analysis is in determining the association between the variables. When a linear association as given by Equation (9.2) is assumed, the association can be tested by the hypothesis for the *slope β_1*.

$$H: \beta_1 = 0$$
$$A: \beta_1 \neq 0 \tag{9.14}$$

If the null hypothesis is accepted, there is no linear association. Later we will test this hypothesis also in terms of the correlation coefficient. Here we give the test statistic for Hypothesis (9.14) using the regression estimate.

Decision rule: Reject H when $\hat{\beta}_1 > k_2$ or $\hat{\beta}_1 < k_1$. Since $\hat{\beta}_1$ is normally distributed with a mean of β_1, which is zero under the null hypothesis, and a variance of $\sigma^2/\sum(X_i - \bar{X})^2$, the distribution of $(\hat{\beta}_1 - \beta_1)\sqrt{\sum(X_i - \bar{X})^2}/\hat{\sigma}$ has t_{n-2} distribution. This fact has already been used in giving confidence intervals for β_1. For a given level of significance, k_1 and k_2 can be determined from the tables. These are:

$$k_1 = -t_{n-2,1-\alpha/2}\frac{\hat{\sigma}}{\sqrt{\sum(X_i - \bar{X})^2}}$$

$$k_2 = t_{n-2,1-\alpha/2}\frac{\hat{\sigma}}{\sqrt{\sum(X_i - \bar{X})^2}} \tag{9.15}$$

Similarly, tests for one-sided alternatives can be given. For testing at level α:

$$H: \beta_1 = 0$$
$$A: \beta_1 < 0$$

Decision rule: Reject H when $\hat{\beta}_1 < k$ where

$$k = -t_{n-2,1-\alpha}\frac{\hat{\sigma}}{\sqrt{\sum(X_i - \bar{X})^2}} \tag{9.16}$$

For testing at level α:

$$H: \beta_1 = 0$$
$$A: \beta_1 > 0$$

Decision rule: Reject H if $\hat{\beta}_1 > k$ where

$$k = t_{n-2,1-\alpha} \frac{\hat{\sigma}}{\sqrt{\sum (X_i - \bar{X})^2}} \tag{9.17}$$

We also give the test for the hypothesis for the intercept β_0, where β_{01} is some given number:

$$H: \beta_0 = \beta_{01}$$
$$A: \beta_0 \neq \beta_{01} \tag{9.18}$$

A special case of interest is when $\beta_{01} = 0$, since we may want to know whether the regression line passes through the origin. The intercept in this case is zero.

Decision rule: Reject H when $\hat{\beta}_0 > k_2$ or $\hat{\beta}_0 < k_1$, k_1, k_2 are determined as before by the following:

$$k_1 = \beta_{01} - t_{n-2,1-\alpha/2}\,\hat{\sigma} \sqrt{\frac{\sum X_i^2}{n \sum (X_i - \bar{X})^2}}$$

$$k_2 = \beta_{01} + t_{n-2,1-\alpha/2}\,\hat{\sigma} \sqrt{\frac{\sum X_i^2}{n \sum (X_i - \bar{X})^2}} \tag{9.19}$$

For one-sided hypotheses, the results are similar.

Tests for the variance σ^2 are obtained as before, since $(n-2)\hat{\sigma}^2/\sigma^2$ has χ^2_{n-2}. Suppose we want to test:

$$H: \sigma^2 = \sigma_0^2$$
$$A: \sigma^2 \neq \sigma_0^2 \tag{9.20}$$

Decision rule: Reject H when $\hat{\sigma}^2 > k_2$ or $\hat{\sigma}^2 < k_1$ where

$$k_1 = \chi^2_{n-2,\alpha/2} \frac{\sigma_0^2}{n-2}$$

$$k_2 = \chi^2_{n-2,1-\alpha/2} \frac{\sigma_0^2}{n-2} \tag{9.21}$$

For one-sided hypotheses, the computations are similar.

Example 9.6: To test the hypothesis that systolic blood pressure depends linearly on age, we test hypotheses (9.14). We have from Examples 9.3 and 9.5, at level .05, using Equation (9.15):

$$k_1 = -2.571 \sqrt{\frac{4.83}{2800}} = -.107$$

$$k_2 = .107$$

Since $\hat{\beta}_1 = .543$, we reject H and claim that there is linear dependence at level .05. Note also that the confidence interval given in Example 9.5 does not include zero and hence the hypothesis is not accepted, giving the same conclusion as in the above case.

Exercises

11. Find the 90% confidence interval for the slope of the regression line in Exercise 3. Test the hypothesis that the slope is zero at level .05.
12. Find the .95-level confidence interval for the intercept of the regression line in Exercise 4. Test the hypothesis that the intercept is 5 at level .05.
13. For the data in Example 9.2, find the .95-level confidence interval for the variance. Test the hypothesis that the slope of the regression line is zero at level .10.
14. For the regression of length of life of tires on car speeds in Exercise 3, test whether the length of life depends on speed at level .05.
15. The final grades (y) in a course and the midterm grades (x) are given for a random sample of ten students. Fit a linear regression line to the data. Test the hypothesis that final grades do not depend on midterm grades at level .05. Give the .95-level confidence interval for the slope.

x	75	93	78	62	37	53	42	85	58	70
y	65	85	80	55	52	67	58	75	62	70

16. For Exercise 15, give the .95-level confidence intervals for:

    (a) intercept
    (b) variance

17. Test the hypothesis that the intercept in Exercise 15 is zero at level .05.

Prediction

One of the major applications of the linear regression model is for the prediction of a future value. The investment policy of a firm may depend on the predicted values of certain stocks, a treatment regimen for a patient may depend on the predicted value of his blood pressure, and the admission of a student to an academic program may depend on his predicted performance in the future.

We are concerned with two different quantities to be predicted. One is the mean value of Y when X equals x_0, and the other is the single value of Y when $X = x_0$. The predicted value for $X = x_0$ is:

$$\hat{y}_0 = \hat{\beta}_0 + \hat{\beta}_1 x_0 \qquad (9.22)$$

Knowing the probability distributions of $\hat{\beta}_0$ and $\hat{\beta}_1$, we can find the probability distribution of $\hat{y}_0$. It can be shown that $\hat{y}_0$ is also normally distributed with a mean of

$$E(\hat{y}_0) = \beta_0 + \beta_1 x_0 \tag{9.23}$$

and a variance of

$$\text{Var}(\hat{y}_0) = \sigma^2 \left(\frac{1}{n} + \frac{(x_0 - \bar{x})^2}{s_x^2} \right)$$

Let its estimate be given by s_0^2. That is:

$$s_0^2 = \hat{\sigma}^2 \left(\frac{1}{n} + \frac{(x_0 - \bar{x})^2}{s_x^2} \right) \tag{9.24}$$

The Confidence Interval for $\beta_0 + \beta_1 x_0$

The $(1 - \alpha)$-level confidence interval for $\beta_0 + \beta_1 x_0$ can now be found as before and is given by:

$$(\hat{y}_0 - t_{n-2, 1-\alpha/2} s_0, \hat{y}_0 + t_{n-2, 1-a/2} s_0) \tag{9.25}$$

as shown in Figure 9.8 for various values of x_0.

For the prediction interval for a single Y at x_0, the variance is increased and equals:

$$\sigma^2 \left[1 + \frac{1}{n} + \frac{(x_0 - \bar{x})^2}{s_x^2} \right]$$

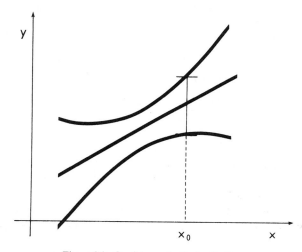

Figure 9.8. Confidence limits for the line.

The prediction interval can be similarly obtained, using the estimate:

$$s_1^2 = \hat{\sigma}^2\left[1 + \frac{1}{n} + \frac{(x_0 - \bar{x})^2}{s_x^2}\right] \tag{9.26}$$

The $(1 - \alpha)$-level prediction interval for a *single* Y is given by:

$$(\hat{y}_0 - t_{n-2,\alpha/2}s_1, \hat{y}_0 + t_{n-2,1-\alpha/2}s_1) \tag{9.27}$$

Example 9.7: The regression equation in Example 9.1 is:

$$y = 5.14 + 1.27x$$

Suppose we want to predict the height of an infant with weight $x_0 = 10$ pounds. Then the predicted height is:

$$\hat{y}_0 = 5.14 + 1.27(10) = 17.84$$

Also, the estimate of $\hat{\sigma}^2$ is 10.91, so that

$$s_0^2 = 10.91\left[\frac{1}{8} + \frac{(10 - 12)^2}{162}\right] = 1.63$$

and

$$s_1^2 = 10.91\left[1 + \frac{1}{8} + \frac{(10 - 12)^2}{162}\right] = 12.54$$

The 95% confidence interval for the mean height of 10-pound infants is:

$$= (17.84 - 2.447(1.277), 17.84 + 2.447(1.277))$$
$$= (17.84 - 3.12, 17.84 + 3.12)$$
$$= (14.72, 20.96)$$

The 95% prediction interval for a single height of a 10-pound infant is:

$$= (17.84 - 2.447(3.542), 17.84 + 2.447(3.542))$$
$$= (17.84 - 8.67, 17.84 + 8.67)$$
$$= (9.17, 26.51)$$

Exercises

18. In Exercise 4, predict the marriage rate for the year 1965. Find the 95% confidence intervals for the mean marriage rate for the year 1965.
19. In Exercise 3, what is the predicted length of life of tires for a speed of 43 miles per hour? Find the .90 prediction interval for the life of a single tire.
20. The body weight (x) and thyroid weight (y) are given for eight rat fetuses below. Find the regression of y on x and predict the thyroid weight of a rat fetus whose body weight is 2.50 g. Give a 95% confidence interval for the

mean thyroid weight for a rat whose body weight is 1.50 g. Give a .95
prediction interval for the thyroid weight for a body weight equal to 1.50 g.

Body Weight (g) x	Thyroid weight (mg) y
1.41	0.55
1.43	0.56
1.20	0.40
1.29	0.46
2.49	1.19
2.67	0.95
1.58	0.60
1.75	0.62

Source: Euchi, Y.; Yamamoto, M.; and Lee, M., Thyroid response in
fetuses of calorie-restricted pregnant rats given Goctrogen, *Proc. Soc.
Exp. Bio. Med.*, 1983 *173*, 436–440.

21. In Exercise 6, what is the preejection period of a subject whose heart
rate is 84? Give the 90% prediction interval for a single period for the
above heart rate.

22. In a regression problem with $n = 10$, we are given the data:

$$\bar{X} = 50 \qquad\qquad \bar{Y} = 75 \qquad\qquad \sum(x_i - \bar{X})^2 = 1600$$

$$\sum(y_1 - \bar{Y})^2 = 1800 \qquad\qquad \sum(x_i - \bar{Y})(y_i - \bar{Y}) = 800$$

(a) Find the linear regression equation
(b) Find the 95% confidence interval for a mean of Y given that $x = 45$
(c) Find the 95% prediction interval for a single Y given that $x = 45$

The Correlation Coefficient

The correlation coefficient is a measure of an association between two random
variables. It measures linear association and has been defined by the ratio of
the covariance to the product of the standard deviations of the two variables.
That is, ρ, the population correlation coefficient between random variables X
and Y, is given by:

$$\rho = \frac{\text{Cov}(X, Y)}{\sqrt{\text{Var } X \text{ Var } Y}} \tag{9.28}$$

Suppose a random sample $(X_1, Y_1), (X_2, Y_2), \ldots, (X_n, Y_n)$ is given. We have seen earlier that the variance of X is estimated by:

$$\frac{\sum(X_i - \bar{X})^2}{n-1}$$

The variance of Y is estimated by:

$$\frac{\sum(Y_i - \bar{Y})^2}{n-1}$$

We can similarly estimate the covariance by:

$$\frac{\sum(X_i - \bar{X})(Y_i - \bar{Y})}{n-1}$$

The estimate of ρ is given by the sample correlation coefficient r:

$$r = \frac{\sum(X_i - \bar{X})(Y_i - \bar{Y})}{\sqrt{\sum(X_i - \bar{X})^2 \sum(Y_i - \bar{Y})^2}} \tag{9.29}$$

$$= \frac{n\sum X_i Y_i - \sum X_i \sum Y_i}{\sqrt{[n\sum X_i^2 - (\sum X_i)^2][n\sum Y_i^2 - (\sum Y_i)^2]}} \tag{9.30}$$

The correlation coefficient r was proposed by Karl Pearson and is also known as Pearson's Correlation Coefficient. The values of the sample correlation coefficient vary between -1 and 1. Also note that the dependent and independent variables are both assumed to be random here, whereas in the linear regression model the independent variable is assumed to be given and hence is assumed to be nonrandom. This is the difference between correlation and regression. (The connections between regression and correlations will be discussed next.) When regression of Y on X is linear, then the slope of the regression line $\hat{\beta}_1$ can be expressed in terms of r. That is,

$$\hat{\beta}_1 = \frac{s_y}{s_x} r \tag{9.31}$$

so that one can be obtained from the other. Also a test based on r can be reduced to a test based on $\hat{\beta}_1$ and vice versa. An important problem is to test the hypothesis that the correlation coefficient ρ is zero. That is, test that two random variables are uncorrelated by testing $\rho = 0$. This is the same as testing $\beta = 0$ when X, Y are jointly normal:

$$H: \rho = 0$$
$$A: \rho \neq 0 \tag{9.32}$$

The test statistic used is r.

Decision rule: Reject H when $r < k_1$ or $r > k_2$.

The constants k_1 and k_2 are determined from the distribution of r under the null hypothesis. This distribution is tabulated in Table VIII in the Appendix. Although the distribution of the sample correlation coefficient is not symmetric, we use the same convention as before for convenience to assign equal probabilities in tails for the two-sided test.

The Coefficient of Determination

The square of the correlation coefficient r^2 is called the *coefficient of determination*. It can be shown that,

$$\sum(Y_i - \bar{Y})^2 = \sum(Y_i - \hat{Y}_i)^2 + \sum(\hat{Y}_i - \bar{Y})^2$$

where $\hat{Y}_i$ is the predicted Y at X_i. The partitioning of the sum of squares of deviations of Y_i from $\bar{Y}$ as a sum of two sums of squares is suggestive. The part $\sum(\hat{Y}_i - \bar{Y})^2$ explains the regression, and the other part explains the random variation. The ratio of $\sum(\hat{Y}_i - \bar{Y})^2$ to the total sum of squares is denoted by r^2. That is:

$$r^2 = \frac{\sum(\hat{Y}_i - \bar{Y})^2}{\sum(Y_i - \bar{Y})^2}$$

It can be seen that r^2 also equals:

$$r^2 = \hat{\beta}_1 \frac{\sum(X_i - \bar{X})(Y_i - \bar{Y})}{\sum(Y_i - \bar{Y})^2}$$

Like r, it is a good measure of the strength of the relationship between X and Y.

Transformation

One transformation of the sample correlation coefficient has t-distribution with $n - 2$ degrees of freedom:

$$t_{n-2} = \sqrt{\frac{r^2(n-2)}{1-r^2}} \tag{9.33}$$

In testing the hypothesis that $\rho = 0$, we can also use the t-statistic or F-statistic, since t_{n-2}^2 has an F-distribution with 1 and $n - 2$ degrees of freedom.

The *distribution for large n*: For large n, another approximation is commonly used. It is well known that the distribution of $\frac{1}{2}\log(1 + r)/(1 - r)$ is normal with a mean of $\frac{1}{2}\log(1 + \rho/1 - \rho)$ and a variance of $1/(n - 3)$. This approximation can be used for testing hypotheses or obtaining confidence intervals for ρ.

Example 9.8: For a sample of size 39, the correlation coefficient $r = .48$. Test the hypothesis that the population correlation coefficient $\rho = .4$ at level .05.

Test the hypothesis:

$$H: \rho = .4$$
$$A: \rho \neq .4$$

The test statistic is:

$$Z = \frac{\frac{1}{2}\log\frac{1+r}{1-r} - \frac{1}{2}\log\frac{1+\rho}{1-\rho}}{\sqrt{\frac{1}{n-3}}}$$

Decision Rule: Reject H if $Z > k_2$ or $Z < k_1$ where k_1, k_2 are determined from standard normal distribution.

We have here:

$$Z = \sqrt{36}\left(\frac{1}{2}\log\frac{1+.48}{1-.48} - \frac{1}{2}\log\frac{1+.4}{1-.4}\right)$$

$$= 3(\log 2.846 - \log 2.333)$$

$$= 3\log 1.2199 = 3(.199) = .597$$

For $\alpha = .05$, we have $k_1 = -1.96$, and $k_2 = 1.96$. Hence we accept H.

Example 9.9: For Example 9.3, the sample correlation coefficient is given by:

$$r = \frac{1520}{\sqrt{(2800)(849.71)}} = .985$$

Testing the hypothesis that X and Y are uncorrelated, we use the statistic in Equation (9.33):

$$t_6 = .985\sqrt{\frac{5}{1-.970}} = 12.72$$

At the $\alpha = .05$ level, the hypothesis is rejected, since $t_{6,.975} = 2.447$.

The Rank Correlation Coefficient

When the data are given in ordinal scale or ranks, another measure of association, called the *rank correlation coefficient*, can be used. The calculations for rank correlation coefficient are the same as those for correlation coefficient r, except that the Xs and Ys are replaced by their ranks. A further simplification is obtained by the following. Let $X_1, \ldots, X_n$ be ranked among themselves and let R_1 be the rank of X_1, R_2 the rank of X_2 and so on, let R_n be the rank of X_n. Similarly, let $Y_1, \ldots, Y_n$ be ranked among themselves and let

$R'_1, \ldots, R'_n$ be their ranks. Let $R_i - R'_i = d_i$, $i = 1, \ldots, n$. Then the rank correlation coefficient r is given by:

$$r_s = 1 - \frac{6 \sum d_i^2}{n(n^2 - 1)} \tag{9.34}$$

Note that r_s is also known as the *Spearman Correlation Coefficient*. The probability distribution of r_s has been tabulated for small values of n in Table X in the Appendix.

The *distribution for large n*: For large n, the distribution of r_s is normal with a mean of zero and a variance of $1/(n-1)$ under the null hypothesis.

Example 9.10: Two radiologists rank X-rays for six patients according to the severity of disease:

Radiologists	1	2	3	4	5	6
A	3	5	4	2	1	6
B	2	4	3	6	5	1
d_i	1	1	1	-4	-4	5

$$\sum d_i^2 = 60$$

$$r_s = 1 - \frac{60(6)}{6(36-1)} = -.714$$

Exercises

23. Let the systolic blood pressure of white females (Y) be linearly dependent on age (X). A sample of seven is given:

X	20	30	40	50	60	70	80
Y	111	115	122	132	145	158	156

(a) Find the correlation coefficent between X and Y
(b) Test the hypothesis that $\rho = 0$ at level .05
(c) Find the rank correlation

24. The heights and weights of 10 subjects about 60 yeas old, are given below:

height (inches)	65	63	68	70	62	67	71	66	69	60
weight (pounds)	155	157	150	160	165	150	152	150	145	140

(a) Find the correlation coefficient between height and weight
(b) Test the hypothesis that the correlation is zero at level .05
(c) Find the rank correlation coefficient between height and weight

25. The number of Americans unemployed and employed by the Armed Forces is given for a period of eight years. Find the correlation coefficient between them and test the hypothesis that they are uncorrelated at level .10. Figures are in millions:

Unemployed	Size of Armed Forces
2.3	1.6
2.3	1.4
3.7	1.6
3.3.	1.6
2.1	3.1
1.9	3.6
1.9	3.5
3.6	3.4

26. Two trace metals, zinc (Y) and cadmium (X), are measured in fish from a certain river. It is known that log Y and log X are linearly related. Find the correlation coefficient between $u = \log Y$ and $v = \log X$. The numbers are in micrograms per gram:

Y	X
78	0.2
6	0.1
15	0.7
68	0.6
1	0.1
250	1.5
52	0.2
69	0.9
57	3.0
71	1.9

Source: Paul, A. C., and Pillai, K. C., Trace metals in a tropical river environment-speciation and biological transfer, *Water, Air and Soil Pullution*, 1983, *19*, 75–86.

Multiple Linear Regression

In the previous sections we considered the linear dependence of one variable on another. However, there are many practical situations where several variables may simultaneously affect a given variable. For example, the weight of a subject may depend on food intake and the amount of exercise. The price of a stock on a given day may depend on its price on the previous day and the price of a related stock. The prediction of the dependent variables such as weight or price in terms of several independent variables is extremely important in scientific, industrial, and business applications.

Let Y be the dependent variable that depends linearly on several variables $X_1, X_2, \ldots, X_k$. For simplicity, we consider the case of only two independent variables in the following model, where β_0, β_1, and β_2 are unknown parameters.

$$Y = \beta_0 + \beta_1 X_1 + \beta_2 X_2 + e \qquad (9.35)$$

We assume further that e is the random error with a mean of zero and an unknown variance of σ^2. As usual, we shall assume that e is normally distributed for purposes of testing the hypotheses and confidence-interval estimation.

Suppose a random sample of n observations on (Y, X_1, X_2) is given and we observe the following data matrix:

$$
\begin{matrix}
Y_1 & X_{11} & X_{21} \\
Y_2 & X_{12} & X_{22} \\
\vdots & \vdots & \vdots \\
Y_n & X_{1n} & X_{2n}
\end{matrix}
$$

The method of least squares provides estimates of β_0, β_1, and β_2 as in the case of simple linear regression. We minimize:

$$\sum_{i=1}^{n} (Y_i - \beta_0 - \beta_1 X_{1i} - \beta_2 X_{2i})^2 \qquad (9.36)$$

Mathematical arguments provide the minimum in terms of the solutions of the following simultaneous linear equations known as normal equations:

$$
\begin{aligned}
n\beta_0 + \beta_1 \sum X_{1i} + \beta_2 \sum X_{2i} &= \sum Y_i \\
\beta_0 \sum X_{1i} + \beta_1 \sum X_{1i}^2 + \beta_2 \sum X_{1i}X_{2i} &= \sum Y_i X_{1i} \\
\beta_0 \sum X_{2i} + \beta_1 \sum X_{1i}X_{2i} + \beta_2 \sum X_{2i}^2 &= \sum Y_i X_{2i}
\end{aligned}
\qquad (9.37)
$$

The solutions of these equations give the estimates of β_0, β_1, and β_2. We do not obtain the explicit solutions of these equations here.

For the model involving three independent variables

$$Y = \beta_0 + \beta_1 X_1 + \beta_2 X_2 + \beta_3 X_3 + e$$

we have the following normal equations:

$$n\beta_0 + \beta_1 \sum X_{1i} + \beta_2 \sum X_{2i} + \beta_3 \sum_{3i} = \sum Y_i$$
$$\beta_0 \sum X_{1i} + \beta_1 \sum X_{1i}^2 + \beta_2 \sum X_{1i}X_{2i} + \beta_3 \sum X_{1i}X_{3i} = \sum X_{1i}Y_i$$
$$\beta_0 \sum X_{2i} + \beta_1 \sum X_{1i}X_{2i} + \beta_2 \sum X_{2i}^2 + \beta_3 \sum X_{2i}X_{3i} = \sum X_{2i}Y_i$$
$$\beta_0 \sum X_{3i} + \beta_1 \sum X_{1i}X_{3i} + \beta_2 \sum X_{2i}X_{3i} + \beta_3 \sum X_{3i}^2 = \sum X_{3i}Y_i$$

These equations can be extended for the general multiple regression model:

$$Y = \beta_0 + \beta_1 X_1 + \beta_2 X_2 + \ldots + \beta_p X_p + e$$

In matrix notation, many of the above calculations can be expressed simply (see books on regression analysis at the introductory level; for example, Neter and Wasserman [1974]).

The tests of hypotheses about the parameters of the model will be considered in Chapter 10.

Example 9.11: Consider the following data on Y, X_1, and X_2 for $n = 5$.

Y	X_1	X_2
2	0	0
4	-1	1
1	1	-1
5	0	1
2	1	0

To fit the linear model in Equation (9.35) for the above data, we have:

$$\sum X_{1i} = 1 \qquad \sum X_{2i} = 1 \qquad \sum Y_i = 14$$
$$\sum X_{1i}^2 = 3 \qquad \sum X_{2i}^2 = 3 \qquad \sum Y_i^2 = 50$$
$$\sum X_{1i}X_{2i} = -2 \qquad \sum Y_i X_{1i} = -1 \qquad \sum Y_i X_{2i} = 8$$

The normal equations from (9.37) are given:

$$5\beta_0 + \beta_1 + \beta_2 = 14$$
$$\beta_0 + 3\beta_1 - 2\beta_2 = -1$$
$$\beta_0 - 2\beta_1 + 3\beta_2 = 8$$

These equations lead to the least-square solution:

$$\hat{\beta}_0 = \tfrac{7}{3}$$
$$\hat{\beta}_1 = \tfrac{4}{15}$$
$$\hat{\beta}_2 = \tfrac{31}{15}$$

The estimated linear relation among $\hat{Y}$, X_1, and X_2 is given by the following equation, where $\hat{Y}$ denotes the predicted value of Y:

$$\hat{Y} = \tfrac{7}{3} + \tfrac{4}{15}X_1 + \tfrac{31}{15}X_2$$

Such a linear relation describes a plane in three dimensions in the same way as a linear relation describes a straight line in two dimensions. The predicted relation is simplified to be:

$$15\hat{Y} = 35 + 4X_1 + 31X_2$$

The observed and the corresponding predicted values are given in Table 9.1.

Table 9.1

Observed Y_i	Predicted $\hat{Y}_i$	Residual $Y_i - \hat{Y}_i$	Residual Square $(Y_i - \hat{Y}_i)^2$
2	7/3	−1/3	1/9
4	62/15	−2/15	4/225
1	8/15	7/15	49/225
5	66/15	9/15	81/225
2	39/15	−9/15	81/225
		Total	16/15

The sum of squares of the residuals is 16/15. The computations are given in fractions to get correct answers. To compare the above linear model with other models, the residual sum of squares will be a useful measure.

Example 9.12: Let D be the infant mortality rate (deaths per 1,000 live births), N be the number of nurses per 10,000 persons, and P be the number of physicians per 10,000 persons in an underdeveloped country. The data on six countries are given below:

D	N	P
46	3.7	2.71
121.4	6.2	1.97
129.0	7.3	1.56
70.3	5.8	3.0
73.9	9.7	4.63
44.9	36.5	10.82

Source: Flegg, A. T., Inequality of income, illiteracy and medical care as determinants of infant mortality in underdeveloped countries, *Population Studies*, 1983, *36*, 441–458.

We want to obtain a multiple regression equation with a dependent variable of $Y = \log_e D$ and independent variables of $X_1 = \log_e N$ and $X_2 = \log_e P$. The transformed data are:

Y	X_1	X_2
3.83	1.31	1.0
4.80	1.82	.68
4.86	1.99	.44
4.25	1.76	1.10
4.30	2.72	1.53
3.80	3.60	2.38

To obtain the least-squares estimate, we have:

$$\sum X_{1i} = 13.20 \qquad \sum X_{2i} = 7.13 \qquad \sum Y_i = 25.84$$
$$\sum X_{1i}^2 = 32.4446 \qquad \sum X_{1i} X_{2i} = 18.0888 \qquad \sum Y_i X_{1i} = 56.2807$$
$$\sum X_{1i} X_{2i} = 18.0888 \qquad \sum X_{2i}^2 = 10.8713 \qquad \sum Y_i X_{2i} = 29.5304$$

The least square equations are:

$$6\beta_0 + 13.20\beta_1 + 7.13\beta_2 = 25.84$$
$$13.20\beta_0 + 32.4446\beta_1 + 18.0888\beta_2 = 56.2807$$
$$7.13\beta_0 + 18.0888\beta_1 + 10.8713\beta_2 = 29.5304$$

The estimates are obtained as:

$$\hat{\beta}_0 = 4.2710$$
$$\hat{\beta}_1 = .6125$$
$$\hat{\beta}_2 = -1.1040$$

The regression equation is:

$$\hat{Y} = 4.2710 + .6125 X_1 - 1.1040 X_2$$

The residuals are:

Y_i	$\hat{Y}_i$	$Y_i - \hat{Y}$	$(Y_i - \hat{Y}_i)^2$
3.83	3.97	$-.14$	.0196
4.80	4.64	.16	.0256
4.86	5.00	$-.14$	.0196
4.25	4.13	.12	.0144
4.30	4.25	.05	.0025
3.80	3.85	$-.05$	.0025
		Total	.0842

Exercises

27. For the following data, fit a regression line of Y in terms of X_1 and X_2. Obtain the residuals:

Y	X_1	X_2
7	1	2
2	0	1
5	1	3
10	2	3
2	1	0

28. For the following data on Y, X_1, X_2, and X_3, fit the regression line $Y = \beta_0 + \beta_1 X_1 + \beta_2 X_2 + \beta_3 X_3 + e$ and obtain the residuals:

Y	X_1	X_2	X_3
6	0	1	2
5	1	0	2
1	1	0	0
5	2	0	1
7	1	1	1
8	1	2	1

29. It is believed that the salinity (Y) of sea water depends on its depth (X_1) and temperature (X_2). Fit a linear regression to the following data and calculate the residuals:

Y (%)	X_1 m	X_2 (°C)
30.2	1	2.79
32.4	12	0.86
33.1	33	− 1.66
34.4	396	1.76
32.8	14	− 0.14
32.5	23	− 1.41

Source: Effect of sea ice meltwater on the alkalinity of sea water by Jones, E. P.; Coote, A. R.; and Levy, E. M., *Journal of Marine Research*, 1983, *41*, 43–52.

30. Suppose the infant mortality rate D in Example 9.12 depends linearly on the number of nurses and physicians and the percentage of women who are illiterate. The data on ten countries are given in terms of their logs. Fit the regression of Y on X_1, X_2, and X_3 and calculate the residuals:

Y	X_1	X_2	X_3
5.03	-2.30	.27	4.51
4.96	$-.51$	-2.04	4.47
4.90	.69	73	4.39
4.96	.18	$-.46$	4.30
3.83	1.31	1.00	3.45
4.80	1.82	.68	4.02
4.86	1.99	.44	4.23
4.25	1.76	1.10	2.82
4.30	2.72	1.53	3.06
3.80	3.60	2.38	3.08

Source: Flegg, A. T., Inequality of income, illiteracy and medical care as determinants of infant mortality in underdeveloped countries, *Population Studies*, 1983, 36 441–458.

Nonlinear Regression

The relationship among two variables may not be linear, and we may be required to fit other mathematical functions to the data. There are many possibilities. We consider first polynomial regression. Consider that the relationship between the peak left ventricle systolic pressure (Y) and the frontal plane QRS-T angle (X) in the study of electrocardiograms is quadratic. It is given by:

$$Y = \beta_0 + \beta_1 X + \beta_2 X^2 + e \qquad (9.38)$$

The model given above is a multiple linear regression model, since by substituting $X = X_1$ and $X^2 = X_2$ we can transform the model in Equation (9.38) as:

$$Y = \beta_0 + \beta_1 X_1 + \beta_2 X_2 + e$$

This is the same model as in Equation (9.35).

The predicted value of Y from the estimated equation is denoted by $\hat{Y}$. A measure of the goodness of fit of the model is taken by the sum of squares of the residuals $Y_i - \hat{Y}_i$ for all i. That is,

$$\sum (Y_i - \hat{Y})^2 \tag{9.39}$$

gives an index for the fit of a model. In choosing among various models, sometimes that in Equation (9.39) is used as an index. We consider an example before mentioning some other nonlinear models.

Example 9.13: Suppose now we fit the following model to the same data of Example 9.11:

$$Y = \beta_0 + \beta_1 X + \beta_2 X_1^2 + e$$

We have the following data:

Y	X_1	$X_2 = X_1^2$
2	0	0
4	-1	1
1	1	1
5	0	0
2	1	1

Now we have:

$$\sum X_{1i} = 1 \qquad \sum X_{2i} = 3 \qquad \sum Y_i = 14$$
$$\sum X_{1i}^2 = 3 \qquad \sum X_{2i}^2 = 3 \qquad \sum Y_i^2 = 50$$
$$\sum X_{1i} X_{2i} = 1 \qquad \sum Y_i X_{1i} = -1 \qquad \sum Y_i X_{2i} = 7$$

The normal equations are given by:

$$5\beta_0 + \beta_1 + 3\beta_2 = 14$$
$$\beta_0 + 3\beta_1 + \beta_2 = -1$$
$$3\beta_0 + \beta_1 + 3\beta_2 = 7$$

The least-square solutions are given by:

$$\hat{\beta}_0 = \tfrac{7}{2}$$
$$\hat{\beta}_1 = -\tfrac{5}{4}$$
$$\hat{\beta}_2 = -\tfrac{3}{4}$$

So the predicted relation is given by:

$$\hat{Y} = \tfrac{7}{2} - \tfrac{5}{4}X - \tfrac{3}{4}X^2$$

The residual sum of squares in this case turns out to be 5.

A comparison of the residual sum of squares with the model in Example 9.11

recommends discarding the above parabolic model in favor of the linear model.

Other Relationships

In some nonlinear situations, the models can be transformed to linear models.
    (I) Consider the model

$$Y = e\beta_0 \exp \beta_1 X \tag{9.40}$$

where e is now a multiplicative error. By taking the logarithm, we have:

$$\log Y = \log \beta_0 + \beta_1 X + \log e$$

This equation is linear in the parameters, since it reduces to

$$Z = \beta_0' + \beta_1 X + \eta \tag{9.41}$$

where $Z = \log Y$, $\beta_0' = \log \beta_0$, and $\eta = \log e$. Equation (9.41) is fitted to the transformed data. The assumptions on errors in Equation (9.40) and its transformed version Equation (9.41) should be compatible.
    (II) Consider the model:

$$Y = \beta_0 + \frac{\beta_1}{X} + e \tag{9.42}$$

Again, if we take $1/X = Z$, we have

$$Y = \beta_0 + \beta_1 Z + e \tag{9.43}$$

which is linear.
    (III) Models such as the following are not easily amenable to linear transformation:

$$Y = \beta_0 + e\beta_1 \exp(\beta_2 X) \tag{9.44}$$

We have:

$$\log(Y - \beta_0) = \log \beta_1 + \beta_2 X + \log e \tag{9.45}$$

Such models are fitted with the help of numerical techniques.

Exercises

31. The data on peak systolic pressure (Y), the frontal plane $QRS\text{-}T$ angle (X_1), and the age (X_2) of six subjects is given. Fit the following models:
    (a) $Y = \beta_0 \times \beta_1 X_1 + \beta_2 X_2 + e$
    (b) $Y = \beta_0 + \beta_1 X_1 + \beta_2 X_2^2 + e$

Y (mnHg)	X_1 (degrees)	X_2 (years)
126	30	1.7
158	55	11.5
135	15	4.0
160	50	2.5
175	45	11.0
200	90	12.0

32. For the data in Exercise 31, fit the models:

(a) $Y = \beta_0 + \beta_1 X_1 + \beta_2 X_1^2 + \beta_3 X_2 + e$
(b) $Y = \beta_0 + \beta_1 X_1 + \beta_2 X_1^2 + \beta_3 X_1^3 + e$

33. The times of deaths of guinea pigs (Y) exposed to different levels of X-rays (X) are given by the table:

X-ray dose	200	300	400	500	700	800
Times	7	5	3	5	3	2
of	22	7	7	8	7	3
death	27	20	8	10	8	5
	28	29	10	15	4	
	30	10				

(a) Assume Equation (9.40) and obtain the estimates of β_0 and β_1
(b) Assume Equation (9.42) and estimate β_0 and β_1

The Multiple Correlation Coefficient

An important problem in multiple regression is concerned with testing hypotheses about the adequacy of the linear model. In a way, what is needed is a test for a hypothesis that the predicted and observed values agree if the model is correct. A measure to test this hypothesis is the multiple correlation coefficient. In the case of simple linear regression, the correlation coefficient between the predicted and observed values is the sample correlation coefficient r. In general, the *multiple correlation coefficient* is the maximum of all correlation coefficients between the dependent variable and all possible linear combinations of the independent variables. To distinguish it from an ordinary correlation coefficient, the multiple correlation coefficient is denoted by R.

Suppose we have the model with p independent variables:

$$Y = \beta_0 + \beta_1 X_1 + \cdots + \beta_p X_p + e \qquad (9.46)$$

Suppose $\hat{Y}$ denotes the predicted value of Y based on a random sample of size n. The multiple correlation coefficient is given by

$$R^2 = 1 - \frac{\sum(Y_i - \hat{Y}_i)^2}{\sum(Y_i - \bar{Y})^2} \qquad (9.47)$$

where $\hat{Y}$ is the predicted value of Y_i and $\bar{Y}$ is the average of Y_is. The distribution of R^2 is given in the following.

Under the assumption that the population multiple correlation coefficient is zero, we use the statistic:

$$\frac{R^2(n - p - 1)}{(1 - R^2)p}$$

This has an F-distribution with p and $n - p - 1$ degree of freedom.

We reject the hypothesis that the model is appropriate if R^2 is too small. The test can be implemented by the tables of F-distribution.

The study of the linear dependence on various variables is sometimes done with the help of R^2. As one goes on adding variables to the model, the value of R^2 keeps on increasing. Similarly, one discards variables if the value of R^2 does not reduce too fast. These stepwise procedures for finding the best prediction equation are numerical. Computer programs for such procedures are available.

R^2 measures the proportion of total variation about $\bar{Y}$ explained by regression. It is often expressed as a percentage. If $R^2 = 0.8$, we say that 80% of the total variation has been explained by the regression equation.

Example 9.14: For data in Example 9.11, we know that $\sum(Y_i - \hat{Y}_i)^2 = \frac{16}{15} = 1.07$. Also since $\sum Y_i^2 = 50$ and $\sum Y_i = 14$, $\sum(Y_i - \bar{Y})^2 = 50 - 39.2 = 10.8$.

$$R^2 = 1 - \frac{1.07}{10.8} = 1 - .0988 = .9012$$

That is, 90% of the variation in the data is explained by the regression.

Example 9.15: The multiple regression in Example 9.12 gives the total sum of squares of the residuals:

$$\sum(Y_i - \hat{Y}_i)^2 = .0842$$

Also

$$\sum(Y_i - \bar{Y})^2 = 112.321 - \frac{(25.84)^2}{6}$$

$$= 1.037$$

Hence,

$$R^2 = 1 - \frac{.0842}{1.037} = .919$$

That is, a 92% variation in the data is explained by the regression equation and only about 8% is an unexplained random variation, showing that the fit is good.

Exercises

34. Find the multiple correlation coefficient in Exercise 27.
35. Find the multiple correlation coefficient in Exercise 28. How much variation in the data is explained by the regression equation?
36. Using the multiple correlation coefficient, find the amount of variation in the data explained by the regression in Exercise 35.
37. For the following data, fit the regression, obtain the residuals, and calculate the multiple correlation coefficient:

Y	X_1	X_2
5.13	− 2.3	.3
5.06	1.2	− 2.1
5.00	.7	.8
3.93	.1	− .5
4.86	1.3	1.0
4.40	1.8	.7

Chapter Exercises

1. For the following data, obtain the simple linear regression of Y on X. Test the hypothesis that the slope of the regression is zero at level .05:

Y	X
3.1	2
4.2	1
3.5	3
4.5	5
3.1	4
4.7	7

2. (a) Find the 95% confidence intervals for the intercept and slope of the linear regression in Exercise 1

   (b) Find the correlation coefficient r
   (c) Find the Spearman correlation coefficient

3. The following calculations are given for a data set for $n = 15$ $\bar{X} = 5$, $\bar{Y} = 2$:

$$\sum(X_i - \bar{X})^2 = 50, \sum(Y_i - \bar{Y})^2 = 30, \sum(X_i - \bar{X})(Y_i - \bar{Y}) = 150$$

   (a) Find the regression line
   (b) Give the estimate of variance σ^2
   (c) Give the 95% confidence interval for the mean when $X = 7$
   (d) Give the 95% confidence interval for the single value when $X = 7$
   (e) Give the 90% confidence interval for the slope

4. The following calculations are given for the multiple regression of Y on X_1 and X_2:

$$n = 10, \sum Y_i = 30, \sum Y_i X_{1i} = 50, \sum Y_i X_{2i} = 60$$
$$\sum X_{1i} = 15, \sum X_{1i}^2 = 30, \sum X_{2i} = 5, \sum X_{2i}^2 = 15$$
$$\sum X_{1i} X_{2i} = 20$$

   (a) Write the normal equations and find $\hat{\beta}_0, \hat{\beta}_1, \hat{\beta}_2$
   (b) Find the multiple correlation coefficient if we know that $\sum(Y_i - \bar{Y})^2 = 40$ and $\sum(Y_i - \hat{Y}_i)^2 = 30$.

5. For a ten-year period, the exports and gross national product of a country (in millions of dollars) are given. Find the regression of exports on the gross national product and predict the exports when the gross national product is $250 million. Find the correlation coefficient between these variables, and test the hypothesis that they are uncorrelated at level .05.

Exports	Gross National Product
27	180.1
32.1	192.1
30.1	204.5
32.1	208.0
37.2	215.7
39.1	220.5
41.6	224.2
40.5	260.1
50.1	276.1
55.2	280.0

6. The average lifetime income per year is given for ten randomly chosen individuals with their education (number of years of schooling) and an index of their family wealth. Give a regression equation that predicts the average lifetime income for this group in terms of the other variables:

Income (in thousands of dollars)	Years of schooling	Wealth of family index
17	12	2
19	12	2
22	12	2
32	16	3
15	12	2
10	6	1
25	16	3
42	18	4
35	12	4
80	16	6

7. In the exhausts of a factory are found zinc, lead, and molybdenum. Thirteen samples of vegetable dry matter at various locations are collected and the amount of trace metals measured (in parts per million for dry matter in grams). The data are:

Z (Y)	Pb (X_1)	Mo (X_2)	Distance from plant (X_3)
124	7	2.6	0.8
72	10	1.9	1.5
60	8	2.1	2
70	7	2.2	2
68	8	2.0	2.5
67	7	2.3	2.5
50	10	0.9	3
86	6	2.1	3.5
66	9	2.0	3.5
68	6	1.6	3.5
48	9.5	1.0	4.5
71	9	1.6	4.5
63	7.5	1.1	6

(a) Fit the multiple regression equation of Y on X_1, X_2, and X_3
(b) Find the correlation coefficient between (Y, X_1), (Y, X_2), and (Y, X_3)
(c) Find the rank correlation coefficient between (Y, X_3)
(d) Find the residuals and obtain the multiple correlation coefficient. How much of the variation in the data is explained by the regression?

Summary

The *regression function* is the mean of the conditional distribution of a random variable given another random variable. In jointly normally distributed random variables, the regression turns out to be linear. The important parameters in a *linear regression* model are the *intercept* and the *slope*. They are estimated by the method of *least squares*, which minimizes the sum of squares of observations from their expected values. The sum of squares of vertical deviations of observations from their assumed values are squared and minimized in the case of linear regression. The *residuals* measure the deviation of observed values from their values predicted by the regression equation. Plotting the *residuals* and the *scatter plot* of observations gives an important insight into the data and the model. The regression equation can be used to *predict* values of the *dependent* variable in terms of the known value of the independent variables.

The *correlation coefficient* can be calculated when both variables are random. The *Spearman correlation coefficient* or the *rank correlation coefficient* is used when the random variables are ranks. The *multiple* linear regression involves more than one dependent variable. Models in *nonlinear regression* can sometimes be transformed to those in linear regression and the least-squares method applied in finding estimates of the parameters. The *multiple correlation coefficient* gives the measure of goodness of fit of the multiple linear regression to the data and gives the percentage of the variation in the data explained by regression.

References

Draper, N. R., and Smith, Harry. *Applied Regression Analysis.* New York: John Wiley & Sons, 1968.

Graybill, Franklin, A. *An Introduction to Linear Statistical Models.* New York: McGraw-Hill, 1961.

Neter, John, and Wasserman, William. *Applied Linear Statistical Models.* Homewood, Ill.: Richard D. Irwin, 1974.

Williams, E. J. *Regression Analysis.* New York: John Wiley & Sons, 1959.

chapter ten

Analysis of Variance

In solving applied problems, we are often concerned with the comparison of two groups. The hypothesis that the means of two groups are equal answers such important questions as whether the mean of a treatment group is the same as that of the control group. In Chapter 7 we considered tests of the equality of means of two normal populations. The analysis of variance technique is used in testing hypotheses whether the means of several normal populations are the same. Obviously we cannot compare two groups at a time, since the overall significance level of the test is not available in terms of the significance levels of individual tests. Such pairwise comparisons may also result in contradictory tests showing that the Group I mean is the same as the Group II mean and the Group II mean is the same as the Group III mean, but the Group I mean is not the same as the Group III mean. This difficulty is avoided by the *analysis-of-variance* test. Analysis of variance is one of the most frequently used techniques in application.

The model for the analysis of variance assumes that the observation has a certain number of components—some constant components and a random component. These are called *fixed-effect models*, since the parameters to be estimated from the models are fixed. There are also analysis-of-variance models where all the components of an observation are assumed to be random. Such models are known as *variance-component models*. We consider both kinds of models in this chapter.

The test is based on partitioning the sum of squares of the deviations of observations from the mean that is used in estimating the overall variation. This is partitioned into several components, which also estimate the variance under various hypotheses, and the test of equality of means is made by comparing the estimates of variances. The basic assumption in all these analyses is that the variances of all the populations are the same.

One-Way Analysis of Variance

Suppose we are given data on k groups. The first observation of the first group is Y_{11}, the second observation of the first group is Y_{12}, and so on, to the nth are denoted by $Y_{13}, \ldots, Y_{1n}$. Similarly, the observations in Group II are denoted by $Y_{21}, Y_{22}, \ldots, Y_{2n}$ and in Group k by $Y_{k1}, Y_{k2}, \ldots Y_{kn}$. That is, the observations are given by a matrix of $Y_{ij}, i = 1, 2, \ldots, k$ and $j = 1, 2, \ldots, n$. Table 10.1 shows this matrix of observations in the actual form.

When we say that Group I is a random sample from a normal population with mean μ_1 and variance σ^2, we mean:

$$E(Y_{1j}) = \mu_1 \text{ and } \text{Var}(Y_{1j}) = \sigma^2, \text{ for all } j$$

That is,

$$Y_{1j} = \mu_1 + e_{1j}$$

where μ_1 is the parameter (fixed constant) and e_{1j} is random with mean 0 and variance σ^2.

Similarly,

$$Y_{2j} = \mu_2 + e_{2j}$$
$$Y_{3j} = \mu_3 + e_{3j}$$
$$\vdots \qquad \vdots \qquad \vdots$$
$$Y_{kj} = \mu_k + e_{kj}$$

all having the same assumptions on $e_{1j}, e_{2j}, \ldots, e_{kj}$.

In other words, we can write

$$Y_{ij} = \mu_i + e_{ij} \qquad (10.1)$$
$$i = 1, 2, \ldots, k, \qquad j = 1, 2, \ldots, n$$

and e_{ij} are normal with mean 0 and variance σ^2. Since Y_{ij} are independent, e_{ij} are also independent.

Alternatively, we can write Equation (10.1) as

$$Y_{ij} = \mu + (\mu_i - \mu) + e_{ij}$$

where $\mu = (1/k)\sum \mu_i$, the average of μ_is.

Table 10.1   Data Matrix

Group I	Y_{11}	Y_{12}	$Y_{13} \cdots Y_{1n}$
Group II	Y_{21}	Y_{22}	$Y_{23} \cdots Y_{2n}$
$\vdots$	$\vdots$	$\vdots$	$\vdots \quad \vdots$
Group k	Y_{k1}	Y_{k2}	$Y_{k3} \cdots Y_{kn}$

Let $\mu - \mu_i = \tau_i$, called *treatment effect*. Then $\sum \tau_i = 0$, since $\sum(\mu - \mu_i) = 0$. Equation (10.1) can now be written as

$$Y_{ij} = \mu + \tau_i + e_{ij} \tag{10.2}$$
$$i = 1, 2, \ldots, k, \qquad j = 1, 2, \ldots, n$$

with the assumptions:

(i) μ is the common mean
(ii) τ_i is the treatment effect of the ith group with $\sum \tau_i = 0$ (10.3)
(iii) e_{ij} are independent, normally distributed random errors with mean 0 and variance σ^2.

For example, consider three groups, each having four observations. $k = 3$, $n = 4$, and:

$$Y_{11} = \mu + \tau_1 + e_{11}$$
$$Y_{12} = \mu + \tau_1 + e_{12}$$
$$Y_{13} = \mu + \tau_1 + e_{13}$$
$$Y_{14} = \mu + \tau_1 + e_{14}$$
$$Y_{21} = \mu + \tau_2 + e_{21}$$
$$Y_{22} = \mu + \tau_2 + e_{22}$$
$$Y_{23} = \mu + \tau_2 + e_{23}$$
$$Y_{24} = \mu + \tau_2 + e_{24}$$
$$Y_{31} = \mu + \tau_3 + e_{31}$$
$$Y_{32} = \mu + \tau_3 + e_{32}$$
$$Y_{33} = \mu + \tau_3 + e_{33}$$
$$Y_{34} = \mu + \tau_3 + e_{34}$$

We can write the above equations also as:

$$Y_{11} = \mu + 1 \cdot \tau_1 + 0\tau_2 + 0\tau_3 + e_{11}$$
$$Y_{12} = \mu + 1\tau_1 + 0\tau_2 + 0\tau_3 + e_{12}$$
$$\vdots \qquad\qquad \vdots$$
$$Y_{21} = \mu + 0\tau_1 + 1\tau_2 + 0\tau_3 + e_{21}$$
$$\vdots$$
$$Y_{31} = \mu + 0\tau_1 + 0\tau_2 + 1 \cdot \tau_3 + e_{31}$$
$$\vdots$$

That is, for the general model in (10.2), with coefficient of τ_i to be one and other coefficients 0:

$$Y_{ij} = \mu + 0\tau_1 + 0\tau_2 + \cdots + 1 \cdot \tau_i + \cdots + 0\tau_k + e_{ij}$$

This is a special case of the multiple regression model discussed in the previous chapter:

$$Y = \beta_0 + \beta_1 x_1 + \beta_2 x_2 + \cdots + \beta_k x_k + e$$

The major difference is that in the analysis-of-variance model Xs are replaced by ones and zeros. In a sense, the analysis-of-variance model is a special regression model where observations on the independent variables $X_1, \ldots, X_k$ are given by ones and zeros.

Estimation of Parameters

As in the case of regression analysis, the estimates of the parameters of the linear analysis-of-variance model can be obtained by the least-squares method. That is, we minimize

$$\sum_i \sum_j (Y_{ij} - \mu - \tau_i)^2$$

Subject to constraints

$$\sum_i \tau_i = 0$$

The solutions of the normal equations provide the following estimates:

$$\hat{\mu} = \bar{Y}_{..} = \frac{1}{nk} \sum_i \sum_j Y_{ij} \tag{10.4}$$

$$\hat{\mu} + \hat{\tau}_i = \bar{Y}_{i.} = \frac{1}{n} \sum_j Y_{ij}, \quad i = 1, 2, \ldots, k \tag{10.5}$$

So that: $\hat{\tau}_i = \bar{Y}_{i.} - \bar{Y}_{..}$

Tests of Hypotheses

The basic purpose here is to test the hypothesis that the means of the k groups—that is $\mu + \tau_1, \mu + \tau_2, \ldots$—are equal. We are testing:

$$H: \tau_1 = \tau_2 = \cdots = \tau_k = 0$$
$$A: \text{at least one of } \tau_i \text{ is not zero} \tag{10.6}$$

The test statistic is obtained by comparing the estimate of σ^2 under the null hypothesis and the alternative hypothesis.

Now the overall variance when the means are all equal is $\sum \sum (Y_{ij} - \bar{Y}_{..})^2 / (nk - 1)$. We call the numerator as the sum of squares for the total. We partition the sum of squares for the total into two separate components, since the cross-product term is zero:

$$SS_T = \sum\sum(Y_{ij} - \bar{Y}_{..})^2 = \sum\sum(Y_{ij} - \bar{Y}_{i.} + \bar{Y}_{i.} - \bar{Y}_{..})^2$$
$$= \sum_i\sum_j(Y_{ij} - Y_{i.})^2 + \sum_i\sum_j(\bar{Y}_{i.} - \bar{Y}_{..})^2$$

or

$$SS_T = \sum_i\sum_j(Y_{ij} - \bar{Y}_{i.})^2 + n\sum_i(\bar{Y}_{i.} - \bar{Y}_{..})^2 \qquad (10.7)$$

Let us denote:

Within the sums of squares $= SS_W = \sum\sum(Y_{ij} - \bar{Y}_{i.})^2$

Between the sums of squares $= SS_B = n\sum_i(\bar{Y}_{i.} - \bar{Y}_{..})^2$

The corresponding estimates of variance can be denoted:

Within mean squares $= MS_W = \dfrac{SS_W}{nk - k}$

Between mean squares $= MS_B = \dfrac{SS_B}{k - 1}$

The variance σ^2 is estimated by MS_W. That is:

$$\hat{\sigma}^2 = MS_W \qquad (10.8)$$

Since $\bar{Y}_{i.}$ is the average of the normally distributed random variables, $\bar{Y}_{i.}$ is normally distributed with a mean of $\mu + \tau_i$ and a variance of $\dfrac{\sigma^2}{n}$. Hence,

$$\frac{n\sum(\bar{Y}_{i.} - \bar{Y}_{..})}{\sigma^2}$$

is distributed as a chi-square with $k - 1$ degrees of freedom. That is, $\dfrac{SS_B}{\sigma^2}$ has a chi-square distribution with $k - 1$ degrees of freedom. Similarly, $\dfrac{SS_W}{\sigma^2}$ has a chi-square distribution with $nk - k$ degrees of freedom. Also, SS_B and SS_W are independent. The test statistic, then, can be obtained in terms of the ratio:

$$F_{k-1,nk-k} = \frac{SS_B/(k-1)}{SS_W/(nk-k)} = \frac{MS_B}{MS_W}$$

Decision Rule: Reject H when

$$F_{k-1,nk-k} > a$$

The constant a can be determined by the tables of the F-distribution. This F-test is the *analysis of variance* test.

The simplicity of this test arises from the fact that it can be described in terms of a table. The table contains the partitioned sum of squares, mean squares, and F ratio. The table is called an *analysis-of-variance table*, or ANOVA table. The headings in the table are sources of variation, degrees of freedom (DF), sum of squares (SS), mean squares (MS), and F ratio. Sometimes the expected mean squares $(E(MS))$ column is also added. The above test is given in Table 10.2.

For computational simplicity, first we calculate the sums of squares for the total and between the sum of squares and then obtain within the sum of squares by subtraction.

Also, we use the simplified formula:

$$SS_T = \sum \sum Y_{ij}^2 - \frac{(\sum \sum Y_{ij})^2}{nk}$$

We call $\dfrac{(\text{Total})^2}{nk}$ the *correction factor (CF)*.

Since $\bar{Y}_{i.} = $ the average of the ith group $= \dfrac{T_i}{n}$ where $T_i = $ the total of the ith group, we have,

$$SS_B = n \sum \bar{Y}_{i.}^2 - \frac{n}{k}(\sum \bar{Y}_{i.})^2$$

$$= \frac{\sum T_{i.}^2}{n} - \frac{(\sum T_{i.})^2}{nk}$$

$$= \frac{\sum T_{i.}^2}{n} - CF$$

and $\quad S_W = SS_T - SS_B$

Table 10.2   One-Way Analysis of Variance

Source of Variation	DF	SS	MS	F
Between groups	$k-1$	$n\sum_i (\bar{Y}_{i.} - \bar{Y}_{..})^2$	MS_B	MS_B/MS_W
Within groups	$k(n-1)$	$\sum_i \sum_j (Y_{ij} - \bar{Y}_{i.})^2$	MS_W	
Total	$nk-1$	$\sum_i \sum_j (Y_{ij} - \bar{Y}_{..})^2$		

Note that after the null hypothesis of the equality of means is rejected, the problem is to find which means differ and which do not. *Multiple comparison* procedures are needed to pursue this problem, and we consider them later in the chapter.

Example 10.1: Consider the following data for iron absorption in three groups of subjects—a control group and two treatment groups:

Control group:	99,	98,	98,	97,	93
Treatment group I:	121,	131,	151,	154,	153
Treatment group II:	116,	147,	156,	158,	138

Here:

$$T_1 = 485, \quad T_2 = 710, \quad T_3 = 715$$
$$k = 3, \quad n = 5$$
$$\sum\sum Y_{ij}^2 = 252{,}204$$
$$\sum\sum Y_{ij} = 1910$$
$$CF = 1910^2/15 = 243{,}206.67$$
$$SS_T = 252{,}204 - 243{,}206.67 = 8{,}997.33$$
$$SS_B = \tfrac{1}{5}[485^2 + 710^2 + 715^2] - CF$$
$$= 250{,}110 - 243{,}206.67 = 6{,}903.33$$
$$SS_W = 8{,}997.33 - 6{,}903.33 = 2{,}094.00$$

The estimates of the mean and treatment effects are:

$$\hat{\mu} = 1910/15 = 127.33$$
$$\hat{t}_1 = 97 - 127.33 = -30.33$$
$$\hat{t}_2 = 142 - 127.33 = 14.67$$
$$\hat{t}_3 = 143 - 127.33 = 15.67$$

Note that $\hat{t}_1 + \hat{t}_2 + \hat{t}_3 = 0$ (except to the extent of round-off errors). The analysis of variance is given in Table 10.3.

Table 10.3 Analysis of Variance

Source of Variation	DF	SS	MS	F
Between groups	2	6,903.33	3,451.67	19.78
Within groups	12	2,094.00	174.5	
Total	14	8,997.33		

Since $F_{2,12,.95} = 3.89$, we reject the hypothesis of the equality of means at level .05.

The usefulness of the analysis-of-variance test arises not only from its simplicity but also from its property of *robustness*. A procedure is said to be robust if it is valid under mild departures from assumptions. This vague notion of robustness has recently been extensively studied and made more precise.

Studies of analysis-of-variance tests have revealed that they are robust under mild departures from the assumption of normality, as well as under the mild departures from homogeneity of variances (for reference, see Scheffé [1959]).

Unequal Group Size

Suppose there are an unequal number of observations in k groups, say $n_1, n_2, \ldots, n_k$. The model is the same, except that in place of $\sum \tau_i = 0$, we need:

$$\sum n_i \tau_i = 0 \qquad (10.9)$$

The estimates of μ, τ_i are given as before by Equations (10.4) and (10.5). Remember that the group average $\bar{Y}_{i.}$ is now based on n_i observations. The analysis-of-variance table is given in Table 10.4, and the test of the hypothesis H is given by the F ratio:

Table 10.4 One-Way Analysis of Variances for Unequal Group Sizes

Source of Variation	DF	SS	MS	F
Between groups	$k-1$	$\sum_i n_i (\bar{Y}_{i.} - \bar{Y}_{..})^2$	MS_B	MS_{B/MS_W}
Within groups	$\sum_i (n_i - 1)$	$\sum_i \sum_j (Y_{ij} - \bar{Y}_{i.})^2$	MS_W	
Total	$\sum_i n_i - 1$	$\sum_i \sum_j (Y_{ij} - \bar{Y}_{..})^2$		

Example 10.2: Four groups are to be compared:

Group I: 10, 17, 15
Group II: 14, 12
Group III: 21, 18, 24
Group IV: 23, 29

$$n_1 = 3, \quad n_2 = 2, \quad n_3 = 3, \quad \text{and} \quad n_4 = 2$$
$$k = 4$$
$$T_1 = 42, \quad T_2 = 26, \quad T_3 = 63, \quad \text{and} \quad T_4 = 52$$
$$\text{Total} = 183$$
$$CF = 183^2/10 = 3348.9$$

$$SS_T = 3665 - 3348.9 = 316.1$$

$$SS_B = \frac{42^2}{3} + \frac{26^2}{2} + \frac{63^2}{3} + \frac{52^2}{2} - 3348.9$$

$$= 3601 - 3348.9 = 252.1$$

$$SS_W = 316.1 - 252.1 = 64$$

Estimates of the mean and effects are:

$$\hat{\mu} = 18.3$$
$$\hat{t}_1 = 14 - 18.3 = -4.3$$
$$\hat{t}_2 = 13 - 18.3 = -5.3$$
$$\hat{t}_3 = 21 - 18.3 = 2.7$$
$$\hat{t}_4 = 26 - 18.3 = 7.7$$

Note that the condition in Equation (10.9) is satisfied, since:

$$3\hat{t}_1 + 2\hat{t}_2 + 3\hat{t}_3 + 2\hat{t}_4 = -12.9 - 10.6 + 8.1 + 15.4 = 0$$

The analysis of variance is given in Table 10.5.

Table 10.5 Analysis of Variance

Source of Variation	DF	SS	MS	F
Between groups	3	252.1	84.033	7.88
Within groups	6	64	10.667	
Total	9	316.1		

Since $F_{3,6,.95} = 4.76$, we reject the hypothesis of the equality of means at level .05.

Exercises

1. Give the analysis of variance test for the following data with four groups, each having three observations:
   (a) What is the model?
   (b) What are the estimates of the effects?
   (c) Test the hypothesis of the equality of means at level .05

   Group I: 7, 9, 5
   Group II: 8, 10, 6
   Group III: 9, 8, 13
   Group IV: 11, 15, 16

2. The hourly production rate on three machines is given below for an item. Test whether the machines have the same mean production rate:

   (a) Give the model
   (b) Estimate the effects
   (c) Test the hypothesis of the equality of the mean hourly production by the machines at level .05

   Machine I: 120, 132, 83, 85, 75
   Machine II: 101, 104, 95, 32, 63
   Machine III: 151, 150, 132, 143, 135

3. Three different teaching methods were used to teach statistics in large sections at Ohio State University. The average scores of four standardized tests given to the three classes that were chosen at random for the experiment are given below. See if the methods are the same:

   Method I: 75, 83, 47, 68
   Method II: 82, 57, 91, 75
   Method III: 86, 60, 80, 92

   (a) Give the model
   (b) Estimate the parameters of the model
   (c) Test the hypothesis that the three methods are the same at level .05

4. Four different methods of advertising were used to promote a department store's sales over several months. Weekly sales totals are given below. Are the methods the same (sales in 100,000 dollars)?

   Newspaper advertising: 7, 8, 5
   Radio advertising: 5, 8, 3
   TV advertising: 8, 4, 4
   Special advertising: 10, 7, 15

   (a) Give the model
   (b) Give estimates of the parameters of the model
   (c) Test the hypothesis that the four different methods of advertising result in the same mean amount of sales

5. Mortgage rates in three states were compared. Test the hypothesis that mortgage rates are the same at level .05, giving the model and the estimates of the parameters of the model:

   California rates: 13.1, 12.5, 13.5, 14.0
   Nevada rates: 12.1, 13.5, 14.5, 12.5
   Ohio rates: 12.1, 11.5, 11.1, 10.5

6. Three different insect sprays were used to see if they are are equally effective

in increasing the production of apples in a certain orchard. The following data are given in terms of bushels per plant. Test the hypothesis that the sprays are equally effective at level .05, giving the model and the estimates of parameters of the model:

Spray I: 40, 20, 35, 25
Spray II: 10, 12, 5, 8, 7
Spray III: 32, 17, 10

7. To compare the mean blood pressure (systolic) of subjects after the administration of drugs "to reduce blood pressure," three drugs were compared. The increase or decrease was noted:

$$n_1 = 3, \ n_2 = 5, \ n_3 = 4, \ \sum\sum Y_{ij} = 14,$$
$$T_1 = 30, \ T_2 = -20, \ T_3 = 4, \ \sum\sum Y_{ij}^2 = 104$$

(a) Give the model
(b) Estimate the parameters of the model
(c) Test the hypothesis that there is no difference between the drugs at level .10

Two-Way Analysis of Variance

One-way analysis of variance may not suffice when an experimenter must study two or more factors simultaneously. In finding the effect of three different drugs, each manufactured by four pharmaceutical companies, we have two-way variability introduced by different drugs and different companies. The effect of weight gain of four feeds given to three different species of animals introduces the same problem. These are experiments with a two-way classification. We assume that the *factor* have several *levels*. First we consider the case of a two-factor experiment with one observation per cell. Later, experiments with more than one observation per cell will be considered. The data for a two-factor experiment with one observation per cell are given by a matrix, $Y_{ij}, i = 1, 2, \ldots, t, j = 1, 2, \ldots, b$, where factor A has t levels and factor B has b levels. The model for the study of such an experiment is:

$$Y_{ij} = \mu + \tau_i + \beta_j + e_{ij} \tag{10.10}$$

Note that $i = 1, 2, \ldots, t, \ j = 1, 2, \ldots, b$.
Assumptions for Equation (10.10) are:
   (i) τ_i is the effect for the ith level of factor A with $\sum \tau_i = 0$
   (ii) β_j is the effect for the jth level of factor B with $\sum \beta_j = 0$
   (iii) e_{ij} are independently normally distributed with means 0 and variance σ^2.

The effects in Equation (10.10) are said to be *additive*. The hypotheses of interest are:

$$H_1: \tau_1 = \tau_2 = \cdots = \tau_t = 0$$

$$A_1: \text{at least one equality is violated} \tag{10.11}$$

and

$$H_2: \beta_1 = \beta_2 = \cdots = \beta_b = 0$$

$$A_2: \text{at least one equality is violated} \tag{10.12}$$

Estimation of Treatment Effects

As we have seen in the one-way model:

$$\hat{\mu} = \bar{Y}_{..}$$

$$\hat{\tau}_i = \bar{Y}_{i.} - \bar{Y}_{..}, \quad i = 1, 2, \ldots, t$$

$$\hat{\beta}_j = \bar{Y}_{.j} - \bar{Y}_{..}, \quad j = 1, 2, \ldots, b$$

where

$$\bar{Y}_{..} = \frac{1}{bt} \sum \sum Y_{ij}$$

$$\bar{Y}_{i.} = \frac{1}{b} \sum_j Y_{ij}$$

$$\bar{Y}_{.j} = \frac{1}{t} \sum_i Y_{ij}$$

To obtain the test statistic for hypotheses (10.11) and (10.12), we partition the sum of squares as before as the cross-product terms vanish:

$$\sum_i \sum_j (Y_{ij} - \bar{Y}_{..})^2 = \sum_i \sum_j (Y_{ij} - \bar{Y}_{i.} - \bar{Y}_{.j} + \bar{Y}_{..} + \bar{Y}_{i.} - \bar{Y}_{..} + \bar{Y}_{.j} - \bar{Y}_{..})^2$$

$$= \sum_i \sum_j (Y_{ij} - \bar{Y}_{i.} - \bar{Y}_{.j} + \bar{Y}_{..})^2 + \sum_i \sum_j (\bar{Y}_{i.} - \bar{Y}_{..})^2$$

$$+ \sum_i \sum_j (\bar{Y}_{.j} - \bar{Y}_{..})^3$$

So that:

$$\sum_i \sum_j (Y_{ij} - \bar{Y}_{..})^2 = \sum_i \sum_j (Y_{ij} - \bar{Y}_{i.} - \bar{Y}_{.j} + \bar{Y}_{..})^2 + b \sum_i (\bar{Y}_{i.} - \bar{Y}_{..})^2 + t \sum_j (\bar{Y}_{.j} - \bar{Y}_{..})^2$$

Again, as before:

$$SS_A = \text{Sum of squares for factor } A = b \sum_i (\bar{Y}_{i.} - \bar{Y}_{..})^2$$

$$SS_B = \text{Sum of squares for factor } B = t \sum_j (\bar{Y}_{.j} - \bar{Y}_{..})^2$$

$$SS_E = \text{Sum of squares for error} = \sum_i \sum_j (Y_{ij} - \bar{Y}_{i.} - \bar{Y}_{.j} + \bar{Y}_{..})^2$$

The corresponding mean squares are denoted by MS_A, MS_B and MS_E.

Decision rule: For Hypothesis (10.11), we reject H_1 if $F = MS_A/MS_E > k$, where F has $(t-1)$ and $(t-1)(b-1)$ degrees of freedom.

For hypothesis (10.12), we reject H_2 if $F = \dfrac{MS_B}{MS_E} > k$, where F has $(b-1)$ and $(t-1)(b-1)$ degrees of freedom.

These calculations are given in Table 10.6.

Table 10.6 Two-Way Analysis of Variance

Source of Variation	DF	SS	MS	F
Factor A	$t-1$	$b\sum_i (\bar{Y}_{i.} - \bar{Y}_{..})^2$	MS_A	MS_A/MS_E
Factor B	$b-1$	$t\sum_j (\bar{Y}_{.j} - \bar{Y}_{..})^2$	MS_B	MS_B/MS_E
Error	$(t-1)(b-1)$	$\sum_i \sum_j (Y_{ij} - \bar{Y}_{i.} - \bar{Y}_{.j} + \bar{Y}_{..})^2$	MS_E	
Total	$tb-1$	$\sum_i \sum_j (Y_{ij} - \bar{Y}_{..})^2$		

Example 10.3: Three different drugs (Factor A) were tested for their effectiveness in relieving pain. Four categories of patients—children, young adults, adults, and old persons (Factor B)—were used for measuring the hours of relief they received from these drugs. Test the hypothesis that all the drugs have the same effectiveness, according to the data:

Factor B

Factor A	Children	Young Adults	Adults	Old Persons
Drug 1	2	4	3	7
Drug 2	1	3	3	5
Drug 3	7	9	5	6

We assume that τ_i, $i = 1, 2, 3$ is the effect of Drug i and β_j, $j = 1, 2, 3, 4$ measures the effect of factor B. We find:

$$\bar{Y}_{..} = 4.583$$
$$\bar{Y}_{1.} = 4.0 \quad \bar{Y}_{2.} = 3.0 \quad \bar{Y}_{3.} = 6.75$$
$$\bar{Y}_{.1} = 3.333, \quad \bar{Y}_{.2} = 5.333, \quad \bar{Y}_{.3} = 3.667, \quad \bar{Y}_{.4} = 6.0$$

The estimates of effects are:

$$\hat{\mu} = 4.583$$
$$\hat{\tau}_1 = 4 - 4.583 = -.583$$
$$\hat{\tau}_2 = 3 - 4.583 = -1.583$$
$$\hat{\tau}_3 = 6.75 - 4.583 = 2.167$$
$$\hat{\beta}_1 = 3.333 - 4.583 = -1.250$$
$$\hat{\beta}_2 = 5.333 - 4.583 = .750$$
$$\hat{\beta}_3 = 3.667 - 4.583 = -.916$$
$$\hat{\beta}_4 = 6.0 - 4.583 = 1.417$$

For the analysis-of-variance table, we have the calculations:

$$\text{Correction Factor} = (55)^2/12 = 252.083$$
$$SS_A = (16^2 + 12^2 + 27^2)/4 - 252.083$$
$$= 30.167$$
$$SS_B = (10^2 + 16^2 + 11^2 + 18^2)/3 - 252.083$$
$$= 14.917$$
$$SS_T = 313 - 252.083 = 60.917$$
$$SS_E = 60.917 - 30.167 - 14.917$$
$$= 15.833$$

Other computations are completed in Table 10.7.

Table 10.7 Analysis of Variance

Source of Variation	DF	SS	MS	F
Factor A	2	30.167	15.083	5.72
Factor B	3	14.917	4.972	1.88
Error	6	15.833	2.639	
Total	11	60.917		

The test of the hypothesis that there is no difference between the various drugs—that is, $\tau_i = 0$ for all i—is given with the help of the F-ratio. In this case, we have $F_{2,6,.95} = 5.14$ and we reject the null hypothesis H_1. Since $F_{3,6,.95} = 4.76$, we do not reject the null hypothesis H_2 at level .05.

Exercises

8. In an experiment comparing three different teaching methods for statistics at the high-school level, four high schools were chosen and students

were chosen at random to be assigned to three different sections for the three teaching methods. The average scores for each class on a standardized test are:

Teaching Methods	Schools			
	1	2	3	4
I	73	61	72	79
II	85	76	86	89
III	64	55	58	66

(a) Give the model
(b) Test if there is any difference among the schools at level .10
(c) Test if there is any difference among teaching methods at level .05

9. The data on production of four machines, each having three operators in different shifts, are given below. Determine if the mean production per machine is the same, and whether the mean production for each operator is the same:

Machines	Operators		
	1	2	3
I	7	9	5
II	6	3	4
III	8	8	7
IV	5	4	8

(a) Give the model with assumptions
(b) Test the hypothesis that there is no difference among machine production at level .05
(c) Test the hypothesis that there is no difference among operators at level .1

10. Data on the nicotine content in filter pads determined by four different laboratories (A, B, C, D) is given below. Test the hypothesis that there is no difference among the laboratories:

Laboratories	Samples				
	1	2	3	4	5
A	.161	.192	.373	.663	.692
B	.157	.193	.366	.660	.688
C	.159	.193	.373	.643	.671
D	.294	.206	.405	.630	.701

Source: Wagner, J. R., and Thaggard, N. A., Gas liquid chromatographic determination of nicotine contained on Cambridge filter pads, *J. Assoc. Off. Anal. Chem.* 1979, 62, 229–236.

(a) Give the model
(b) Test the hypothesis that there is no difference between laboratories at level .05
(c) Why can we not use a one-way analysis of variance for this problem? Test the hypothesis that the sample means are the same at level .1.

Two-Way Analysis of Variance with Interaction

The model of a two-way analysis of variance considered in the last section assumes that the effects of factor A and B are *additive*; that is, there is no joint effect of both the factors. In experimental terminology, the joint effect

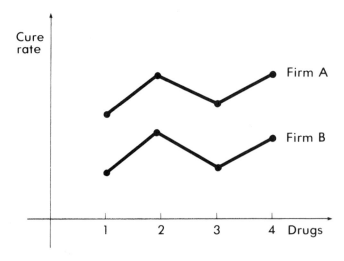

Figure 10.1. No interaction.

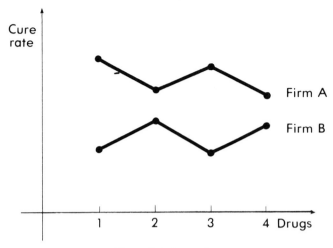

Figure 10.2. Interaction.

of two factors is known as the *interaction* between the factors. Consider that four different drugs to cure headaches are manufactured by two different pharmaceutical firms. If there is no interaction between the drugs and the firms, we expect that the mean cure rate for the two firms will be similar, as shown in Figure 10.1. However, if the cure rates are different for the different drugs of the two firms, we have interaction, as shown in Figure 10.2.

Interaction is an important concept, and its precise nature will become clear when we study factorial experiments in Chapter 11. Here we add another term to the model to take care of the joint effect of both factors.

Suppose we have t levels of factor A and b levels of factor B, as before. Assume that there are c (greater than one) observations in each cell. We denote by Y_{ijk} the kth observation on the ith level of factor A and the jth level of factor B.

We assume that:

$$Y_{ijk} = \mu + \tau_i + \beta_j + \gamma_{ij} + e_{ijk} \qquad (10.13)$$
$$i = 1, 2, \ldots, t$$
$$j = 1, 2, \ldots, b$$
$$k = 1, 2, \ldots, c$$

The assumptions of the model are:

(i) $\sum_i \tau_i = 0, \ \sum_j \beta_j = 0$

(ii) The *interaction* effect is γ_{ij} is such that $\sum_i \gamma_{ij} = \sum_j \gamma_{ij} = 0$

(iii) e_{ijk} are normally and independently distributed with means 0 and variance σ^2

The test of the hypotheses of interest in addition to H_1 and H_2 as given in Equations (10.11) and (10.12) is:

$$H_3: \gamma_{ij} = 0 \text{ for all } i \text{ and } j$$
$$A_3: \text{at least one equality is violated}$$

Estimates of the parameters in the model are given below.

$$\hat{\mu} = \bar{Y}_{...} = \frac{1}{tbc} \sum_i \sum_j \sum_k Y_{ijk}$$

$$\hat{\tau}_i = \bar{Y}_{i..} - \bar{Y}_{...}, \quad i = 1, 2, \ldots, t$$
$$\hat{\beta}_j = \bar{Y}_{.j.} - \bar{Y}_{...}, \quad j = 1, 2, \ldots, b$$
$$\hat{\gamma}_{ij} = \bar{Y}_{ij.} - \bar{Y}_{i..} - \bar{Y}_{.j.} + \bar{Y}_{...}, \quad \text{for all } i \text{ and } j$$

where

$$\bar{Y}_{i..} = \frac{1}{bc} \sum_j \sum_k Y_{ijk}$$

$$\bar{Y}_{.j.} = \frac{1}{tc} \sum_i \sum_k Y_{ijk}$$

$$\bar{Y}_{ij.} = \frac{1}{c} \sum_k Y_{ijk}$$

As before, we partition the sum of squares for the total in terms of the sums of squares for factors A and B for interaction and for error. We can write:

$$\sum_i \sum_j \sum_k (Y_{ijk} - \bar{Y}_{...})^2 = \sum_i \sum_j \sum_k (Y_{ijk} - \bar{Y}_{ij.})^2$$
$$+ c \sum_i \sum_j (\bar{Y}_{ij.} - \bar{Y}_{i..} - \bar{Y}_{.j.} + \bar{Y}_{...})^2$$
$$+ bc \sum_i (\bar{Y}_{i..} - \bar{Y}_{...})^2 + tc \sum_j (\bar{Y}_{.j.} - \bar{Y}_{...})^2$$
$$= SS_E + SS_{AB} + SS_A + SS_B$$

The corresponding mean squares are denoted by MS_E, MS_{AB}, MS_A, and MS_B.

Decision rule: We reject the hypotheses H_1, H_2, and H_3 based on the F ratios as given in Table 10.8.

The test for the hypothesis that $\tau_i = 0$ for all i is given by the ratio MS_A/MS_E, which has an F-distribution with $t - 1$ and $tb(c - 1)$ degrees of freedom. The test for the hypothesis that $\beta_j = 0$ for all j is given by the ratio MS_B/MS_E, which has an F-distribution with $b - 1$ and $tb(c - 1)$ degrees of freedom. The tests for interactions $\gamma_{ij} = 0$ for i and j are given by the ratio MS_{AB}/MS_E, which has an F-distribution with $(t - 1)(b - 1)$ and $tb(c - 1)$ degrees of freedom.

When the interactions are not significant, the model assumes that factors

Table 10.8 Two-Way Analysis of Variance with Interaction

Source of Variation	DF	SS	MS	F
Factor A	$t-1$	$bc\sum_{k}(\bar{Y}_{i..} - \bar{Y}_{...})^2$	MS_A	MS_A/MS_E
Factor B	$b-1$	$tc\sum_{k}(\bar{Y}_{.j.} - \bar{Y}_{...})^2$	MS_B	MS_B/MS_E
Interaction	$(t-1)(b-1)$	$c\sum_{i}\sum_{j}(\bar{Y}_{ij.} - \bar{Y}_{i..} - \bar{Y}_{.j.} + \bar{Y}_{...})^2$	MS_{AB}	MS_{AB}/MS_E
Error	$tb(c-1)$	$\sum_{i}\sum_{j}\sum_{k}(Y_{ijk} - \bar{Y}_{ij.})^2$	MS_E	
Total	$tbc-1$	$\sum_{i}\sum_{j}\sum_{k}(Y_{ijk} - \bar{Y}_{...})^2$		

A and B have additive effects. When they are significant, we say that there is *nonadditivity*. Notice that when $c = 1$—that is, there is only one observation per cell—the degrees of freedom for the error sum of squares is zero, and in this case we cannot test the hypothesis for factors A and B as well as for the interactions. However, a test of nonadditivity can be developed by assuming a special form of the interaction effects, even when there is one observation per cell. Tukey (1949) gives a test for this situation. Consider the following model with usual assumptions:

$$Y_{ijk} = \mu + \tau_i + \beta_j + \gamma\tau_i\beta_j + e_{ijk} \tag{10.14}$$

When we test the hypothesis that $\gamma = 0$ and we reject the hypothesis, then we can say that there is nonadditivity. This test can be made even when the number of observations in each cell is one. (For details, see Tukey [1949]).

Example 10.4: Let $t = 3$, $b = 2$, and $c = 2$. The table gives the data:

Factor A	Factor B	
	I	II
I	2, 4	3, 7
II	1, 3	3, 5
III	7, 9	5, 6

The observation Y_{ijk} in the above notation is given as:

$$
\begin{array}{lll}
Y_{111} = 2 & Y_{211} = 1 & Y_{311} = 7 \\
Y_{112} = 4 & Y_{212} = 3 & Y_{312} = 9 \\
Y_{121} = 3 & Y_{221} = 3 & Y_{321} = 5 \\
Y_{122} = 7 & Y_{222} = 5 & Y_{322} = 6
\end{array}
$$

Notice that we have:

$$\bar{Y}_{11.} = 3 \qquad \bar{Y}_{12.} = 5$$
$$\bar{Y}_{21.} = 2 \qquad \bar{Y}_{22.} = 4$$
$$\bar{Y}_{31.} = 8 \qquad \bar{Y}_{32.} = 5.5$$
$$\bar{Y}_{1..} = 4$$
$$\bar{Y}_{2..} = 3$$
$$\bar{Y}_{3..} = 6.75$$
$$\bar{Y}_{.1.} = \tfrac{13}{3} = 4.33$$
$$\bar{Y}_{.2.} = 14.5/3 = 4.83$$
$$\bar{Y}_{...} = 55/12 = 4.58$$

$$CF = (55)^2/12 = 252.08$$
$$SS_T = 313 - 252.08 = 60.92$$
$$SS_A = [16^2 + 12^2 + 27^2]/4 - 252.08$$
$$= 30.17$$
$$SS_B = [26^2 + 29^2]/6 - 252.08$$
$$= 0.75$$
$$SS_{AB} = \tfrac{1}{2}(6^2 + 4^2 + 16^2 + 10^2 + 8^2 + 11^2)$$
$$- 30.17 - 0.75 - 252.08$$
$$= 13.50$$
$$SS_E = SS_T - SS_A - SS_B - SS_{AB}$$
$$= 60.92 - 30.17 - 0.75 - 13.50$$
$$= 16.50$$

The above calculations are arranged in Table 10.9.

Table 10.9 Analysis of Variance

Source of Variation	DF	SS	MS	F
Factor A	2	30.17	15.08	5.48
Factor B	1	0.75	0.75	0.27
Interaction	2	13.05	6.75	2.45
Error	6	16.50	2.75	
Total	11	60.92		

The tabulated F values are then compared to the values in the table. Since we find that $F_{2,6,.95} = 5.14$, we have Factor A significant at the 5% level. Again, since $F_{1,6,.95} = 5.99$ and $F_{2,6,.95} = 5.14$, we find that Factor B and the interactions are not significant at the 5% level. To see further which of the levels of Factor A are different, one finds the simultaneous confidence intervals discussed later.

Exercises

11. Two drugs and a placebo are given to athletes, and their performance is measured under two different exercises in an experiment. Discuss the model and analyze the experiment, giving the analysis-of-variance table:

	Placebo	Drug A	Drug B
Exercise I	50, 30	60, 40	50, 40
Exercise II	70, 65	30, 50	50, 45

12. To test the contention that music increases the milk yield of cows, an experimenter obtained the following data on 18 cows under various conditions. Analyze the data and suggest the model, testing the various hypotheses you propose:

Cows \ Music	None	Soft	Loud
Holstein	3, 7, 8	4, 6, 9	5, 9, 10
Jersey	2, 5, 6	3, 7, 8	8, 9, 4

13. In order to see whether there was a difference among the laboratories for estimating the amount of lead in samples of urine, three identical samples were sent to three different laboratories. Each laboratory was asked to use two different methods of estimation. The experiment was repeated twice. Suggest a model for the analysis of the data, and give the analysis-of-variance table for testing the various hypotheses.

Laboratories

	1	2	3
Method I	2, 5	1, 7	3, 5
Method II	15, 30	20, 15	7, 10

Analysis of Covariance

In many experiments, certain variables are known to affect the response but are beyond the control of the investigator. Sometimes controlling a variable may be too expensive. A common practice in such cases is to observe the variables simultaneously with the observation of response. For example, when we study the effect of a vitamin on the weight of a subject, age may be an important factor that cannot be controlled. Similarly, time may be an important variable in certain experiments and not controllable. The variable we measure with the response simultaneously is a *covariable*. Let Y_{ij} be the response and let X_{ij} be the covariable. For simplicity, we assume that Y_{ij} depends on X_{ij} linearly. Consider a one-way classification for the model:

$$Y_{ij} = \mu + \tau_i + \beta X_{ij} + e_{ij} \qquad (10.15)$$

$$i = 1, 2, \dots, t \quad \text{and} \quad j = 1, 2, \dots, n$$

The assumptions are:

(i) $\sum \tau_i = 0$
(ii) X_{ij} are known constants, independent of τ_i
(iii) β is the unknown regression parameter
(iv) e_{ij} are independently normally distributed with mean 0 and variance σ^2

The hypotheses we test here are:

$$H_1 : \tau_i = 0 \text{ for all } i$$

$$A_1 : \text{at least one of the equalities is violated}$$

and

$$H_2 : \beta = 0$$

$$A_2 : \beta \neq 0$$

Estimation of Parameters

The least-squares method is used as in regression analysis. The normal equations are:

$$n\mu + n\tau_i + \beta \sum_j x_{ij} = \sum_j Y_{ij}, \quad \text{for all } i$$

$$\mu \sum_i \sum_j X_{ij} + \sum_i \sum_j \tau_i X_{ij} + \beta \sum_i \sum_j X_{ij}^2 = \sum_i \sum_j Y_{ij} X_{ij}$$

$$nt\mu + n \sum_i \tau_i + \beta \sum_i \sum_j X_{ij} = \sum_i \sum_j Y_{ij}$$

The estimate of the regression coefficient is:

$$\hat{\beta} = \frac{\sum\sum(Y_{ij} - \bar{Y}_{..})(X_{ij} - \bar{X}_{..})}{\sum\sum(X_{ij} - \bar{X}_{..})^2}$$

$$\hat{\mu} = \bar{Y}_{..} - \hat{\beta}\bar{X}_{..}$$

$$\hat{\tau}_i = \bar{Y}_{i.} - \bar{Y}_{..} - \hat{\beta}(\bar{X}_{i.} - \bar{X}_{..})$$

Testing Hypotheses

The test statistic here essentially adjusts the sum of squares after the effect of the regression on X_{ij} is removed. Let us partition the sum of squares of the deviations of Y_{ij} from $\bar{Y}_{..}$ as well as X_{ij} from $\bar{X}_{..}$ in the same way as before:

$$\sum\sum(Y_{ij} - \bar{Y}_{..})^2 = \sum\sum(Y_{ij} - \bar{Y}_{i.})^2 + n\sum(\bar{Y}_{i.} - \bar{Y}_{..})^2$$

We use the notation for the above identity as:

$$S_{yy} = T_{yy} + E_{yy}$$

Similarly,

$$\sum\sum(X_{ij} - \bar{X}_{..})^2 = \sum\sum(X_{ij} - \bar{X}_{i.})^2 + n\sum(\bar{X}_{i.} - \bar{X}_{..})$$

$$S_{xx} = T_{xx} + E_{xx}$$

and

$$\sum\sum(Y_{ij} - \bar{Y}_{..})(X_{ij} - \bar{X}_{..}) = \sum(X_{ij} - \bar{X}_{i.})(Y_{ij} - \bar{Y}_{i.})$$
$$+ n\sum(\bar{X}_{i.} - \bar{X}_{..})(\bar{Y}_{i.} - \bar{Y}_{..})$$

$$S_{xy} = T_{xy} + E_{xy}$$

The tests of hypotheses H_1 and H_2 are given in Table 10.10 with the help of F ratios:

Table 10.10   Analysis of Covariance

Source of Variation	DF	SS	MS	F
Regression	1	E_{xy}^2/E_{xx}	MS_R	MS_R/MS_E
Treatments	$t-1$	By subtraction	MS_T	MS_T/MS_E
Error	$t(b-1)-1$	$E_{yy} - \dfrac{E_{xy}^2}{E_{xx}}$	MS_E	
Total	$tb-1$	$S_{yy} - S_{xy}^2/S_{xx}$		

The test for the hypothesis H_1 that the treatment effects are the same is based on $F = MS_T/MS_E$, where F has an F-distribution with $(t-1)$ and

$(t(b-1)-1)$ degrees of freedom. The test for the hypothesis H_2 that the regression coefficient is zero is based on $F = MS_R/MS_E$, where F has an F-distribution with 1 and $(t(b-1)-1)$ degrees of freedom.

Example 10.5: Suppose the serum levels of iron (Y) are dependent on the age (X) of the subject. Three groups are studied and the following data are available:

I	Y	99	98	98	97	93
	X	35	30	25	30	25
II	Y	112	131	151	163	153
	X	40	45	45	50	45
III	Y	116	147	156	158	138
	X	40	45	50	55	30

Test the hypothesis that there is no difference among the three groups. Also test the hypothesis that the serum level of iron does not depend on age.

$$S_{xx} = 25,100 - \frac{600^2}{15} = 1,100$$

$$S_{xy} = 79,110 - \frac{1910 \times 600}{15} = 2,710$$

$$S_{yy} = 252,960 - \frac{1910^2}{15} = 9,753.3$$

$$T_{xx} = \frac{123,050}{5} - 24,000 = 610$$

$$T_{xy} = \frac{392,225}{5} - 76,400 = 2045$$

$$T_{yy} = \frac{1,250,550}{5} - \frac{1910^2}{15} = 6,903.3$$

By substraction, we get:

$$E_{xx} = 490$$
$$E_{xy} = 665$$
$$E_{yy} = 2850$$

$$\frac{E_{xy}^2}{E_{xx}} = 902.5$$

$$E_{yy} - \frac{E_{xy}^2}{E_{xx}} = 1{,}947.5$$

$$\frac{S_{xy}^2}{S_{xx}} = 6676.5$$

$$S_{yy} = \frac{S_{xy}^2}{S_{xx}} = 3076.8$$

These calculations give us the analysis of covariance as shown in Table 10.11.

Table 10.11 Analysis of Covariance

Due to	DF	SS	MS	F
Regression	1	902.5	902.5	5.10
Treatments	2	226.8	113.4	.64
Error	11	1947.5	177.05	
Total	14	3076.8		

Since $F_{1,11,.95} = 4.84$, we reject the hypothesis that $\beta = 0$ at level .05. However, with the covariate assumption, the treatments are not significant at level .05, since $F_{2,11,.95} = 3.98$.

Exercises

14. The weight gains (y) in mice fed on four different kinds of feed, A, B, C, and D, are given in the following table; x denotes the amount of feed eaten. Test the hypothesis that the four kinds of feed are the same. Are the gains in weight affected by the quantity of food eaten?

	A		B		C		D	
x	y	x	y	x	y	x	y
16	48	29	14	99	21	47	22
28	52	45	36	52	34	20	14
14	52	15	10	83	21	71	50
48	58	21	22	63	12	36	40

15. The following data are given as the initial and final weights of guinea pigs in an experiment to study the effect of vitamins A, B, and C on their wight.

(a) Using initial weight as the covariable, test the hypothesis that there is no difference among the vitamins at level .05.
(b) Test the hypothesis that the final weight does not depend on initial weight at level .1.
(c) Using the final-initial weight difference as the response, perform a one-way analysis of variance for the data at level 05.

A	initial	23	17	20
	final	43	38	39
B	initial	18	19	23
	final	47	29	52
C	initial	20	19	25
	final	59	49	65

16. Three different methods were used to extract cadmium from dust. The data on cadmium content (ppm) and the size of particles (μ) are given for the three methods. Are the methods different? Let $y = $ cadmium content and $x = $ size:

Methods	y	x	y	x	y	x
I	55	15	135	75	110	125
II	17	40	36	106	18	500
III	5	15	10	125	8	250

(a) What is your model? Give assumptions.
(b) Test at level .05 the hypothesis that there is no difference between the methods.
(c) Test the hypothesis that the cadmium content does not depend on particle size at level .10.

The Variance-Component Model

The analysis of data through the one- and two-way classification models considered so far has assumed that the effects τ_i, β_j are unknown parameters in the model. But in some experiments such an assumption is not valid. The yield of milk from a cow depends upon its genetic composition, so a model of milk yield should include this contribution. That is, the effect is not a parameter, but a random variable. When the models contain random variables as effects, we call them *random-effect models* or *variance-component models*. When the experimental subjects are chosen from a super population, we also use variance-component models.

For a one way-analysis of variance with a random effect, we assume:

$$Y_{ij} = \mu + a_i + e_{ij}, \quad i = 1, 2, \ldots, t, \quad j = 1, 2, \ldots, n, \qquad (10.16)$$

(i) μ is a constant parameter
(ii) a_i are random effects with mean zero and variance σ_a^2.
(iii) e_{ij} are independently normally distributed random errors with mean zero and variance σ_e^2

The hypothesis that there are no treatment effects in the model is tested by $\sigma_a^2 = 0$. Although the model in Equation (10.16) is the same as the model of a one-way analysis of variance, we use different notation, a_i in place of τ_i, to show that we have random effects and not fixed effects. We test the hypothesis:

$$H: \sigma_a^2 = 0$$
$$A: \sigma_a^2 \neq 0 \qquad (10.17)$$

The test statistic is the same as in the case of the fixed-effect model: That is, the F statistic given by Table 10.2. It can further be shown, with the above assumptions, that:

$$E(\text{Between } MS) = \sigma_e^2 + n\sigma_a^2$$
$$\text{and} \quad E(\text{Within } MS) = \sigma_e^2$$

When the null hypothesis is true,

$$F = \frac{\text{Between } MS}{\text{Within } MS}$$

has F-distribution with $t - 1$ and $t(n - 1)$ degrees of freedom. The test is given in Table 10.12.

Table 10.12 Analysis of Variance

Source of Variation	DF	SS	MS	F	E(MS)
Between	$t-1$	$n\sum(Y_{i.} - Y_{..}^2)$	MS_B	$\dfrac{MS_B}{MS_W}$	$\sigma_e^2 + n\sigma_a^2$
Within	$t(n-1)$	By subtraction	MS_W		σ_e^2
Total	$tn-1$	$\sum\sum(Y_{ij} - \bar{Y}_{..})^2$			

Estimation of σ_a^2 and σ_e^2

As before, we estimate σ_e^2 by the mean square for error (within mean square):

$$\hat{\sigma}_e^2 = MS_W$$

and

$$\hat{\sigma}_a^2 = \frac{MS_B - MS_W}{n} \tag{10.18}$$

Care is needed in the use of the estimate given in Equation (10.18). When MS_W is larger than MS_B, the estimate may be negative. In that case, the convention is to estimate σ_a^2 by zero.

Random-effect models are used in genetics and psychological research. They are especially relevant in studies where animal or human subjects are involved. Extensions to two-way or multiway analysis of variance follows along the lines of fixed-effect models. For reference, see Searle (1971).

Example 10.6: Consider Example 10.1 and apply the random-effect model to the data. Test the hypothesis that $\sigma_a^2 = 0$ is the same as given by the F ratio in Table 10.3. The estimate of σ_a^2 is given by:

$$\sigma_a^2 = (3,451.67 - 174.5)/5$$
$$= 655.43$$

The magnitude of this estimate confirms the result obtained.

Exercises

17. In Exercise 2, use the random-effect model and explain the result of analysis by estimating σ_a^2.
18. For Exercise 7, give a random-effect model. What are your assumptions? Estimate the variance of the random effect.
19. In order to investigate the effect of carbon monoxide on thinking, an experiment was performed involving human subjects. They were subjected to three levels of carbon monoxide (non toxic levels), and the same arithmetic test was given to five subjects at each level of exposure.

    (a) Use the component-of-variance model to test if there is any effect of exposure to carbon monoxide at level .05:
    (b) Give the model and assumptions.

Control	Level 1	Level 2	Level 3
87	45	80	87
35	90	75	92
83	40	85	70
65	82	89	88
42	87	91	82

Nonparametric Tests

When the assumption of normality of errors in the analysis-of-variance model does not clearly hold, the data either should be transformed so that they satisfy the assumptions of the model or nonparametric methods should be used. In Chapter 8 we discussed procedures to compare two populations. Some of the procedures for comparing more than two populations—for example, testing the hypothesis of the equality of medians of c populations—have been discussed in contingency-table analysis. We give below two more c-sample procedures. The Kruskal-Wallis test for one-way classification can be regarded as a generalization of the Wilcoxon-Mann-Whitney procedure. Similarly, Friedman's test for two-way classification is also a generalization of the Wilcoxon-Mann-Whitney test for c samples. Several of these procedures are discussed by Conover (1980).

Kruskal-Wallis Test

Suppose c treatments are assigned at random to groups or individuals, and we want to test the hypothesis of the equality of the distributions of these groups. We have a one-way classification model with $n_1, n_2, \ldots n_c$ number of observations in the c group. The model is the same as (10.2), but we do not have any assumption of normality of errors.

The Kruskal-Wallis statistic, K, is given by

$$K = \frac{12}{N(N+1)} \sum_{i=1}^{c} \left[R_i - \frac{n_i}{2}(N+1) \right]^2 \qquad (10.19)$$

where $N = n_1 + n_2 + \cdots + n_c$ and $R_1, R_2, \ldots, R_c$ are the sum of the ranks of the c groups when the combined sample is ranked. When the group sizes are equal, $n_i = n$, so that:

$$K = \frac{12}{nN(N+1)} \sum R_i^2 - 3(N+1) \qquad (10.20)$$

The distribution of K for small values of n_i is given by Kraft and van Eeden (1968). Table XVI in the Appendix gives a few values. For large n_i, K is approximately distributed as chi-square with $c-1$ degrees of freedom.

Example 10.7: The clinical trials for evaluating the treatments in Hodgkins' disease consist of three treatments; X-ray therapy (I), chemotherapy (II), and X-Ray therapy combined with chemotherapy (III). The times of survival in months is given for 12 patients.

Treatment I: 17, 25, 36, 18
Treatment II: 15, 30, 23, 16
Treatment III: 35, 42, 75, 45

Replacing the combined sample with their ranks, we have:

Treatment I: 3, 6, 9, 4
Treatment II: 1, 7, 5, 2
Treatment III: 8, 10, 12, 11

$R_1 = 22$, $R_2 = 15$, $R_3 = 41$, so that:

$$K = \frac{12}{4 \times 12 \times 13}(22^2 + 15^2 + 41^2) - 39$$

$$= \frac{2390}{52} - 39 = 6.96$$

We reject the hypothesis of the equality of treatments, since the significance level of this statistic is .019 (from Table XVI).

The Friedman Test

For a two-way classification, one nonparametric test is the Friedman Test. As before, let factor A have t levels and factor B have b levels. Suppose there is one observation per cell. We replace the observations by their ranks for levels of factor B in the sample. Then the hypothesis for the equality of levels of factor A is tested by Friedman's statistic,

$$Q = \frac{12}{bt(t + 1)}\sum_{i=1}^{t} R_i^2 - 3b(t + 1) \tag{10.21}$$

where R_i is the sum of ranks of the ith level of factor A. The probability distribution of Q for small t and b is given in Table XVII in the Appendix. For large b and t, Q is approximately chi-square with $(t - 1)$ degrees of freedom.

Example 10.8: Suppose patients are observed in four hospitals, A, B, C, and D under clinical trials with treatments I, II, and III. Survival times are given in months:

Hospitals	Treatments		
	I	II	III
A	27	25	45
B	35	40	52
C	46	33	85
D	28	26	55

We want to test the hypothesis of the equality of treatments. The hospitals act as rows and the treatments as columns. Let the rows now be ranked so that we have:

	I	II	III
	2	1	3
	1	2	3
	2	1	3
	2	1	3
	$R_1 = 7$	$R_2 = 5$	$R_3 = 12$

The Friedman statistic is:

$$J = \frac{12}{4 \times 3 \times 4}[7^2 + 5^2 + 12^2] - 3 \times 4 \times 4$$

$$= \frac{218}{4} - 48 = 6.5$$

The hypothesis of the equality of treatments is rejected at .05 level of significance using Table XVII in the Appendix.

Exercises

20. Analyze the data in Exercise 9 by using nonparametric methods.
21. The clotting time of plasma under four different methods of treatment is given below. Five subjects were used to obtain plasma and the samples were assigned at random to treatments. Analyze the experiments using nonparametric techniques.

Subjects　　　Treatments	I	II	III	IV
1	8.4	7.5	9.6	9.3
2	12.8	15.2	17.1	16.2
3	9.8	7.8	8.7	12.7
4	8.6	8.3	10.2	8.9
5	7.9	8.2	9.1	11.5

22. Analyze the experiment discussed in Exercise 8 without assuming normality.
23. Analyze the data in Exercise 10 using nonparametric tests. Compare your results with the analysis-of-variance tests.
24. For Exercise 5, use nonparametric tests for testing the hypothesis that mortgage rates are the same in three states. Compare your result with the analysis of variance.

25. The pathology of tissues obtained during biopsies on four patients for diagnosing a certain disease is ranked from 1 (worst) to 4 (best) for four organs. Test the hypothesis that the biopsies are the same for the patients.

Biopsied Organ

	liver	spleen	stomach	kidney
Patient 1	4	3	1	2
2	3	4	2	1
3	4	1	3	2
4	3	4	2	1

Test for Linearity of Regression

Analysis-of-variance tests are applicable in many other contexts. We consider here the test for the linearity of regression. The linear regression model assumes that observation Y for a given X is normally and independently distributed with a mean linear in X and a variance independent of X. We assume that for every value of X there are several values of Y available. If there is only one value of Y for every value of X, we are unable to test the hyothesis.

Let there be n_1 values of Y at X_1 given by $Y_{11}, Y_{12}, \ldots, Y_{1n_1}$. Similarly, let $Y_{21}, Y_{22}, \ldots Y_{2n_2}$ be the values at X_2, $Y_{31}, Y_{32}, \ldots, Y_{3n_3}$ at X_3, and so on:

$$Y_{c1}, Y_{c2}, \ldots, Y_{cn_c} \quad \text{at} \quad X_c$$

Let $n_1 + n_2 + \cdots + n_c = n$. There are n pairs of values. A typical data set is shown in Figure 10.3.

The hypothesis of linearity essentially tests the hypothesis that the means of Ys are in a straight line. This is similar to testing the hypothesis that the means of k population are the same. Test:

$$H: E(Y|x) = \beta_0 + \beta_1 x$$
$$A: E(Y|x) \neq \beta_0 + \beta_1 x$$

The test statistic is based on partitioning the sum of squares of the deviations of observations from predicted values, in terms of the sums of squares of the deviations of predicted values of linear regression from the group averages, and the sum of squares of the deviations of group averages from the predicted values at given Xs.

Let the following be the predicted value of Y at X_j in terms of the estimates $\hat{\beta}_0$ and $\hat{\beta}_1$:

$$\hat{Y}_j = \hat{\beta}_0 + \hat{\beta}_1 X_j, \quad j = 1, 2, \ldots, c$$

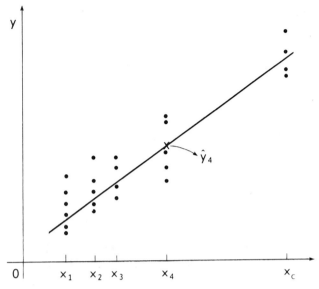

Figure 10.3. Linear regression.

Then it can be shown that:

$$\sum_i \sum_j (Y_{ij} - \hat{Y}_j)^2 = \sum_i \sum_j (Y_{ij} - \bar{Y}_j)^2 + \sum_j n_j(\bar{Y}_j - \hat{Y}_j)^2$$

or $\qquad\qquad SS_E = SS_{PE} + SS_{LF}$

$\sum_i \sum_j (Y_{ij} - \bar{Y}_j)^2$ is generally known as the *sum of squares for pure error* and
is denoted by SS_{PE}.
$\sum_i \sum_j n_j(\bar{Y}_j - \hat{Y}_j)^2$ is generally known as the *sum of squares for lack of fit*
and is denoted by SS_{LF}.

Pure error mean squares $= MS_{PE} = \dfrac{SS_{PE}}{n - c}$

Lack of fit mean squares $= MS_{LF} = \dfrac{SS_{LF}}{c - 2}$

Then the hypothesis of linearity is tested by:
Decision rule: Reject H if

$$F = \frac{MS_{LF}}{MS_{PE}} > k$$

where k is determined from the F-distribution with $(c - 2)$, $(n - c)$ degrees
of freedom. The calculations are given in Table 10.13.

Table 10.13   Analysis-of-Variance Test of Linearity

Source of Variation	DF	SS	MS	F
Lack of fit	$c-2$	SS_{LF}	MS_{LF}	$\dfrac{MS_{LF}}{MS_{PE}}$
Pure error	$n-c$	SS_{PE}	MS_{PE}	
Total	$n-2$	SS_E		

For computational convenience, we can get SS_E by subtracting the sum of squares due to regression, that is, $\hat{\beta}_1^2 \sum(X_i - \bar{X})^2$ from the total sum of squares, or:

$$SS_E + \hat{\beta}_1^2 \sum(X_i - \bar{X})^2 = SS_T = \sum_i \sum_j (Y_{ij} - \bar{Y}_{..})^2$$

Example 10.9: The following data are given on the systolic blood pressure (Y) and age (X) on 12 subjects. We first fit the linear regresion of Y on X.

X	20	30	40	50	60	70
Y	120	128	130	134	131	127
	125	132	124	132		140
						129

$$n_1 = 2, \quad n_2 = 2, \quad n_3 = 2, \quad n_4 = 2, \quad n_5 = 1, \quad n_6 = 3$$

$$\sum X_i = (2 \times 20) + (2 \times 30) + (2 \times 40) + (2 \times 50) + (1 \times 60) + (3 \times 70)$$

$$= 550, \qquad\qquad \bar{X} = 45.833$$

$$\sum Y_i = 1552 \qquad\qquad \bar{Y} = 129.333$$

$$\sum X_i^2 = 29{,}100 \qquad\qquad \sum(X_i - \bar{X})^2 = 3891.6667$$

$$\sum\sum Y_{ij}^2 = 201{,}020 \qquad\qquad \sum\sum(Y_{ij} - \bar{Y}_{..})^2 = 294.6667$$

$$\sum\sum X_i Y_{ij} = 71{,}740 \qquad\qquad \sum\sum(Y_{ij} - \bar{Y}_{..})(X_i - \bar{X}) = 606.6667$$

$$\hat{\beta}_0 = 122.18843 \qquad\qquad \hat{\beta}_1 = .15589$$

The regression equation is:

$$Y = 122.18843 + .15589X$$

SS for regression $= (.15589)^2 \times 3891.6667 = 94.5725$
$SS_E = 294.6667 - 94.5725 = 200.0942$

$$SS_{PE} = 201{,}020 - \left(\frac{245^2}{2} + \frac{260^2}{2} + \frac{254^2}{2} + \frac{266^2}{2} + \frac{131^2}{1} + \frac{396^2}{3} \right)$$

$$= 201{,}020 - 200{,}881.5 = 138.5$$

Therefore, $SS_{LF} = 200.0942 - 138.5 = 61.5942$. The analysis of variance is given in Table 10.14.

Table 10.14 Analysis-of-Variance Test for Linearity

Source of Variation	DF	SS	MS	F
Lack of fit	4	61.5942	15.3986	.667
Pure error	6	138.5	23.08	
Total	10	200.0942		

Since $F_{4,6,.95} = 4.53$, we do not reject the hypothesis of linearity at level .05.

Exercises

26. The square root of the stopping distance in feet (Y) for an automobile traveling at speed (X) in miles per hour is given for a certain automobile.

X	20	30	40	50
Y	4	5	7	9
	5	6	8	
		5	5	

(a) Obtain the linear regression of Y on X
(b) Test for linearity at level .05

27. For the following data, fit the linear regression of Y on X and give a test of linearity at level .1.

X	1	3	4	7	9
Y	2.1	3.2	4.1	5.3	6.8
	2.4	2.8	3.5		7.1
	1.8	3.1			
		2.5			

Multiple Comparison Methods

When the hypothesis of the equality of means of several populations is rejected by the analysis of variance, an unanswered question is: Which ones are the same and which ones are different? Procedures of *multiple comparison* and *simultaneous confidence intervals* attempt to answer these questions. Confidence intervals for regression parameters were obtained separately in

Chapter 9 in the simple linear-regression model. The joint confidence intervals can be obtained by Bonferroni inequality.

Bonferroni Simultaneous Confidence Intervals

We have seen from Equation (9.11) that $(1 - \alpha)$-level confidence intervals for β_0 are given by:

$$\hat{\beta}_0 \pm t_{n-2, 1-\alpha/2} \frac{\hat{\sigma} \sum x_i^2}{\sqrt{n \sum (x_i - \bar{x})^2}} \qquad (10.22)$$

The $(1 - \alpha)$-level confidence intervals for β_1 from Equation (9.12), are given by:

$$\hat{\beta}_1 \pm t_{n-2, 1-\alpha/2} \frac{\hat{\sigma}}{\sqrt{\sum (x_i - \bar{x})^2}}, \qquad (10.23)$$

Now let A be the event that β_0 is not in the confidence interval (with $\bar{A}$ the event that it is in the interval) and B be the event that β_1 is not in the confidence interval. Note that $P(A) = \alpha$ and $P(B) = \alpha$. Then $A \cap B$ is the event that both β_0 and β_1 are not in the confidence intervals.

$$P(\bar{A} \cap \bar{B}) = P(\widetilde{A \cup B}) = 1 - P(A) - P(B) + P(A \cap B)$$
$$\geq 1 - P(A) - P(B) = 1 - 2\alpha$$

So the joint confidence is at least $1 - 2\alpha$. For example, if each interval is at the .95 confidence level, then the joint statement can be made at level .90.

In the same way, the joint confidence level of $1 - \gamma$ can be obtained if each is obtained at $(1 - \gamma/2)$ level for two intervals and $(1 - \gamma/k)$-level, if we have k intervals. The Bonferroni method gives the simultaneous confidence intervals in this way for any number of parameters.

Example 10.10: Consider the 95% confidence intervals for β_0 and β_1 in Example 9.5. The rectangular region in two dimensions given by ($102.85 \leq \beta_0 \leq 114.35$) and ($.436 \leq \beta_1 \leq .650$) gives 90% joint-confidence intervals for (β_0, β_1).

Tukey's Method of Multiple Comparisons

When the analysis-of-variance test has rejected the null hypothesis that the means are equal—for example, in one-way classification—the next step is to find which means are different. Consider, for simplicity, three group means, μ_1, μ_2, and μ_3. Tukey's method is to provide joint confidence intervals for all possible pairwise differences, such as $\mu_1 - \mu_2$, $\mu_2 - \mu_3$, and $\mu_3 - \mu_1$. The confidence interval provides a further determination as to which means are unequal. If zero is included in the confidence interval for $\mu_1 - \mu_2$, then we assume that $\mu_1 = \mu_2$ at that level of confidence.

Consider means $\mu_1, \mu_2, \ldots, \mu_k$ of k groups and let the number of observations in each group be n. Suppose the group averages are $\bar{Y}_{1.}, \bar{Y}_{2.}, \ldots, \bar{Y}_{k.}$. Then the $(1 - \alpha)$-level simultaneous confidence intervals for pairs of differences are given by

$$\left(\bar{Y}_{i.} - \bar{Y}_{j.} - \frac{\hat{\sigma}}{\sqrt{n}} q_{k,nk-k,1-\alpha}, \bar{Y}_{i.} - \bar{Y}_{j.} + \frac{\hat{\sigma}}{\sqrt{n}} q_{k,nk-k,1-\alpha} \right) \qquad (10.24)$$

where $q_{k,nk-k,1-\alpha}$ is the $(1 - \alpha)$th percentile of the "studentized range" and is tabulated in Table IX. σ^2 is estimated by the mean square for error. In a one-way analysis-of-variance table, it is MS_W.

Example 10.11: Consider Example 10.1 with $k = 3, n = 5$. We are interested in finding the 95%-level simultaneous confidence intervals for pairwise differences $\mu_1 - \mu_2$, $\mu_2 - \mu_3$, and $\mu_3 - \mu_1$.

The group averages are:

$$\bar{Y}_{1.} = 97, \quad \bar{Y}_{2.} = 142, \quad \bar{Y}_{3.} = 143$$

$$\hat{\sigma} = \frac{\sqrt{174.5}}{5} = 5.91$$

$$q_{3,12,.95} = 3.77$$

The simultaneous confidence interval at level .95 is given by:

$$\bar{Y}_{i.} - \bar{Y}_{j.} \pm 5.91(3.77)$$

or

$$\bar{Y}_{i.} - \bar{Y}_{j.} \pm 22.28$$

So we have:

$$22.72 < \mu_2 - \mu_1 < 67.28$$
$$23.72 < \mu_3 - \mu_1 < 68.28$$
$$-21.28 < \mu_3 - \mu_2 < 23.28$$

Since zero is contained in the confidence interval for $\mu_3 - \mu_2$, we conclude that μ_2 and μ_3 are equal at level .05. However μ_1 and μ_2 and μ_1 and μ_3 are different.

Scheffé's Method of Multiple Comparisons

When there are an unequal number of observations in the groups or we want more general comparisons, we use Scheffé's method of multiple comparisons. We define *contrast* among means as

$$L = \sum c_i \mu_i$$

where

$$\sum c_i = 0$$

Comparisons such as $\mu_1 - 2\mu_2 + \mu_3$, $\mu_2 - 2\mu_1 + \mu_3$, $\mu_1 - 3\mu_2 + 2\mu_3$ are contrasts. They include pairwise differences $\mu_1 - \mu_2$, and so on.

A contrast can be estimated by:

$$\hat{L} = \sum c_i \bar{Y}_i$$

The variance of $\hat{L}$ is $\sigma^2 \sum (c_i^2/n_i)$ and is estimated by $\hat{Var}(\hat{L}) = \hat{\sigma}^2 \sum (c_i^2/n_i)$. Let $N = n_1 + n_2 + \cdots + n_k$.

Then $(1 - \alpha)$-level simultaneous confidence intervals for all contrasts are given by:

$$[\hat{L} - S\sqrt{\hat{Var}(\hat{L})}, \hat{L} + S\sqrt{\hat{Var}(\hat{L})}]$$

$S^2 = (k - 1) F_{k-1, N-k, 1-\alpha}$ and $F_{k-1, N-k, 1-\alpha}$ is the $(1 - \alpha)$ percentile of F-distribution with $(k - 1)$ and $(N - k)$ degrees of freedom.

Example 10.12: Suppose three groups of subjects are used to test the effectiveness of drugs A, B, and C to reduce blood pressure. Let the mean reductions be μ_1, μ_2, and μ_3, respectively. Let $n_1 = 3$, $n_2 = 5$, and $n_3 = 4$. It is known that $\bar{Y}_{1.} = 10$, $\bar{Y}_{2.} = -4$, and $\bar{Y}_{3.} = 1$ from the data and $\sigma^2 = 6$. Suppose the contrast of interest is:

$$L = \mu_1 + 2\mu_2 - 3\mu_3$$

Here $c_1 = 1, c_2 = 2, c_3 = -3$. That is, we are interested in comparing the effect of Drug A combined with twice the effect of Drug B with three times the effect of Drug C:

$$\hat{L} = 10 + 2(-4) - 3(1) = -1$$

and

$$\hat{V}(\hat{L}) = 6[\tfrac{1}{3} + \tfrac{4}{5} + \tfrac{9}{4}] = 20.3$$

$$S^2 = (3 - 1)F_{2,11,.95} = 2 \times 3.98 = 7.96$$

The .95-level of Scheffé's confidence interval is:

$$(-1 - 12.71, -1 + 12.71) = (-13.71, 11.71)$$

Since zero is included in this interval, we regard the contrast to be zero at level .05.

Exercises

28. Using Tukey's method of multiple comparisons, find the joint confidence intervals at level .95 for means the data in Exercise 2.
29. Give Bonferroni's joint confidence intervals at level .95 for the mean sales in Exercise 4.

20. In Exercise 6, find the simultaneous confidence intervals at level .90 for the contrast $2\mu_1 + \mu_2 - 3\mu_3$. Compare these intervals with the corresponding intervals obtained by using the estimate of the variance of this contrast by the t-statistic, but only for this specific contrast.

Chapter Exercises

1. The following data give yields of a certain chemical obtained by the use of three different methods of manufacture.

Method I	Method II	Method III
40	50	28
45	49	32
47	54	29
39	57	31
		32

(a) Test the hypothesis that three methods are equivalent at level .05.
(b) Find the joint confidence intervals for the mean productions at level .90 using Bonferroni's method.
(c) Use Scheffé's method to find the simultaneous confidence intervals at level .95 for all contrasts. Give limits for $\mu_1 - 2\mu_2 + \mu_3$.

2. In a study to investigate the effect of accumulation of cadmium and zinc in barley grown in sludge-treated soils, the following data are given for cadmium for one case.

Treatment	Year				
	1	2	3	4	5
0	.04	.01	.03	.04	.04
1	.08	.11	.09	.13	.12
2	.10	.11	.09	.16	.16
3	.14	.38	.34	.30	.24

Source: Change, A. C.; Page, A. L.; Warneke, J. E.; Rasketo, M. R.; and Jones, T. E., "Accumulations of cadmium and zinc in barley grown on sludge-treated soils, *J. Environ. Qual.*, 1983, *12*, 391–397.

(a) Apply a two-way analysis of variance to test the hypothesis of the equality of means of cadmium for various years

(b) Give the model and ANOVA table

(c) Give the 95% Bonferroni intervals at level .95 for the treatment means

3. To find the effect of Mount St. Helen's volcanic ash on plant growth and mineral uptake, an experiment was conducted. The following data are given for chard (in $g\,kg^{-1}$).

	N	P	K
no treatment	34	3.7	36
50 ash	30	3.6	39
100 ash	39	3.7	45

(a) Using a two-way analysis-of-variance model, test the hypothesis that there are no differences among treatments at level .05

(b) Give the model and assumptions

   Source: Cochran, V. L.; Bezdicek, D. F.; Elliott, L. F.; and Papendick, R. I., The effect of Mount St Helens' volcanic ash on plant growth and mineral uptake, J. Environ. Qual., 1983, 12, 415–417.

4. The following table gives data for the responses (Y) at certain levels of a drug (X).

X	.02	.08	3.2
	70	85	100
	75	60	83
	65	79	115
		87	120
		120	

(a) Fit a linear regression to the data.

(b) Give the confidence intervals at level .95 separately for the intercept and the slope.

(c) What is the joint confidence level for both using Bonferroni's method?

5. The following data give the "median" family income of Hispanic, Black, and White families during 1972–81 in constant dollars.

(a) Test the hypothesis that the "median" family income of the three groups is the same at level .05

(b) Test the hypothesis that the median family income has remained the same over the years at level .05

(c) Give the analysis-of-variance model and your assumptions

Year	Hispanic	Black	White
1972	17,790	14,992	25,107
1973	17,836	14,877	25,777
1974	17,594	14,765	24,728
1975	16,140	14,835	24,110
1976	16,390	14,766	24,823
1977	17,141	14,352	25,124
1978	17,518	15,166	25,606
1979	18,255	14,590	25,689
1980	16,242	13,989	24,176
1981	16,401	13,266	23,517

Source: Bureau of the Census, Money, Income and Poverty Status of Families of Persons in the United States, 1981.

6. The three methods of birth control used in China over the years 1971–77 have been IUD insertion (I), sterilization (II), and induced abortions (III). The percentages of I and II are given in the following:

   (a) Test the hypothesis that the two methods have the same mean percentages at level .05
   (b) Test the hypothesis that there is no difference among the years at level .05
   (c) What is the model used?
   (d) What assumptions are being made?

	IUD Insertions	Sterilizations
1971	47.4	22.8
1972	51.7	21.3
1973	58.2	20.4
1974	59.1	17.5
1975	60.3	21.4
1976	51.9	18.8
1977	55.0	22.2

Source: Zhang Lizhong, Birth control, late marriage, and decline in the national increase rate of population, *Population and Economics* (Renkou Yu Jing ji), 1980, *1*, 35–39.

Summary

A comparison of the means of several normal populations, each having the same variance, is made through the technique of *analysis of variance*. The *linear model* represents the observation in various components of variation, including the common mean and *treatment effect* and *error*. A one-way analysis of variance generalizes to a two-way analysis of variance when the observation is assumed to have components involving several effects. The technique of analysis of variance requires the partitioning of the sum of squares of deviations about the average in several independent component sums of squares. The *analysis-of-covariance* model assumes dependence on a variable that cannot be controlled. The *variance-component model* assumes that the model components are random variables and not constants. The tests turn out to be the same as in the case of the *fixed-effect* model. The *linearity of regression* model can be tested by an analysis-of-variance test when we partition the total variation in terms of the variation about the fitted line and pure error component. *Multiple comparison* methods are needed to decide which means are different once the hypothesis of the equality of means is rejected. *Simultaneous confidence intervals* provide the required methods. They include those by Tukey, Scheffé, and Bonferroni.

References

Canover, W. J. *Practical Nonparametric Statistics*, 2nd ed. New York: John Wiley & Sons, 1980.
Fisher, R. A. *Statistical Methods for Research Workers*. New York: Hafner Publishing Co., 1958.
Kraft, C. H., and van Eeden, C. *A Nonparametric Introduction to Statistics*. New York: Macmillan, 1968.
Miller, Rupert G. *Simultaneous Statistical Inference*. New York: McGraw-Hill, 1966.
Scheffé, H. *Analysis of Variance*. New York: John Wiley & Sons, 1959.
Searle, S. R. *Linear Models*. New York: John Wiley & Sons, 1971.
Tukey, John W. One degree of freedom for nonadditivity, *Biometrics*, 1949, 5, 232–242.

chapter eleven

The Design of Experiments

Scientists are faced with the problem of designing experiments that provide observations without bias. When we select a group of animals for experimentation, a group of patients from a hospital for clinical study, or a set of machines in a factory for time and motion studies, we want the *experimental units* to be such that every individual in the population to be studied has the same chance of being selected. When the experimental material is not homogeneous, such as a plot of land with different fertility characteristics or animals with different body weights, the results of the experiments may be confounded with these differences in experimental units. To enforce homogeneity of the material, the material may be further divided into smaller groups that are more homogeneous within themselves.

The theory of the design of experiments is concerned with developing procedures so as to remove any variability in the experimental material and reduce the *experimental error*. Observational studies where members of a population are observed for certain purpose form a part of *sample survey design* which will be discussed in Volume II. Here we are concerned with controlled experiments where responses to *treatments*, such as methods, procedures, or stimuli, are tested by the experiment. The repetition of the experiment under similar conditions is often needed to estimate the error of the response. This process is known as *replication*. The more observations we have, the more precision we have in estimating mean response, since the variance of mean is inversely proportional to sample size, and an increase in sample size reduces its variance.

To reduce the inherent variability in the experimental material, Sir R. A. Fisher introduced the concept of *randomization* in experimentation in 1920s. Randomization requires the use of a chance device in the selection or allocation of treatment to experimental units. We have already discussed the method of getting a random sample with the use of random numbers. Again, for experimental purposes, randomization can be accomplished by the use of random numbers. The original use of randomization and the development of designs was made for agricultural experiments at the Rothamstead Experimental Station in Great Britain. Today, the use of the modern design of experiments is made extensively in medicine, biology, technology, and business.

The *scientific method* entails the proposal of a hypothesis and its verification by experimentation or observation. We discuss a few situations in this chapter where the basic notions of the scientific method are introduced.

Often the theory requires the proposal of a linear model, and tests of hypothesis are made for the parameters of the model by analysis of variance, discussed in Chapter 10.

Completely Randomized Designs

Suppose three different fertilizers, including a control, are being studied to see their effect on tomato production. Twelve pots with the same soil mixture are available for experiment. It is assumed that the tomatoes will be grown in the same environment (in a hothouse), and other experimental conditions will remain uniform as much as possible. In this case, we assign each treatment to four randomly chosen pots and use the total tomato production on each pot as the response. Then we have three treatments with four observations on each treatment. The assumptions for good experimentation are that one of the treatments be a control (or a placebo) so that treatments can be compared with a control.

Inclusion of a control in comparative studies is important because it allows us to see the effect when no treatment is present under a given set of experimental conditions. Many studies in practice suffer from a lack of appropriate controls. The observation Y_{ij} from this experiment is the same as in one-way classification with an equal number of observations in each group. The model for a completely randomized design having t treatments, each having n obserations, is:

$$Y_{ij} = \mu + \tau_i + e_{ij}$$
$$i = 1, 2, \ldots, t, \quad j = 1, 2, \ldots, n$$

The assumptions are the same as before:

(i) τ_i are treatment effects with $\sum \tau_i = 0$
(ii) e_{ij}s are errors independently normally distributed with mean zero and

variance σ^2. We test the hypothesis:

$$H: \tau_i = 0, \quad \text{for all } i$$

$$A: \text{at least one } \tau_i \text{ is not zero}$$

The test is performed by analysis of variance as given in Table 10.2. In place of designating the sum of squares between groups, we now call them the *treatment sum of squares*, and the sum of squares within groups is designated as the *error sum of squares*.

When the number of observations for treatments is unequal, the model for the experiment is the same as a one-way classification with an unequal number of observation per cell. The analysis-of-variance test is gives in Table 10.3 for this case.

Example 11.1: A pharmaceutical company has developed two new drugs (A and B) to treat dust allergy. A clinical trial is conducted by a clinic using drugs A and B as well as control C, using patients as experimental units. The patients are children from the same neighborhood. Assuming that the patients with a dust allergy are homogeneous, a completely randomized design is used. The number of patients assigned for A, B, and C may not be the same here. The data obtained can be treated as a one-way classification with an unequal number of observations for testing the effectiveness of drugs A and B.

Example 11.2: Let A and B be the fertilizers, with C as the control, used in the tomato experiment. The data of the completely randomized design for tomato plants are given as the yield (pounds) per pot. Using the linear model for the yield, we are interested in testing the hypothesis that the effect of fertilizer is zero.

	Treatments		
	A	B	C
	7.5	8.3	6.5
	6.3	7.8	6.0
	5.8	8.9	7.0
	6.5	6.8	6.8
Totals	26.1	31.8	26.3

Correction factor $= (84.2)^2/12 = 590.8033$

Treatment sum of squares $= \frac{1}{4}[26.1^2 + 31.8^2 + 26.3^2] - 590.8033$

$$= 5.2317$$

Total sum of squares $= 600.5 - 590.8033 = 9.6967$

Error sum of squares $= 9.6967 - 5.2317 = 4.465$

Analysis-of-Variance Table

Source of Variation	DF	SS	MS	F
Treatments	2	5.2317	2.6159	5.273
Error	9	4.4650	0.4961	
Total	11	9.6967		

Since $F_{2,9,.95} = 4.26$, we reject the hypothesis of the equality of treatment effects at level .05. The three means are for:

Treatment $A = 6.525$
Treatment $B = 7.95$
Control $C = 6.575$

Notice that treatment B is significantly different from control, and Treatment A is not significantly different from control at level .05 using multiple comparison tests.

Exercises

1. Four different methods of making steel rods are tested by Worthington Steel Company, Ohio, on the same material. The data obtained on the discrepancy from the desired diameters (mm) by the four methods are given below. A completely randomized design has been used.

	Methods		
A	B	C	D
1.1	.6	1.4	.4
1.3	.7	1.2	.5
.9	.8	.8	.6
.8	.5	1.5	.3
.7	.4	1.3	.2

(a) Give the model and assumptions to test the hypothesis that the four methods are the same
(b) Perform an analysis-of-variance test at level .05 to test the hypothesis of the equality of the four methods
(c) If there is a significant difference among the methods, use multiple comparisons, giving the methods that are really different using a .95-level simultaneous confidence intervals

2. Cakes can be baked at three different temperatures (T_1, T_2, T_3). A completely randomized design was used to bake 12 cakes, and their fluffiness was measured, using an index with the data:

Temperatures

T_1	T_2	T_3
2.3	4.5	7.3
7.5	6.5	4.4
4.5	2.8	6.3
	5.4	4.6
		5.1

(a) Give the model and assumptions
(b) Test the hypothesis that the fluffiness at the three temperatures is the same at level .05

3. Fifteen students were randomly assigned to three different sections of calculus using regular teaching method (A), using hand calculators, (B) and using microcomputers (C). The scores on a standard test after a semester of instruction are given:

Teaching Methods

A	B	C
84	65	75
79	89	95
83	82	92
65	78	89
87	75	98

(a) What is the model for a completely randomized design in this experiment? What are your assumptions?
(b) Test the hypothesis that the methods are the same at a .1 level of significance.
(c) If the methods are significantly different, give the methods that are different from the regular method. Use a .95 level of confidence.

Randomized Complete Block Designs

When there is heterogeneity in the experimental material, we cannot use a completely randomized design. We divide the experimental material in various smaller subsets, called *blocks*, in such a way that variability within a block is small. For example, if an agricultural piece of land has a fertility gradient in the x-direction, we divide it in blocks by lines in the y-direction so that between blocks there is high variability in fertility,

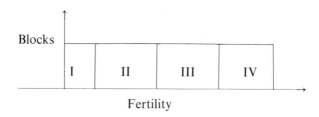

but within the blocks there is little variability. Similarly, if experimentation is needed on human subjects to measure the effects of a drug, we may divide them according to age, the blocks being the various age groups, such as children, adults, older adults, and seniors. The subdivision of each block into *experimental units* is made and each unit is called a *plot*. Plots may be patients, animals, test tubes, tissues, machines, or even small plots of land. *Randomized block designs* are the experimental procedures in which treatments are assigned to plots within each block at random. When the number of plots in a block is the same as the number of treatments, we have *randomized complete block designs*; otherwise, we have incomplete blocks designs.

   If we are studying t treatments, we must have t plots in each block. The assignment of treatments at random in each block means that all the $t!$ different arrangements of treatments are given the same chance. This can be done by using the table of random numbers. Tables of random permutations can also be used to make assignments in blocks. Extensive tables are given by Moses and Oakford (1963).

   The model of a randomized complete block design is the same as the model for a two-way classification in analysis of variance. Let Y_{ij} denote the response on ith treatment in jth block, $i = 1, 2 \ldots t, j = 1, 2 \ldots b$. We assume that

$$Y_{ij} = \mu + \tau_i + \beta_j + e_{ij}$$

where τ_i is the *treatment effect* and β_j is the *block effect*. As usual, we assume:

(i) $\sum \tau_i = \sum \beta_j = 0$
(ii) e_{ij}s are independently normally distributed errors with means zero and variance σ^2

The interest here is in testing the hypothesis:

$$H_1 : \tau_i = 0 \quad \text{for all } i$$
$$A_1 : \text{at least one } \tau_i \text{ is not zero}$$

It is not usual to test the hypothesis that the block effects are zero. However, to see if blocking has any effect, we can test the hypothesis:

$$H_2: \beta_j = 0 \quad \text{for all } j$$
$$A_2: \text{at least one } \beta_j \text{ is not zero}$$

The test for the hypotheses can be made by the analysis of variance as given in Table 10.5. We shall denote one of the factors by *blocks* and the other factor by *treatments* in the analysis-of-variance table for the randomized block design.

Example 11.3: A randomized complete block design was used to compare the production by four machine operators ($A, B, C,$ and D). Three kinds of machines are used by operators in random order. Let the assignment be:

	Machines	
I	II	III
A	A	B
B	C	A
C	B	D
D	D	C

Here the "blocks" are machines and operators are "treatments." The data obtained from the design are given (in terms of screws made per hour by the machine in hundreds).

		Machine			Total
		I	II	III	
	A	12	15	18	45
Operators	B	11	16	27	54
	C	18	17	29	64
	D	20	19	18	57
Total		61	67	92	220

Correction factor $= (220)^2/12 = 4033.3333$

Operators sum of squares $= \frac{1}{3}(45^2 + 54^2 + 64^2 + 57^2) - 4033.3333$
$$= 62.0$$

Machines sum of squares $= \frac{1}{4}(61^2 + 67^2 + 92^2) - 4033.3333$
$$= 135.1667$$

Total sum of squares $= 4338 - 4033.3333 = 304.6667$

Error sum of squares $= 304.6667 - 135.1667 - 62 = 107.5$

Analysis-of-Variance Table

Source of Variation	DF	SS	MS	F
Machines	2	135.1667	67.5833	3.7721
Operators	3	62.0	20.6667	1.1535
Error	6	107.5	17.9167	
Total	11	304.6667		

The test of the hypothesis that there is no difference among the operators is not rejected at level .05, since $F_{3,6,.95} = 4.76$. To test the hypothesis of the equality of machine effects, we find that $F_{2,6,.95} = 5.14$, and we also do not reject the hypothesis. That is, blocking by different machines was not useful, and a completely randomized design may be appropriate in future.

Example 11.4: Four varieties of barley were compared in a randomized complete block experiment. The results of the experiment are given:

Block	Variety A	B	C	D	Totals
I	95	68	84	64	311
II	94	59	86	68	307
III	89	70	74	65	298
IV	100	79	84	80	343
Totals	378	276	328	277	1259

Correction factor $= (1259)^2/16 = 99{,}067.563$

Total sum of squares $= 101{,}357 - 99{,}067.563 = 2289.437$

Blocks sum of squares $= \frac{1}{4}[311^2 + 307^2 + 298^2 + 343^2] - CF$
$$= 288.187$$

Treatment sum of squares $= \frac{1}{4}[378^2 + 276^2 + 328^2 + 277^2] - CF$
$$= 1775.687$$

Error sum of squares $= 2289.437 - 288.187 - 1775.687$
$$= 225.563$$

Analysis-of-Variance Table

Source of Variation	DF	SS	MS	F
Blocks	3	288.187	96.0623	3.8329
Treatments	3	1775.687	591.8957	23.6167
Error	9	225.563	25.0626	
Total	15	2289.437		

To test the hypothesis of the equality of treatment effects, we see that $F_{3,9,.95} = 3.86$, and hence we reject it at level .05. However, the hypothesis of the equality of block effects is not rejected at level .05.

Exercises

4. Three varieties of corn were tested in a randomized block design on four blocks, and the heights of the plants were measured. Is there a significant difference among the varieties at level .05?

		Varieties		
		I	II	III
Blocks	1	50.1	54.2	57.1
	2	51.2	58.3	60.1
	3	47.1	55.1	56.2
	4	56.1	58.7	59.1

5. Hemoglobin content (in grams %) was measured in five dogs who were tested for three drugs (A, B, and C) for their effect on hemoglobin. Test the hypothesis that the drugs have the same effect at level .05.

	Drugs		
	A	B	C
Dog 1	12.1	9.5	12.2
Dog 2	11.9	8.4	13.1
Dog 3	10.3	10.3	13.3
Dog 4	9.4	11.2	11.8
Dog 5	8.5	11.5	12.5

6. The number of units produced by four machines over a period of five days was recorded. Regard the days as blocks. For the following data, test whether there is any difference among machine production at level .05. If there is a significant differences use multiple comparisons at level .95 to find which machines are different.

	Machines			
	I	II	III	IV
Day 1	50	62	71	61
Day 2	40	51	43	81
Day 3	51	52	82	92
Day 4	45	61	78	85
Day 5	43	63	72	78

Latin Square Designs

When the experimental units are such that there is two-way variability, we shall not be able to control it by randomized block designs. Suppose three drugs are to be tested on human subjects who have different ages and weights. So we need to divide the subjects not only with respect to age but also with respect to weight. Nine subjects need to have three treatments for one replication of the experiment. Let the treatments (drugs) be denoted by A, B, and C. A *Latin square design* is such that there is one, and only one, treatment for each age and weight class. For example, in the above case, the assignment is:

		Age Class		
		1	2	3
	1	A	B	C
Weight Class	2	B	C	A
	3	C	A	B

The two classifications can be regarded as rows and columns. To achieve randomness in the assignment of treatments by Latin squares, we can do one of two things:

(a) Choose a Latin square at random from a catalog of Latin squares of given order. A sample of order 2 and 3 is given in Table 11.1, for example. Other tables are available.

Table 11.1   Latin squares of order 2 and 3

```
    A   B                 B   A
    B   A                 A   B

A   B   C         A   C   B
B   C   A         B   A   C
C   A   B         C   B   A

B   A   C         B   C   A
C   B   A         C   A   B
A   C   B         A   B   C

C   A   B         C   B   A
A   B   C         A   C   B
B   C   A         B   A   C

A   B   C         A   C   B
C   A   B         C   B   A
B   C   A         B   A   C

B   A   C         B   C   A
A   C   B         A   B   C
C   B   A         C   A   B

C   A   B         C   B   A
B   C   A         B   A   C
A   B   C         A   C   B
```

(b) Take a cyclic Latin square and permute the rows or columns, or both, at random. For example, the square

$$
\begin{array}{cccc}
A & B & C & D \\
B & C & D & A \\
C & D & A & B \\
D & A & B & C
\end{array}
$$

can be changed to the following if we replace column 2 by 1

$$
\begin{array}{cccc}
B & A & C & D \\
C & B & D & A \\
D & C & A & B \\
A & D & B & C
\end{array}
$$

Similarly, by interchanging the third row with the second, we have:

$$
\begin{array}{cccc}
B & A & C & D \\
D & C & A & B \\
C & B & D & A \\
A & D & B & C
\end{array}
$$

Let Y_{ijk} be the observation in ith row and jth column on kth treatment in a Latin square. We use the three-way classification analysis-of-variance model for a Latin square:

$$Y_{ijk} = \mu + \alpha_i + \beta_j + \tau_k + e_{ijk}$$
$$i = j = k = 1, 2, \ldots, t \tag{11.1}$$

The assumptions of the model are:

(i) $\sum \alpha_i = \sum \beta_j = \sum \tau_k = 0$
(ii) e_{ijk}s are independently normally distributed errors with means zero and variance σ^2.

The test of the following hypothesis, is obtained by analysis of variance given in Table 11.2:

$$H: \tau_k = 0 \text{ for all } k$$

$$A: \text{at least one } \tau_k \text{ is not zero}$$

The analysis is an extension of a two-way classification with no interaction.

We note that Y has three subscripts, ij and k, but in actuality there is only one value for a given i and j. The dot notation also uses three subscripts. By $\bar{Y}_{1..}$ we mean the average of the treatments of the first row. Similarly, by $\bar{Y}_{.1.}$ we denote the average of treatments for the first column, and by $\bar{Y}_{..1}$ we denote the average of the first treatment. The overall average is denoted by $\bar{Y}_{...}$. The appropriate sum of squares needed in the analysis of variance is given in Table 11.2.

Table 11.2 Analysis of Variance for Latin Square Design

Source of Variation	DF	SS	MS	F
Rows	$t-1$	$t\sum_i (\bar{Y}_{i..} - \bar{Y}_{...})^2$	MS_R	
Columns	$t-1$	$t\sum_j (\bar{Y}_{.j.} - \bar{Y}_{...})^2$	MS_C	
Treatments	$t-1$	$t\sum_{ij} (\bar{Y}_{..k} - \bar{Y}_{...})^2$	MS_T	MS_T/MS_E
Error	$t^2 - 3t + 2$	By subtraction	MS_E	
Total	$t^2 - 1$	$\sum_i \sum_{jk} (Y_{ijk} - \bar{Y}_{...})^2$		

The test of the hypothesis H is made by $F = MS_T/MS_E$, which has F-distribution with $t - 1, t^2 - 3t + 2$ degrees of freedom. Row and column effects can be tested in the same manner as the treatments.

When $t = 3$, that is, when we are analyzing a single Latin square experiment of order 3, the degrees of freedom for error is only 2. To base the estimate of σ^2 on small degrees of freedom does not provide a good test. Therefore, we replicate the experiment n times. Suppose $n = 2$ in the above case. Then we have 18 observations, and degrees of freedom for error are now $17 - 2 - 2 - 2 = 11$.

Source of Variation	DF
Rows	2
Columns	2
Treatments	2
Error	11
Total	17

The degrees of freedom for error can be easily increased by a few replications. In general, similar modification can be made in the analysis-of-variance table. Suppose we have n replicates of a Latin square experiment with t treatments. Then we have the breakdown of degrees of freedom:

Source of Variation	DF
Rows	$t - 1$
Columns	$t - 1$
Treatments	$t - 1$
Error	$nt^2 - 3t + 2$
Total	$nt^2 - 1$

Example 11.5: Three drugs, $(A, B,$ and $C)$ are to be studied for their effect on preventing vomiting and nausea after surgery. Three kinds of patients I, II and III (with heart surgery, appendectomy, and hysterectomy) are treated with the three drugs in three different hospitals (1, 2, and 3). We can use a Latin square design with the hospitals as rows, the patient category as columns, and drugs as treatments.

		Patient Category		
		I	II	III
	1	A	C	B
Hospitals	2	B	A	C
	3	C	B	A

Suppose three replicates of this experiments are available. Then we have:

Source of Variation	DF
Hospitals	2
Patients	2
Treatments	2
Error	20
Total	26

Example 11.6: Peanut microcomputers produced by IBM in its three manufacturing plants during three working shifts were made under three different methods of assembly (A, B, and C). The data on two replications are given below. Find if the methods of assembly are different at level .05.

		Plants			Total
		1	2	3	
Shifts	1	45,50	40,50	35,40	260
	2	75,60	50,60	50,52	347
	3	60,55	45,50	60,75	345
Total		345	295	312	952

The Latin square design used was:

		Plants		
		1	2	3
Shifts	1	B	A	C
	2	C	B	A
	3	A	C	B

Treatments	A	B	C
Totals	307	340	305

Correction factor $= (952)^2/18 = 50350.222$
Total sum of squares $= 52354 - 50350.222 = 2003.778$
Row sum of squares $= \frac{1}{6}[260^2 + 347^2 + 345^2] - CF = 822.1113$
Column sum of squares $= \frac{1}{6}[345^2 + 295^2 + 312^2] - CF = 215.4447$
Treatment sum of squares $= \frac{1}{6}[307^2 + 340^2 + 305^2] - CF = 128.7780$

Analysis of Variance Table

Source of Variation	DF	SS	MS	F
Shifts (rows)	2	822.1113	411.0557	
Plants (columns)	2	215.4447	107.7224	
Methods (treatments)	2	128.7780	64.3890	.846
Error	11	837.444	76.1313	
Total	17	2003.778		

To test the hypothesis of no difference between methods of assembly, we have $F = .846$, which is not significant at level .05 since $F_{2,11,.95} = 3.98$. Thus we do not reject the hypothesis.

Exercises

7. Three replications of a second-order Latin square were made for testing two methods (A, B) of instruction in swimming. There were two instructors (rows) and they used two different swimming pools (columns) according to the following scheme:

$$
\begin{array}{cc}
A & B \\
B & A
\end{array}
$$

The scores on a standardized measure of swimming ability are:

	Replications	
I	II	III
20 25	15 20	14 30
35 19	25 17	35 12

Test the hypothesis if the methods of instruction are the same at level .05.

8. A Latin square of order 3 was used to test three feeds for pigs $(A, B,$ and $C)$. The experiment was repeated twice. Given the following data, test

if the feeds are significantly different at level .05 in increasing the weight of the pigs.

C 121,135	B 105,95	A 145,143
B 100,110	A 135,115	C 110,105
A 125,105	C 115,110	B 105,98

9. Four different laboratories were used to find the lead content of four specimens (A, B, C, and D) with four different methods of analysis. Regard laboratories as rows and methods of analysis as columns. Given the following data, find if the specimens are significantly different at level .05. A Latin square design was used.

Methods

		1	2	3	4
	1	B .19	A .13	C .21	D .31
Laboratories	2	D .36	C .24	A .17	B .17
	3	C .28	B .18	D .30	A .15
	4	A .18	D .35	B .15	C .27

Factorial Experiments

Many experiments involve the study of several factors simultaneously. A drug may be studied at several levels of concentration and at each level with several methods of administration. All these factors can be studied in one experiment. The advantage of studying several factors simultaneously is that we can test hypotheses of the equality of effects at various levels of each factor, as well as the interaction among these factors. When one factor at a time is studied, the information on interaction cannot be obtained. An important contribution of statistics to the design of experiments has been the introduction of factorial experiments.

For simplicity, the case of two factors each at two levels is considered. Denote the factors by a and b and the levels by 0 and 1. This is a 2×2 or 2^2 experiment. The four treatment combinations can be represented as

$$a_0 b_0$$
$$a_0 b_1$$
$$a_1 b_0$$
$$a_1 b_1$$

$a_0 b_0$ gives the treatment where both a and b are at 0 level; $a_0 b_1$ gives a treatment where a is at 0 level, but b is at level 1, and so on. The experiment can be performed using these four treatments in a randomized block design. The object of this section is to study the comparison among treatments and levels and the interactions among them.

Contrasts

A *contrast* is a linear combination of treatments such that the coefficients add to zero. For example, the treatment combination

$$a_0 b_0 + a_0 b_1 - a_1 b_0 - a_1 b_1$$

is a contrast, since $1 + 1 - 1 - 1 = 0$.

However, $a_0 b_0 + a_0 b_1 + a_1 b_0 + a_1 b$

is not a contrast. Many important tests of hypotheses in factorial experiments can be written in terms of contrasts. For example, if we want to test whether factor a has any effect, we need to test the hypothesis that the main effect contrast due to a is zero. One way to arrive at a contrast for the main effect a is to consider the average of the difference $a_1 - a_0$ at both levels b_0 and b_1 of the other factor. We define the main effect by $A = \frac{1}{2}[(a_1 - a_0)b_1 + (a_1 - a_0)b_0]$. That is:

$$A = \tfrac{1}{2}(a_1 - a_0)(b_1 + b_0) \tag{11.1}$$

Similarly, the main effect due to b is:

$$B = \tfrac{1}{2}(b_1 - b_0)(a_1 + a_0) \tag{11.2}$$

The treatment effect that measures the *interaction* between the factors a and b is taken as

$$AB = \tfrac{1}{2}(a_1 - a_0)(b_1 - b_0) \tag{11.3}$$

These effects have an important property: They are *orthogonal* to each other.

Orthogonal Contrasts

Two contrasts

$$c_1 t_1 + c_2 t_2 + c_3 t_3 + c_4 t_4 \qquad \text{with} \quad \sum c_i = 0$$

and

$$d_1 t_1 + d_2 t_2 + d_3 t_3 + d_4 t_4 \qquad \text{with} \qquad \sum d_i = 0$$

among treatments t_1, t_2, t_3, and t_4 are *orthogonal* if

$$c_1 d_1 + c_2 d_2 + c_3 d_3 + c_4 d_4 = 0 \tag{11.4}$$

We can easily verify from Equations (11.1) and (11.2) that the following are orthogonal.

$$2A = a_1 b_1 + a_1 b_0 - a_0 b_1 - a_0 b_0$$
$$2B = a_1 b_1 - a_1 b_0 + a_0 b_1 - a_0 b_0$$

The basic use of the orthogonal property is in breaking the sums of squares for treatments. For example, the sum of squares for treatments having three degrees of freedom in the case of four treatments can be decomposed into three separate sums of squares for orthogonal contrast, each having one degree of freedom. The sums of squares for orthogonal contrasts are independent of each other and have a chi-squared distribution with one degree of freedom when divided by σ^2.

The tests of hypotheses in the case of factorial experiments are basically of the following type, with suitable alternatives:

$$H_1 : A = 0 \quad \text{(main effect } a \text{ is zero)}$$
$$H_2 : B = 0 \quad \text{(main effect } b \text{ is zero)}$$
$$H_3 : AB = 0 \quad \text{(interaction effect is zero)}$$

The tests are given by analysis of variance. Suppose we have n replications of the 2^2 design, giving us $4n$ observations. Let y_{ijk} be the kth observation on the ith factor at the jth level with $i = 0, 1, j = 0, 1$, and $k = 1, 2, \ldots, n$. The model here is,

$$y_{ijk} = \mu + \alpha_i + \beta_j + \gamma_{ij} + e_{ijk} \tag{11.5}$$

with usual assumptions

$$\sum_i \alpha_i = \sum_j \beta_j = \sum_i \gamma_{ij} = \sum_j \gamma_{ij} = 0$$

and e_{ijk} are random errors. This is the same model as (10.13) for the two-way analysis of variance with interaction. The decomposition of the degrees of freedom is given in Table 11.3. The sum of squares and mean squares for the various entries in Table 11.3 are obtained in the same way as in the section "Two-Way Analysis of Variance" in Chapter 10.

Table 11.3	Decomposition of degrees of freedom

Source of Variation	DF
Main effect A	1
Main effect B	1
Interaction AB	1
Error	$4n - 4$
Total	$4n - 1$

The analysis of 2^n factorial designs follows the analysis for the 2^2 design. In this case, the number of main effects is n. The number of interactions among pairs of factors is $\binom{n}{2}$. Similarly, the number of *third-order interactions* is $\binom{n}{3}$ and so on. Notice that the $2^n - 1$ degrees of freedom among 2^n treatments are composed of degrees of freedom for main effects, second-order interactions, third-order interactions, and so on. The degrees of freedom for treatments is $2^n - 1$. The following identity verifies the required decomposition:

$$2^n - 1 = \binom{n}{1} + \binom{n}{2} + \cdots + \binom{n}{n}$$

Factorial experiments with a different number of levels for each factor can be analyzed by the same method. Consider an example where two levels of a drug are to be tested, each with three different methods of formulations, by four hospitals. This is a $2 \times 3 \times 4$ factorial design. The contrasts in such experiments are difficult to interpret.

For a detailed discussion of factorial experiments, see Kempthorne (1952).

Example 11.7: The mean germination time of carrot and celery seeds is obtained in an experiment under two drying rates (fast and slow). The data on three replicates are:

	Drying rate	
	fast (b_0)	slow (b_1)
Carrots (a_0)	1.4, 1.6, 1.5	1.7, 1.9, 2.2
Celery (a_1)	4.2, 4.6, 4.1	4.9, 5.1, 5.3

This is a 2×2 factorial experiment, with factor a being the vegetable variety, $a_0 = $ carrots, $a_1 = $ celery, and with factor b being the drying rates, with $b_0 = $ fast and $b_1 = $ slow.

Here we want to test the hypotheses about the equality of the population means of germination times for factors a and b and to test if the interaction between them is zero.

The sum of squares for factor a is obtained by totaling the germination times for both levels of a and subtracting the correction factors from the sum of squares of the totals divided by the number of observation in each total. That is, the sum of squares for a, SS_A, is given by:

$$SS_A = \tfrac{1}{6}[(\text{total for celery})^2 + (\text{total for carrots})^2] - CF$$

Similarly, the sum of squares for factor b can be obtained. The sum of squares for the interaction between factors a and b is obtained by summing the squares of the totals for treatments a_0b_0, a_0b_1, a_1b_0, and a_1b_1, divided by the number of observations on which the totals are based and subtracting SS_A, SS_B, and CF. The subtotals are given by:

	b_0	b_1	Totals
a_0	4.5	5.8	10.3
a_1	12.9	15.3	28.2
Totals	17.4	21.1	38.5

Correction factor $= (38,5)^2/12 = 123.5208$
$SS_A = \tfrac{1}{6}[10.3^2 + 28.2^2] - CF = 26.7009$
$SS_B = \tfrac{1}{6}[17.4^2 + 21.1^2] - CF = 1.1409$
$SS_{AB} = \tfrac{1}{3}[4.5^2 + 5.8^2 + 12.9^2 + 15.3^2] - SS_A - SS_B - CF = .1007$
SS total $= 151.83 - CF = 28.3092$

Analysis-of-Variance Table

Source of Variation	DF	SS	MS	F
Vegetables	1	26.7009	26.7009	582.9891
Drying times	1	1.1409	1.1409	24.9105
Vegetables × Drying times	1	.1008	.1008	2.2009
Error	8	.3666	.0458	
Total	11	28.3092		

We find that $F_{1,8,.95} = 5.32$, and hence we reject the hypothesis of the equality of means for the levels of vegetables and drying times. However, the interaction is not significant at level .05.

Example 11.8: Two methods of teaching mathematics (factor *a*) are being experimented upon at a state university and two different teachers (factor *b*) are used. The same experiment is repeated three times. We have a 2×2 factorial experiment with three replications. The average class scores are:

		Factor (b_0)	(b_1)
Factor	a_0	71, 65, 83	85, 96, 87
	a_1	65, 55, 50	80, 91, 84

We would like to know if there are differences between methods and teachers and if there is any interaction. The computations to obtain the sum of squares for *A*, *B*, and interaction *AB* are:

Correction factor $(CF) = 912^2/12 = 69,312$

SS_A = Sum of squares for $A = \frac{1}{6}[487^2 + 425^2] - CF = 320.3333$

SS_B = Sum of squares for $B = \frac{1}{6}[389^2 + 523^2] - CF = 1496.3333$

SS_{AB} = Sum of squares for $AB = \frac{1}{3}[219^2 + 268^2 + 170^2 + 255^2] - SS_A$
$$- SS_B - CF$$
$$= 108.0001$$

SS_T = Sum of squares for total $= 71,652 - CF = 2,340$

The analysis-of-variance table gives the test:

Analysis-of-Variance Table

Source of Variation	DF	SS	MS	F
Methods	1	320.3333	320.3333	6.17
Teachers	1	1496.3333	1496.3333	28.82
Methods × Teachers	1	108.0001	108.0001	2.08
Error	8	415.3333	51.9167	
Total	11	2,340		

The hypotheses of the equality of mean levels of both factors are rejected; zero interaction is not rejected at level .05 since $F_{1,8,.95} = 5.32$.

Exercises

10. Write a complete model for a 2^3 factorial experiment with two replications. Give the assumptions of the model. What tests are appropriate? Give the partial table of analysis of variance with the source of variation and degrees of freedom.

11. For counting the radioactivity of a liquid, a factorial experiment was designed to see if the paper, the size of the drop of solution, and its placement on the inner or outer side of the paper had any effect. Consider paper (a), drop size (b), and position of placement (c) at two levels each, and give a model of the experiment. Given the following data on two replicates, test the hypothesis about factors and interactions at level .05. Counts are given in thousands.

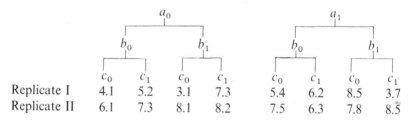

	c_0	c_1	c_0	c_1	c_0	c_1	c_0	c_1
Replicate I	4.1	5.2	3.1	7.3	5.4	6.2	8.5	3.7
Replicate II	6.1	7.3	8.1	8.2	7.5	6.3	7.8	8.5

12. An experiment was designed to find the amount of warping in brass plates. Two temperatures (a) and two laboratories (b) were used. The experiment was repeated three times. Test the hypothesis that the temperature has no effect on warping at level .05. Also test the hypothesis that there is no interaction between temperature and laboratories at level .05.

	Replicates					
	I		II		II	
	a_0	a_1	a_0	a_1	a_0	a_1
b_0	18	24	17	25	20	25
b_1	20	30	18	35	25	35

13. The moisture content in lumber was obtained under two different temperatures (a) in two different kind of kilns (b). Give the model for this 2^2 factorial experiment when there aro four replications. Use a randomized block design, (the block being the time of year). The term

denoting the block effect should be present in the model. Test the hypothesis that there is no effect of temperature or kiln at level .05.

		Percent moisture	
		a_0	a_1
Block I	b_0	8.1	7.4
	b_1	9.2	8.2
Block II	b_0	7.1	8.3
	b_1	8.5	9.5
Block III	b_0	8.2	8.7
	b_1	7.2	6.5
Block IV	b_0	11.1	12.5
	b_1	13.1	9.8

Chapter Exercises

1. The mean germination time of carrot and celery seeds is obtained under two drying rates (fast and slow) and two drying temperatures ($15°C$, $30°C$). The data on two replicates are given:

	Drying Rate	Drying Temperature	
		c_0	c_1
Carrots (a_0)	Fast (b_0)	1.4, 1.6	2.7, 2.9
	Slow (b_1)	1.5, 1.4	2.8, 2.9
Celery (a_1)	Fast (b_0)	4.2, 4.6	5.9, 5.1
	Slow (b_1)	4.9, 4.7	4.5, 5.7

Let the three factors here be a = vegetable variety, b = drying rate, and c = drying temperature, each having two levels. This is a 2^3 experiment having 8 treatments:

(a) Give a model of the experiment
(b) Test the hypothesis that there are no main effects and interactions at level .05
(c) Estimate the 95% confidence interval for the main effect B where

$$B = \tfrac{1}{4}(b_1 - b_0)(c_1 + c_0)(a_1 + a_0)$$

2. Obtain a 5×5 Latin square. Permute two of its rows and three of its

columns at random to obtain a "random" Latin square. Give the table of the source of variation and the degrees of freedom for one replicate of this square.

3. Find five randomized blocks for four treatments (A, B, C, and D) using the table of random numbers. Give the partial table of analysis of variance with the source of variation and the degrees of freedom for this experiment.

4. For a 3×3 Latin square experiment, the following calculations are given:

Total $= 99$
Sum of squares of totals of treatments $= 4{,}017$
Sum of squares of totals of rows $= 3{,}717$
Sum of squares of totals of column $= 3{,}717$
Sum of squares of observations $= 1{,}669$

(a) Complete the analysis-of-variance table
(b) Test the hypothesis that there is no difference among treatments at level .05

Summary

In this chapter we discussed *completely randomized designs*, where the treatments are assigned to experimental units randomly. When the experimental material is heterogeneous and can be divided in *blocks*, we use a *randomized block design*. For taking into account two-way variability in the experimental material, we use *Latin squares*. *Factorial experiments* are needed when several factors are being studied simultaneously, each having several levels. They allow us to test the hypothesis about the difference between levels of factors and the interactions among them.

References

Cochran, W. G., and Cox, G. M. *Experimental Designs*. New York: John Wiley & Sons, 1957.
Davis, Owen L. (Editor). *The Design and Analysis of Industrial Experiments*, 2nd ed. London: Longman Group, 1979.
Duckworth, W. E., *Statistical Techniques in Technological Research*. London: Mathuen & Co., 1968.
Finney, D. J. *An Introduction to the Theory of Experimental Design*. Chicago: University of Chicago Press, 1960.
Hills, Michael. *Statistics for Comparative Studies*. London: Chapman and Hall, 1974.
Kempthorne, Oscar. *The Design and Analysis of Experiments*. New York: John Wiley & Sons, 1952.
Moses, L. E., and Oakford, Robert V. *Tables of Random Permutations*. Palo Alto: Stanford University Press, 1963.
Winer, B. J. *Statistical Principles in Experimental Designs*, 2nd ed. New York: McGraw-Hill, 1971.

chapter twelve

Multivariate Data Analysis

The statistical procedures discussed in previous chapters have mostly been based on univariate responses. We have discussed several procedures that are concerned with more than one variable, such as the correlation coefficient and multiple regression. There are many situations in practice where several responses from a subject are simultaneously observed and decisions are needed on their interrelationships. For example, blood pressure, pulse rate, and weight and height of a subject are known to be related and should not be studied independently of each other. Scores on various subjects in an examination—say in mathematics, English, chemistry, and physics—are related and should be studied together. Problems arising in the study of several responses from a subject simultaneously are discussed in the field of multivariate statistical analysis. The increasing use of multivariate procedures has become possible with the recent introduction of these procedures on computers, such as in *SAS, SPSS,* and *BMDP.*

In this chapter some elementary procedures in multivariate statistics are discussed. The language of multivariate problems is in terms of vectors and matrices, and we use the notation of vectors and matrices. Several books on multivariate statistics have recently been written for practitioners; see, for example, Morrison (1977), Gnanadesikan (1977), Seal (1964), and Srivastava and Carter (1983).

Vector and Matric Notation

A *vector* **X** in p dimensions has p components $X_1, X_2, \ldots, X_p$ and is usually denoted by column vector:

$$\mathbf{X} = \begin{bmatrix} X_1 \\ X_2 \\ \vdots \\ X_p \end{bmatrix}$$

Let X_1 = age, X_2 = height, and X_3 = the weight of an individual; then the vector $\mathbf{X}$ is the column of age, height, and weight. When we write the components in a row rather than in a column, it is called the *transpose* of $\mathbf{X}$, denoted by $\mathbf{X}'$. That is:

$$\mathbf{X}' = [X_1, X_2, \ldots, X_p]$$

Example 12.1: Data on 12 subjects for their age, height, weight, and serum cholesterol and serum triglyceride levels were collected in an experiment and are given in Table 12.1.

Here $p = 5$ and $\mathbf{X}_i$, $i = 1, 2, \ldots, 12$ are independent vectors of observations, such that

$$\mathbf{X}_1 = \begin{bmatrix} 64 \\ 173 \\ 71 \\ 182 \\ 209 \end{bmatrix}, \quad \mathbf{X}_2 = \begin{bmatrix} 61 \\ 185 \\ 92 \\ 182 \\ 88 \end{bmatrix}, \quad \mathbf{X}_3 = \begin{bmatrix} 76 \\ 155 \\ 50 \\ 203 \\ 124 \end{bmatrix}$$

and so on, with

$$\mathbf{X}' = (\text{age, height, weight, serum cholesterol, serum triglycerides}).$$

Table 12.1   Data on Twelve Subjects

Age (year)	Height (cm)	Weight (kg)	Serum Cholesterol (mg/100 ml)	Serum Triglycerides (mg/100 ml)
64	173	71	182	209
61	185	92	182	88
76	155	50	203	124
52	168	67	187	88
82	170	66	245	143
50	168	55	158	80
74	170	74	180	104
44	164	70	284	150
66	170	69	191	73
80	156	69	202	190
51	174	61	198	87
78	170	64	191	114

Source: Krempler, F.; Kostner, G.M.; Haslaner, F.; Bolzano, K.; and Sandhoffer, F., Studies on the role of specific cell surface receptors in the removal of lipoprotein (a) in man, *J. Clin. Invest.*, 1983, *71*, 1431–1441.

A rectangular array of numbers, written in large parentheses, is a *matrix*. That is, matrix $\mathbf{A}$ is an $m \times n$ array of numbers with m rows and n columns. For example:

$$\mathbf{A} = \begin{bmatrix} a_{11} & a_{12} \ldots a_{1n} \\ a_{21} & a_{22} \ldots a_{2n} \\ \vdots & \vdots \quad \vdots \\ a_{m1} & a_{m2} \ldots a_{mn} \end{bmatrix}$$

The *order* of the matrix is $m \times n$.

The data given in Example 12.1 is a matrix of numbers with 12 rows and 5 columns. It is a 12×5 matrix. Notice that a vector $\mathbf{X}$ of p components is a matrix of order $p \times 1$.

The matrix notation is very useful in writing a large array of numbers in condensed form. Various operations on vectors and matrices are collected in the Appendix. We use vector and matrix notation to describe the multivariate problems discussed in this chapter.

Let $\mathbf{X}$ be a p-dimensional random vector. That is, its components are random variables. The vector of expectation of the components will be the expectation of the random vector. That is:

$$\boldsymbol{\mu} = E(\mathbf{X}) = \begin{bmatrix} E(X_1) \\ E(X_2) \\ \vdots \\ E(X_p) \end{bmatrix}$$

To describe the variance-covariance matrix (or simply the covariance matrix, also called the dispersion matrix), we denote

$$\sigma_{11} = \operatorname{Var} X_1$$
$$\sigma_{22} = \operatorname{Var} X_2$$
$$\vdots \qquad \vdots$$
$$\sigma_{pp} = \operatorname{Var} X_p$$

and $\sigma_{ij} = \operatorname{Cov}(X_i, X_j)$, $i \neq j = 1, 2, \ldots, p$.

The covariance matrix of $\mathbf{X}$ is given by:

$$\Sigma = \begin{bmatrix} \sigma_{11} & \sigma_{12} & \sigma_{13} & \cdots & \sigma_{1p} \\ \sigma_{21} & \sigma_{22} & \cdots & & \sigma_{2p} \\ \vdots & \vdots & & & \vdots \\ \sigma_{p1} & \sigma_{p2} & \cdots & & \sigma_{pp} \end{bmatrix}$$

If a *random sample* of N multivariate observations is given, we have:

$$\mathbf{X}_\alpha, \quad \alpha = 1, 2, \ldots, N$$

Now:

$$\mathbf{X}_\alpha = \begin{bmatrix} X_{\alpha 1} \\ X_{\alpha 2} \\ \vdots \\ X_{\alpha p} \end{bmatrix} \qquad \alpha = 1, 2, \ldots, N$$

Hence, the average vector is,

$$\bar{\mathbf{X}} = \begin{bmatrix} \bar{X}_1 \\ \bar{X}_2 \\ \vdots \\ \bar{X}_p \end{bmatrix}$$

where $\bar{X}_i = \sum\limits_{\alpha=1}^{N} X_{\alpha i}/N$. That is, we have averages of the rows of matrix $(\mathbf{X}_1, \mathbf{X}_2, \ldots, \mathbf{X}_N)$.

Similarly, we can denote in case i is not equal to j,

$$S_{ij} = \sum_{\alpha=1}^{N} \sum_{\beta=1}^{N} (X_{\alpha i} - \bar{X}_i)(X_{\beta j} - \bar{X}_j)/(N-1)$$

the sample covariance between the ith and jth components of the vector. S_{ii} will be the variance of X_i,

$$S_{ii} = \sum_\alpha (X_{\alpha i} - \bar{X}_i)^2/(N-1)$$

sample variance of the ith component. Then the matrix of the variance and covariance of the sample is given by:

$$\mathbf{S} = \begin{bmatrix} S_{11} & S_{12} \ldots S_{1p} \\ S_{21} & S_{22} \ldots S_{2p} \\ \vdots & \vdots \quad \vdots \\ S_{p1} & S_{p2} \quad S_{pp} \end{bmatrix}$$

The matrix can also be written in terms of vector multiplication (row-column multiplication):

$$\mathbf{S} = \sum_{\alpha=1}^{N} (\mathbf{X}_\alpha - \bar{\mathbf{X}})(\mathbf{X}_\alpha - \bar{\mathbf{X}})'/(N-1)$$

The *sample correlation coefficient* between X_i and X_j is given by:

$$r_{ij} = \frac{\sum_\alpha \sum_\beta (X_{\alpha i} - \bar{X}_i)(X_{\beta j} - \bar{X}_j)}{\sqrt{\sum_\alpha (X_{\alpha i} - \bar{X}_i)^2 \sum_\beta (X_{\beta j} - \bar{X}_j)^2}}, \quad i, j = 1, 2, \ldots, p$$

Notice that $r_{ii} = 1, i = 1, 2, \ldots, p$. We can write the sample correlation matrix as:

$$\mathbf{R} = \begin{bmatrix} 1 & r_{12} & \cdots & & r_{1p} \\ r_{21} & 1 & r_{23} & \cdots & r_{2p} \\ \vdots & \vdots & \vdots & & \vdots \\ r_{p1} & r_{p2} & \cdots & & 1 \end{bmatrix}$$

The sample correlation matrix is used in many multivariate problems of inference.

Notice that the covariance and correlation matrices are symmetric—that is, the respective rows and columns are the same. Therefore, in computation, only the upper triangular matrix above the diagonal need be evaluated.

Example 12.2: Let the following six observations be available on the vector of weight and serum cholesterol and serum triglyceride levels.

	Weight	CL	ST
	71	182	209
	92	182	88
	50	203	124
	67	187	88
	66	245	143
	55	158	80
Total	401	1157	732

The vector of averages is:

$$\bar{\mathbf{X}} = \begin{bmatrix} 66.8 \\ 192.8 \\ 122.0 \end{bmatrix}$$

The sample covariance matrix is obtained by:

$$S_{11} = \tfrac{1}{5}[71^2 + 92^2 + 50^2 + 67^2 + 66^2 + 55^2 - 401^2/6] = 214.9$$

Similarly,
$$S_{22} = 861.37$$
and
$$S_{33} = 2418.0$$

Now $S_{21} = S_{12} = \tfrac{1}{5}\Big[(71 \times 182) + (92 \times 182) + (50 \times 203)$

$$+ (67 \times 187) + (66 \times 245) + (55 \times 158)$$

$$- \frac{401 \times 1157}{6} \Big] = -24.23$$

and $S_{23} = S_{32} = 440.6$
$S_{31} = S_{13} = -10.6$

So the sample covariance matrix is:

$$S = \begin{bmatrix} 214.97 & -24.23 & -10.6 \\ -24.23 & 861.37 & 440.6 \\ -10.6 & 440.6 & 2418.0 \end{bmatrix}$$

The correlation matrix can now be obtained from **S**. We have:

$$r_{12} = \frac{-24.23}{\sqrt{214.97 \times 861.37}} = -.0563$$

$$r_{13} = \frac{-10.6}{\sqrt{214.97 \times 2418.0}} = -.0147$$

and

$$r_{23} = \frac{440.6}{\sqrt{861.37 \times 2418.0}} = .3053$$

So that:

$$R = \begin{bmatrix} 1 & -.0563 & -.0147 \\ -.0563 & 1 & .3053 \\ -.0147 & .3053 & 1 \end{bmatrix}$$

Exercises

1. The consumer price indexes for the following countries for three selected items are given for the year 1980. Find the average vector and covariance matrix of the sample.

	Food	Clothing	Housing
Australia	292	301	322
Austria	187	178	233
Canada	290	196	243
France	295	267	309
West Germany	159	179	182
Italy	382	437	386
Japan	288	289	227
Netherlands	198	257	244
Sweden	282	197	315
United Kingdom	468	306	422
United States	255	178	263

Source: United States Department of Labor, Bureau of Labor Statistics.

2. The scores of five freshmen in English, mathematics, and physics midterm examinations are given below. Find the average vector and the covariance matrix of the sample.

English	Mathematics	Physics
86	91	75
72	80	55
66	72	70
85	90	91
45	40	65

Graphical Representation of Multivariate Data

Data with two coordinates can be plotted in (x, y)-plane. When data have more than two components, say $x_1, x_2, \ldots, x_p$, we say that we have p dimensions. It is difficult to represent three or more components on graphs, so several ways of representing such points have been devised. An exhaustive list of such representations has been given by Chambers, Cleveland, Kleiner, and Tukey in *Graphical Methods for Data Analysis* (1983). A few of them are given here. Just as a scatter diagram in two dimensions is convenient to describe the relationship among the variables that generated the data, graphical representations of points in p dimensions are needed to shed light on the relationship among multivariate data. For an initial look at multivariate data, graphical representation is highly desirable. If we can see that the data are formed of three clusters, as in Figure 12.1, we can deal with three populations rather than only one. Similarly, maverick observations (outliers) may be easily identified by a graphical representation of the data, as in Figure 12.2.

Scaling

Before we graph the components of vector $\mathbf{X}' = (X_1, X_2, \ldots, X_p)$, it must be standardized in some fashion. That is, all the components, which may have different units of measurements, must be converted to numbers between zero and one. This can be done by subtracting the smallest value of the component from it and dividing it by its range. If the component X_i has range (a_i, b_i), then we transform X_i to Y_i by:

$$Y_i = \frac{X_i - a_i}{b_i - a_i}$$

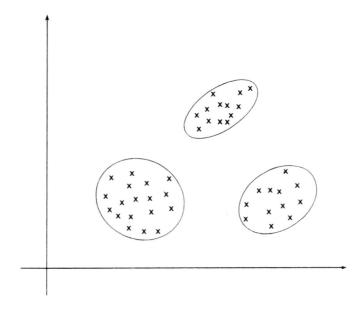

Figure 12.1. Clusters of data.

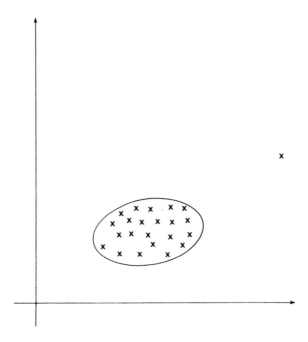

Figure 12.2. Outlier.

The range of Y_i is now from zero to one. Hence, in future discussions of graphical displays, we will use standardized measures on components.

Example 12.3: Consider the data in Table 12.1. For age, the smallest value is 44 and the largest value is 82, so the range is 38. Hence, if $X_1 = 64$, $Y_1 = \dfrac{64 - 44}{38} = .53$. Similarly for height, the smallest values is 155 and the range is $185 - 155 = 30$. Hence, if $X_2 = 173$, $Y_2 = \dfrac{173 - 155}{30} = .6$.

Profiles

In profiles, the components of the vector are represented by vertical lines uniformly spaced along the horizontal line. A typical profile for a five-dimensional vector is given in Figure 12.3. When the vertices of the points in Figure 12.3 are joined by lines, we have a *polygon profile* of the same data.

Four data points of Table 12.1 are graphed in Figure 12.4. The standard values are given in Table 12.2.

Table 12.2 Standardized Values

	Age	Height	Weight	SC	ST
(a)	.53	.6	.5	.19	1.0
(b)	.45	1.0	1.0	.19	.11
(c)	.84	0	0	.36	.38
(d)	.21	.43	.4	.23	.11

When the lengths of components are marked along the rays of a unit circle, we get a *circular profile* and the points are joined by lines. The data in Table 12.2 are given in circular profiles, also called *stars*, in Figure 12.5. Note that the rays should be drawn at equal angles.

Trigonometric Series Form

The vector of observations is represented by a smooth curve where the components are used in defining the curve. The curve for the vector **X** is assumed to be:

$$f(t) = \frac{X_1}{\sqrt{2}} + X_2 \operatorname{Sin} t + X_3 \operatorname{Cos} t + X_4 \operatorname{Sin} 2t + X_5 \operatorname{Cox} 2t + \cdots$$

It is graphed as a function of t. This curve in terms of trigonometric functions is also known as the *Fourier series*. The function can be graphed with the help of a graphic computer procedure such as CALCOM. Details of the procedure are given by Andrews (1972). A typical vector looks like the periodic curve in Figure 12.6.

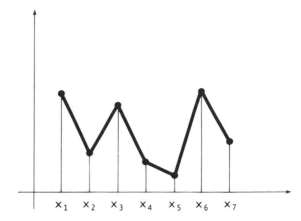

Figure 12.3. Profile and Polygonal profile.

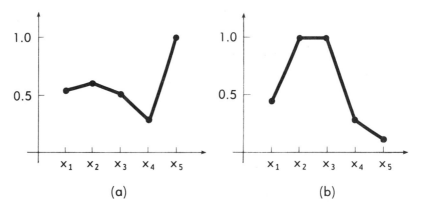

(a)

(b)

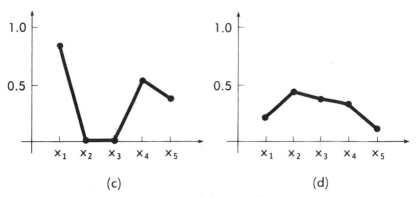

(c)

(d)

Figure 12.4. Polygon Profiles.

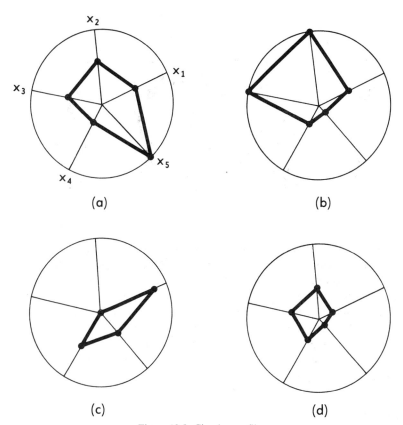

(a)   (b)

(c)   (d)

Figure 12.5. Circular profiles.

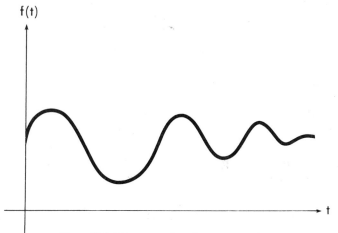

Figure 12.6. Trigonometric series representation.

Figure 12.7. Chernoff face.

Chernoff Faces

The vector in p dimensions is represented by the caricature of a face. The components define various characteristics of the face, such as the length of the nose, the mouth, and the direction of the eyebrows. Of course, various different assignments of the components of the features will provide different faces. A typical face looks like the one in Figure 12.7 as originally proposed by Chernoff (1971). Several recent extensions of the face have been proposed (see, for example, Wainer [1983]). Computer programs for drawing faces are available from Chernoff and Wainer. This representation allows the "data to look at us!" (Wainer).

Exercises

3. Data on telephones for 1978 and for newspapers, radio, and television sets for 1977 are given below for a few selected countries. Find the standardized values and graph the data with circular profiles. Data are given per 1,000 of population.

	Telephones	Newspapers	Radio	TV Sets
Australia	44.	310	1,037	357
Canada	11.2	221	1,043	430
Denmark	56.9	362	825	471
France	37.2	205	330	278
Japan	42.4	546	571	242
Soviet Union	8.0	396	481	217
United Kingdom	41.5	410	716	390
United States	77.0	287	2,048	623

Source: *Information Please Almanac, A* and *W* Publishers, New York, 1983.

4. The following data on ten female macaque monkeys is abstracted from Bliss (1970), p. 297. X_1 = age in years at the emergence of the deciduous canine first molar, X_2 = age at the second molar, and X_3 = age at teeth, and X_4 = age at first permanent molar.

X_1	X_2	X_3	X_4
.11	.14	.38	1.40
.15	.15	.33	1.46
.15	.15	.31	1.30
.17	.13	.32	1.35
.17	.16	.34	1.32
.17	.15	.38	1.31
.15	.15	.44	1.35
.15	.14	.34	1.36
.17	.12	.41	1.28
.16	.16	.42	1.32

(a) Graph the polygonal profiles for the data
(b) Graph the circular profiles for the data
(c) Comment on the similarity of the graphs

Multinomial Distribution

The generalization of binomial distribution leads to multinomial distribution. In Bernoulli trials, each trial has two possible outcomes, whereas in general we assume that there are more than two outcomes, as in the toss of a die or in drawing a card from a deck of cards. Let there be k possible outcomes of a trial with probabilities $p_1, p_2, \ldots, p_k$ such that:

$$p_1 + p_2 + \cdots p_k = 1$$

Let there be n independent trials and let $X_1, X_2, \ldots, X_k$ be the number of occurrences of each outcome with:

$$X_1 + X_2 + \cdots + X_k = n$$

The probability distribution of $(X_1, \ldots, X_k)$ is called the *multinomial distribution*. The probability density function of $\mathbf{X}' = (X_1, X_2, \ldots, X_k)$ is given by

$$\frac{n!}{x_1! x_2! \ldots x_k!} p_1^{x_1} p_2^{x_2} \cdots p_k^{x_k} \tag{12.1}$$

with $p_1 + p_2 + \cdots + p_k = 1$, $x_1 + x_2 + \cdots + x_k = n$, and $x_i \geq 0$. $x_1, x_2, \ldots, x_k$ are integers. Notice that there are really $k - 1$ random variables, since the last one is determined in terms of the $k - 1$ variables. So the dimension of the random vector $\mathbf{X}$ is only $k - 1$. The mean vector is:

$$E(\mathbf{X}) = \begin{bmatrix} np_1 \\ np_2 \\ \vdots \\ np_k \end{bmatrix} \tag{12.2}$$

The covariance matrix is:

$$\begin{bmatrix} np_1(1 - p_1) & -np_1p_2 & \cdots & -np_1p_k \\ -np_2p_1 & np_2(1 - p_2) & \cdots & -np_2p_k \\ \vdots & \vdots & & \vdots \\ -np_1p_k & np_2p_k & \cdots & np_k(1 - p_k) \end{bmatrix} \tag{12.3}$$

There are many other discrete distributions giving multivariate generalizations of such well-known univariate distributions as Poisson or hypergeometric. (See Johnson and Kotz [1972] for further details.)

Example 12.4: In an auto accident, the probabilities of head injuries, leg injuries, and neck injuries are .3, .1, and .4, respectively. The probability of no injury is .2. This is a multinomial trial with $k = 4$ outcomes. In ten accidents, the probability of getting two head injuries, two leg injuries, three neck injuries, and three no injury is:

$$\frac{10!}{2!\,2!\,3!\,3!}(.3)^2(.1)^2(.4)^3(.2)^3$$

$$= 25200 \times (.09)(.01)(.064)(.008)$$

$$= .0116$$

In 100 accidents, the expected number of injuries is given by the vector:

$$E(\mathbf{X}) = \begin{bmatrix} 30 \\ 10 \\ 40 \\ 20 \end{bmatrix}$$

The covariance matrix is given by:

$$\begin{bmatrix} 21 & -3 & -12 & -6 \\ -3 & 9 & -4 & -2 \\ -12 & -4 & 24 & -8 \\ -6 & -2 & -8 & 16 \end{bmatrix}$$

Example 12.5: In a bridge hand (having 13 cards), we need the probability of having 2 aces and 3 kings. This is obtained by considering a multinomial trial with outcomes ace, king, and "other" and probabilities $\frac{1}{13}$, $\frac{1}{13}$, and $\frac{11}{13}$, respectively. Thus the multinomial distribution gives the above probability as:

$$\frac{13!}{2!\,3!\,8!}\left(\frac{1}{13}\right)^2\left(\frac{1}{13}\right)^3\left(\frac{11}{13}\right)^8$$

$$= 12870 \times \frac{11^8}{13^{13}} = .0091$$

The data from multinomial distributions can be obtained in terms of contingency tables, which have been discussed earlier.

Exercises

5. A fair die is tossed independently 20 times. Find the probability that the frequency of dots is as follows:

   Dots: 1 2 3 4 5 6
   Frequency: 2 5 4 3 3 3

6. Five balls are drawn with replacement from an urn containing seven red, eight white, and ten blue balls. What is the probability of obtaining two red, two white, and one blue ball? (*Hint:* This is a multinomial trial with probabilities 7/25, 8/25, and 10/25.)

7. Find the probability that a bridge hand contains three queens but no aces.

8. In a group, 10% of the people have high blood pressure and 20% have high cholesterol levels. A clinic servicing the population screened 100 persons for the presence of these conditions. What is the expected number of persons with high blood pressure? Supposing that the other 70% of the population is normal, what is the covariance matrix of the three groups of people in the sample?

The Inference for Multivariate Normal Distribution

Just like univariate normal distribution, multivariate normal distribution is frequently used. Many statistical procedures have been developed for the multivariate normal. A vector random variable X is said to have multivariate normal distribution if every linear combination of X, that is, $á\,X$, for all vectors of constants a has univariate normal distribution. This property gives the probability density function of X. Let X be the p dimensional vector with the mean vector μ and the covariance matrix $\sum$. Then the probability density function can be completely determined in terms of μ and $\sum$. (For the discussion of the probability density function, see Anderson [1958]).

An interesting property of multivariate normal distribution is that the marginal distributions are also multivariate normal, as are the conditional distributions. Bivariate normal distribution can be visualized to have a bell-shaped surface. The marginal distributions of the bivariate normal are univariate normal distributions. The conditional distribution is also univariate normal.

Let $\mathbf{X}_1, \mathbf{X}_2, \ldots \mathbf{X}_N$ be a random sample of N vectors from a multivariate normal distribution. Then the maximum likelihood estimates of $\boldsymbol{\mu}$ and $\sum$ are given by:

$$\hat{\boldsymbol{\mu}} = \bar{\mathbf{X}} \tag{12.4}$$

$$\hat{\sum} = \frac{N-1}{N} \mathbf{S} \tag{12.5}$$

As in the case of univariate distributions, the unbiased estimate of $\sum$ is $\mathbf{S}$.

Test of Hypothesis for the Mean of
Normal Populations

Based on a sample size N, suppose we want to test the following hypothesis about multivariate normal distribution:

$$H: \boldsymbol{\mu} = \boldsymbol{\mu}_0$$
$$A: \boldsymbol{\mu} \neq \boldsymbol{\mu}_0 \tag{12.6}$$

First, assume that $\sum$ is known. In the case when $p = 1$, the multivariate normal distribution reduces to one-dimensional normal distribution $N(\mu, \sigma^2)$. For testing a hypothesis in the univariate case.

$$H: \mu = \mu_0$$

we use the statistic

$$Y = \frac{\sqrt{N}(\bar{X} - \mu_0)}{\sigma} \tag{12.7}$$

Under the null hypothesis, Y has a standard normal distribution. From Equation (12.7), we find that we could also use Y^2, which, for testing H, is:

$$Y^2 = N(\bar{X} - \mu_0)(\sigma^2)^{-1}(\bar{X} - \mu) \tag{12.8}$$

Y^2 has a chi-squared distribution with one degree of freedom under the null hypothesis, and the test can be easily implemented.

The multivariate case can be studied with the help of a statistic similar to the one in Equation (12.8). Consider the statistic:

$$Y = N(\bar{\mathbf{X}} - \boldsymbol{\mu}_0)' \sum{}^{-1} (\bar{\mathbf{X}} - \boldsymbol{\mu}_0) \tag{12.9}$$

Y has a chi-squared distribution with p degrees of freedom under the null hypothesis. The test for $\boldsymbol{\mu} = \boldsymbol{\mu}_0$ can be carried out with the help of chi-squared tables. When $\sum$ is not known, we need a test analogous to the t-test in the univariate case. This test is based on *Hotelling's T^2* test.

Hotelling's T^2 Test

Hotelling's T^2 statistic is defined by:

$$T_1^2 = N(\bar{\mathbf{X}} - \boldsymbol{\mu}_0)'\mathbf{S}^{-1}(\bar{\mathbf{X}} - \boldsymbol{\mu}_0) \tag{12.10}$$

Its distribution under $\boldsymbol{\mu} = \boldsymbol{\mu}_0$ is given by the following.

Let $\mathbf{X}_1, \mathbf{X}_2, \ldots, \mathbf{X}_N$ be a random sample from p-variate normal distribution. Then under the null hypothesis H, the distribution of $\dfrac{N-p}{p(N-1)} T_1^2$ is an F-distribution with p and $N - p$ degrees of freedom. That is:

$$\frac{N-p}{p(N-1)} T_1^2 \sim F_{p,N-p}$$

The statistics T_1^2 can also be used in constructing confidence regions in p-dimensions for the mean vector of the normal distribution when the covariance is unknown.

Example 12.6: Suppose observations on ten subjects on their cholesterol levels (X_1), blood pressure (X_2), and pulse rate (X_3) are given. The sample mean vector and covariance matrix are given below. We want to test the hypothesis that the mean cholesterol level is 260, the blood pressure is 108, and the pulse rate is 75 at level .05:

$$H: \boldsymbol{\mu} = \begin{bmatrix} 260 \\ 108 \\ 75 \end{bmatrix} = \boldsymbol{\mu}_0$$

$$A: \boldsymbol{\mu} \neq \boldsymbol{\mu}_0$$

The data of the sample gives:

$$\bar{\mathbf{X}} = \begin{bmatrix} 265 \\ 110 \\ 70 \end{bmatrix} \quad \text{and} \quad \mathbf{S} = \begin{bmatrix} 50 & 10 & 15 \\ 10 & 10 & 5 \\ 15 & 5 & 25 \end{bmatrix}$$

To find $\mathbf{S}^{-1}$, we need the determinant of S:

$$
|\mathbf{S}| = \begin{vmatrix} 50 & 10 & 15 \\ 10 & 10 & 5 \\ 15 & 5 & 25 \end{vmatrix}
$$

$$= 50(10 \times 25 - 5 \times 5) - 10(10 \times 25 - 5 \times 15) + 15(10 \times 5 - 10 \times 15)$$
$$= (50 \times 225) - (10 \times 175) - (15 \times 100)$$
$$= 8{,}000$$

and

$$S^{-1} = \frac{1}{8000}\begin{bmatrix} 225 & -175 & -100 \\ -175 & 1025 & -100 \\ -100 & -100 & 400 \end{bmatrix}$$

$$\bar{X} - \mu_0 = \begin{bmatrix} 265 - 260 \\ 110 - 108 \\ 70 - 75 \end{bmatrix} = \begin{bmatrix} 5 \\ 2 \\ -5 \end{bmatrix}$$

$$T_1^2 = \frac{10}{8,000}(5\ 2\ -5)\begin{bmatrix} 225 & -175 & -100 \\ -175 & 1025 & -100 \\ -100 & -100 & 400 \end{bmatrix}\begin{matrix} 5 \\ 2 \\ -5 \end{matrix}$$

$$= \frac{1}{800}(1275\ 1675 - 2700)\begin{bmatrix} 5 \\ 2 \\ -5 \end{bmatrix} = 29.03125$$

Hence, $F = \dfrac{N-p}{p(N-1)}T_1^2 = \dfrac{7}{27} \times 29.03125 = 7.5266$

We reject the hypothesis H at level .05 since $F_{3,7,.95} = 4.35$.

Two Populations

Suppose now we have two multivariate normal populations with different means μ_1 and μ_2, but with the same covariance matrix $\sum$. Let there be samples of sizes N_1 and N_2 from the two populations, respectively.

The test of the hypothesis

$$H: \mu_1 = \mu_2$$
$$A: \mu_1 \neq \mu_2$$

with $\sum$ unknown, is given with the help of

$$T_2^2 = \frac{N_1 N_2}{N_1 + N_2}(\bar{X}_1 - \bar{X}_2)'S^{-1}(\bar{X}_1 - \bar{X}_2)$$

Here $\bar{X}_1$, $\bar{X}_2$ are the sample means and S is the common sample covariance matrix. If S_1 and S_2 are the sample covariance matrices of the two samples. then:

$$S = [(N_1 - 1)S_1 + (N_2 - 1)S_2]/(N_1 + N_2 - 2)$$

The statistic $F = (N_1 + N_2 - p - 1)T_2^2/[p(N_1 + N_2 - 2)]$ has F-distribution with p and $[N_1 + N_2 - p - 1]$ degrees of freedom, and the test of the hypothesis H can be made.

An extension of Hotelling's T^2 test in the case of several multivariate normal populations leads to techniques of Multivariate Analysis of Variance—MANOVA. As usual, we use one-way MANOVA to test hypotheses about the equality of means of several groups where there are several measurements on a given individual. Standard statistical packages such as SPSS and SAS provide numerical computations for multivariate analysis of variance.

Example 12.7: To test the hypothesis that the grade point average (GPA) of males and females in the first two years at a university are the same, data on 252 males and 154 females were obtained with the results:

	Males		Females	
	Mean Vector	Covariance Matrix	Mean Vector	Covariance Matrix
Year I	$\begin{bmatrix} 2.61 \\ 2.63 \end{bmatrix}$	$\begin{bmatrix} 0.261 & 0.181 \\ 0.181 & 0.203 \end{bmatrix}$	$\begin{bmatrix} 2.54 \\ 2.55 \end{bmatrix}$	$\begin{bmatrix} .303 & .206 \\ .206 & .194 \end{bmatrix}$

Source: Srivastava, M. S., and Carter, E. M., 1983, p. 93.

Assuming that GPAs are distributed as bivariate normal, we want to test the hypothesis of the equality of means of the populations, assuming equal covariance matrices. The pooled sample covariance matrix is:

$$\frac{1}{404}\left[251\begin{bmatrix} .261 & .181 \\ .181 & .203 \end{bmatrix} + 153\begin{bmatrix} .303 & .206 \\ .206 & .194 \end{bmatrix}\right]$$

$$\mathbf{S} = \begin{bmatrix} .2769 & .1905 \\ .1905 & .1996 \end{bmatrix}$$

$$\mathbf{S}^{-1} = \begin{bmatrix} 10.5169 & -10.0374 \\ -10.0374 & 14.5894 \end{bmatrix}$$

$$\bar{\mathbf{X}}_1 - \bar{\mathbf{X}}_2 = \begin{bmatrix} 2.61 \\ 2.63 \end{bmatrix} - \begin{bmatrix} 2.54 \\ 2.55 \end{bmatrix} = \begin{bmatrix} .07 \\ .08 \end{bmatrix}$$

$$T_2^2 = \frac{252 \times 154}{252 + 154}(.07\ .08)\mathbf{S}^{-1}\begin{bmatrix} .07 \\ .08 \end{bmatrix} = 3.1052$$

$$F = \frac{252 + 154 - 2 - 1}{2(252 + 154 - 2)} \times 3.1052 = 1.55$$

We do not reject H since $F_{2,403,.95} = 3.0$ at .05 level of significance.

Exercises

9. For the data in Exercise 4, assume that $\mathbf{X}' = (X_1, X_2, X_3, X_4)$ is multivariate normal with mean $\boldsymbol{\mu}$ and covariance $\sum$:

   (a) Find the estimate for $\boldsymbol{\mu}$ and $\sum$
   (b) Test the hypothesis at level .05 using Hotelling's test:

   $$H: \boldsymbol{\mu}' = (.15, .15, .30, 1.30)$$

   $$A: \text{not equal}$$

10. For Exercise 3, find the average vector and sample covariance matrix.
11. For the year 1981, the military expenditures of six European nations and seven Asian nations are given below. Let $X_1 = $ expenditure per person, $X_2 = $ percent of government spending, and $X_3 = $ percent of gross national product. Assume that the vector $\mathbf{X}' = (X_1, X_2, X_3)$ is normally distributed and that the covariance matrix is the same for both European and Asian countries. Test the equality of the mean vector of these countries at level .05.

European Countries

X_1	X_2	X_3
161	3.8	1.2
149	5.1	1.5
86	3.3	1.6
105	12.0	2.9
455	7.7	3.2
154	20.2	1.9

Asian Countries

X_1	X_2	X_3
95	48.3	8.3
7	16.9	3.8
5	12.3	3.4
98	5.0	0.9
74	14.7	11.2
113	36.0	5.7
157	23.0	3.4

Source: International Institute for Strategic Studies, London, as published in *Information Please Almanac*, *A* and *W* Publishers, 1983, p. 412.

12. A study to compare the mean height X_1 and the mean chest circumference (X_2) of two-year-old boys and girls based on 100 boys and 80 girls gave the data:

	Sample Mean Vector	Sample Covariance Matrix
Boys	$\begin{bmatrix} 80 \\ 65 \end{bmatrix}$	$\begin{bmatrix} 150 & 40 \\ 40 & 16 \end{bmatrix}$
Girls	$\begin{bmatrix} 75 \\ 60 \end{bmatrix}$	$\begin{bmatrix} 189 & 42 \\ 42 & 12 \end{bmatrix}$

Test the hypothesis that the mean vectors for boys and girls are equal at level .05, assuming that the vector $\mathbf{X}' = (X_1, X_2)$ is bivariate normal with the same covariance matrix.

Linear Discriminant Functions

In many situations, we want to classify a given observation according to one of several populations. For example, in differential diagnosis, a physician will want to classify a patient according to one of several disease categories. Similarly, with several observations about an air traveler, we can classify him to one of the two possible populations—hijacker type and nonhijacker type. These problems belong to the area of *discriminant analysis*. In its simple form, we have two populations. π_1 is the population having multivariate normal distribution with mean μ_1 and covariance $\sum$. π_2 is the population having multivariate normal distribution with mean μ_2 and covariance $\sum$. An observation $\mathbf{X}$ is given, and the problem is to classify it to π_1 or π_2.

Errors of Misclassification

Just like errors in testing hypotheses, we have two errors of misclassification. Classifying an observation to π_2 when it belongs to π_1 leads to error e_1, and classifying an observation to π_1 when it belongs to π_2 leads to error e_2. In classification rules, these errors play an important role.

Fisher's Linear Discriminant Function

Sir R. A. Fisher introduced a linear discriminant function for classifying an observation in the above case.

$$L(\mathbf{X}) = \mathbf{X}' \sum^{-1} (\mu_1 - \mu_2)$$

Classification Rule:

Classify $\mathbf{X}$ to π_1 if $L(\mathbf{X}) > c\frac{1}{2}(\boldsymbol{\mu}_1 - \boldsymbol{\mu}_2)' \sum^{-1}(\boldsymbol{\mu}_1 + \boldsymbol{\mu}_2)$
Classify $\mathbf{X}$ to π_2 otherwise

Errors of misclassification in this case are equal and are given by

$$e_1 = e_2 = \Phi(-\tfrac{1}{2}\Delta)$$

where

$$\Delta^2 = (\boldsymbol{\mu}_1 - \boldsymbol{\mu}_2)' \sum^{-1}(\boldsymbol{\mu}_1 - \boldsymbol{\mu}_2)$$

is a well-known quantity known as *Mahalanobis distance* and Φ is the cumulative distribution function of the standard normal distribution.

Unknown $\boldsymbol{\mu}_1$, $\boldsymbol{\mu}_2$, *and* $\sum$.

Let $\bar{\mathbf{X}}_1$ and $\bar{\mathbf{X}}_2$ be the sample means from the two populations with $\mathbf{S}$ as the common sample covariance matrix. Then the rule based on the above statistics is in terms of *Anderson's statistic*:

$$T(\mathbf{X}) = (\bar{\mathbf{X}}_1 - \bar{\mathbf{X}}_2)'\mathbf{S}^{-1}\mathbf{X}$$

Classification Rule:

Classify $\mathbf{X}$ to π_1 if

$$T(\mathbf{X}) > \tfrac{1}{2}(\bar{\mathbf{X}}_1 - \bar{\mathbf{X}}_2)'\mathbf{S}^{-1}(\bar{\mathbf{X}}_1 + \bar{\mathbf{X}}_2)$$

and to π_2 otherwise.

The errors of misclassification are involved in this case. For further reference, see Srivastava and Carter (1983). Standard statistical packages such as SAS, SPSS, and BMDP provide computations for obtaining discriminant functions and misclassification errors. Here is a simple example.

Example 12.8: Assume that the high-school grade average (X_1) and ACT scores (X_2) of entering freshmen at Ohio State University have bivariate normal distribution. Two different populations are given: π_1 represents those who graduate after four years and π_2 represents those who drop out after a year. It is given that:

$$\boldsymbol{\mu}_1 = \begin{bmatrix} 3 \\ 65 \end{bmatrix} \quad \boldsymbol{\mu} = \begin{bmatrix} 2.5 \\ 55 \end{bmatrix}$$

$$\sum = \begin{bmatrix} .64 & 4 \\ 4 & 100 \end{bmatrix}$$

Then *Mahalanobis distance* is:

$$\Delta^2 = (.5 \ 10)\begin{bmatrix} .64 & 4 \\ 4 & 100 \end{bmatrix}^{-1}\begin{bmatrix} .5 \\ 10 \end{bmatrix} = 1.0201$$

$$\Delta = 1.01$$

since
$$\Sigma^{-1} = \frac{1}{48}\begin{bmatrix} 100 & -4 \\ -4 & .64 \end{bmatrix}$$

Fisher's discriminant function is:

$$(X_1, X_2)\frac{1}{48}\begin{bmatrix} 100 & -4 \\ -4 & .64 \end{bmatrix}\begin{bmatrix} .5 \\ 10 \end{bmatrix}$$

$$= .2083 X_1 + .0917 X_2$$

$$c = \tfrac{1}{2}(.5 \ 10)\frac{1}{48}\begin{bmatrix} 100 & -4 \\ -4 & .64 \end{bmatrix}\begin{bmatrix} 5.5 \\ 120 \end{bmatrix}$$

$$= \tfrac{1}{96}[10 \ 4.4]\begin{bmatrix} 5.5 \\ 120 \end{bmatrix}$$

$$= 6.0729$$

Classification Rule:
Classify an entering freshman to π_1 if $.2083X_1 + .0917X_2 > 6.0729$ and to π_2 otherwise.

Misclassification Errors: $e_1 = e_2 = \Phi(-.505) = .31$.

Exercises

13. Find the classification rule in samples of 100 each, given the estimate of the mean and covariance matrix for π_1 in Example 12.8 to be

$$\begin{bmatrix} 3.1 \\ 72 \end{bmatrix} \text{ and } \begin{bmatrix} .7 & 4.2 \\ 4.2 & 98 \end{bmatrix}$$

and for π_2

$$\begin{bmatrix} 2.4 \\ 53 \end{bmatrix} \text{ and } \begin{bmatrix} .8 & 3.8 \\ 3.8 & 102 \end{bmatrix}$$

14. The percentage of nitrogen (X_1), chlorine (X_2), and potassium (X_3) in the leaf of plant A has the population mean

$$\mu_1 = \begin{bmatrix} 3 \\ 1 \\ 5 \end{bmatrix}$$

and the covariance matrix $\Sigma = \begin{bmatrix} 9 & 1.0 & 1.0 \\ 1 & 1.1 & 1.5 \\ 1 & 1.5 & 5 \end{bmatrix}$

Plant B has the same covariance matrix but has mean $\mu_2 = \begin{bmatrix} 4 \\ 2 \\ 4 \end{bmatrix}$

Assuming multivariate normality, classify a plant with the following observation and find the errors of misclassification:

$$X = \begin{bmatrix} 3.5 \\ 1 \\ 3.5 \end{bmatrix}$$

Appendix: Determinants and Matrices

Determinants

The system of linear equations arises in many applications and forms an important area of study in mathematics. These equations occur in the study of linear models and multivariate analysis in statistics. The notion of a *determinant* is basic to the study of *matrices*.

A square array of numbers such as $\begin{vmatrix} a_1 & b_1 \\ a_2 & b_2 \end{vmatrix}$ is called a *determinant* and is defined by

$$\begin{vmatrix} a_1 & b_1 \\ a_2 & b_2 \end{vmatrix} = a_1 b_2 - a_2 b_1$$

For example, $\begin{vmatrix} 2 & 3 \\ 5 & 8 \end{vmatrix} = (2 \times 8) - (3 \times 5) = 1$. Similarly, $\begin{vmatrix} 7 & 4 \\ 2 & 5 \end{vmatrix} = (7 \times 5) - (2 \times 4) = 27$. The number of rows (or columns) is called the *order* of the determinant. The determinants of the third and higher orders are expressed in terms of the determinants of an order lower than the one considered. The process of evaluation of a higher-order determinant is exemplified by considering a third-order determinant. The value of the third-order determinant is in terms of second-order determinants using expansions in terms of elements of a row or a column. That is, if we expand the determinant in terms of the elements of the first row, then:

$$D = \begin{vmatrix} a_1 & b_1 & c_1 \\ a_2 & b_2 & c_2 \\ a_3 & b_3 & c_3 \end{vmatrix} = a_1 \begin{vmatrix} b_2 & c_2 \\ b_3 & c_3 \end{vmatrix} - b_1 \begin{vmatrix} a_2 & c_2 \\ a_3 & c_3 \end{vmatrix} + c_1 \begin{vmatrix} a_2 & b_2 \\ a_3 & b_3 \end{vmatrix}$$

$$= a_1(b_2 c_3 - b_3 c_2) - b_1(a_2 c_3 - c_2 a_3) + c_1(a_2 b_3 - b_2 a_3)$$

$$= a_1 b_2 c_3 + a_2 b_3 c_1 + a_3 b_1 c_2 - a_1 b_3 c_2 - a_2 b_1 c_3 - a_3 b_2 c_1$$

The *rule* of expansion in terms of the elements of a row requires that we use positive and negative signs alternately. The coefficients of these elements are the minors of the determinant.

Definition: The *minor* of an element of a determinant of order n is the determinant of order $n - 1$, obtained by suppressing the row and column in which the element occurs. That is, the minors of a_1, b_1, and c_1, respectively, are

$$A_1 = \begin{vmatrix} b_2 & b_3 \\ c_2 & c_3 \end{vmatrix}, \quad B_1 = \begin{vmatrix} a_2 & c_2 \\ a_3 & c_3 \end{vmatrix}, \quad C_1 = \begin{vmatrix} a_2 & b_2 \\ a_3 & b_3 \end{vmatrix}.$$

They are denoted by capital letters for convenience.

Hence, $D = a_1 A_1 - b_1 B_1 + c_1 C_1$.

Interchanging Rows or Columns

A determinant changes sign when any two rows or columns are interchanged. For example, by interchanging the second row with first, we have:

$$\begin{vmatrix} a_1 & b_1 & c_1 \\ a_2 & b_2 & c_2 \\ a_3 & b_3 & c_2 \end{vmatrix} = (-1) \begin{vmatrix} a_2 & b_2 & c_2 \\ a_1 & b_1 & c_1 \\ a_3 & b_3 & c_3 \end{vmatrix}$$

Similarly, by interchanging the third row with the second and then with the first, we have:

$$= (-1)^2 \begin{vmatrix} a_3 & b_3 & c_3 \\ a_1 & b_1 & c_1 \\ a_2 & b_2 & c_2 \end{vmatrix}$$

A Shortcut Method of Evaluating a Third-Order Determinant

Write the first two rows of the determinant after the first three rows. Take the sum of the products of those diagonals that have all three elements as shown below; subtract from it the sum of the product of the other three diagonal terms in the corssward direction.

We then have:

$$D = a_1 b_2 c_3 + a_2 b_3 c_1 + a_3 b_1 c_2$$
$$- a_3 b_2 c_1 - a_1 b_3 c_2 - a_2 b_1 c_3$$

Example: Evaluate the determinant:

$$D = \begin{vmatrix} 2 & 3 & 1 & 3 \\ 4 & 5 & 3 & 1 \\ 5 & 1 & 3 & 2 \\ 1 & 2 & 3 & 1 \end{vmatrix}$$

Using the rule and expanding in terms of the elements of the first row, we have:

$$D = 2 \begin{vmatrix} 5 & 3 & 1 \\ 1 & 3 & 2 \\ 2 & 3 & 1 \end{vmatrix} - 3 \begin{vmatrix} 4 & 3 & 1 \\ 5 & 3 & 2 \\ 1 & 3 & 1 \end{vmatrix}$$

$$+ 2 \begin{vmatrix} 4 & 5 & 1 \\ 5 & 1 & 2 \\ 1 & 2 & 1 \end{vmatrix} - 3 \begin{vmatrix} 4 & 5 & 3 \\ 5 & 1 & 3 \\ 1 & 2 & 3 \end{vmatrix}$$

Now:

$$\begin{vmatrix} 5 & 3 & 1 \\ 1 & 3 & 2 \\ 2 & 3 & 1 \end{vmatrix} = 5 \begin{vmatrix} 3 & 2 \\ 3 & 1 \end{vmatrix} - 3 \begin{vmatrix} 1 & 2 \\ 2 & 1 \end{vmatrix} + 1 \begin{vmatrix} 1 & 3 \\ 2 & 3 \end{vmatrix}$$

$$= [5 \times (-3)] - [3 \times (-3)] + [-3]$$

$$= -9$$

$$\begin{vmatrix} 4 & 3 & 1 \\ 5 & 3 & 2 \\ 1 & 3 & 1 \end{vmatrix} = 4 \begin{vmatrix} 3 & 2 \\ 3 & 1 \end{vmatrix} - 3 \begin{vmatrix} 5 & 2 \\ 1 & 1 \end{vmatrix} + 1 \begin{vmatrix} 5 & 3 \\ 1 & 3 \end{vmatrix}$$

$$= [4 \times (-3)] - (3 \times 3) + (12)$$

$$= -9$$

$$\begin{vmatrix} 4 & 5 & 1 \\ 5 & 1 & 2 \\ 1 & 2 & 1 \end{vmatrix} = 4 \begin{vmatrix} 1 & 2 \\ 2 & 1 \end{vmatrix} - 5 \begin{vmatrix} 5 & 2 \\ 1 & 1 \end{vmatrix} + 1 \begin{vmatrix} 5 & 1 \\ 1 & 2 \end{vmatrix}$$

$$= [4 \times (-3)] - (5 \times 3) + 9$$

$$= -18$$

and

$$\begin{vmatrix} 4 & 5 & 3 \\ 5 & 1 & 3 \\ 1 & 2 & 3 \end{vmatrix} = 4\begin{vmatrix} 1 & 3 \\ 2 & 3 \end{vmatrix} - 5\begin{vmatrix} 5 & 3 \\ 1 & 3 \end{vmatrix} + 3\begin{vmatrix} 5 & 1 \\ 1 & 2 \end{vmatrix}$$

$$= [4 \times (-3)] - (5 \times 12) + (3 \times 9)$$

$$= -45$$

Thus: $D = [2 \times (-9)] - [3 \times (-9)] + [1 \times (-18)] - [3 \times (-45)]$
$$= -18 + 27 - 18 + 135 = 126$$

An important property of a determinant is that if two of its columns or rows are identical it is zero. This fact can be seen by noticing that an interchange of the rows or columns will require that the determinant be multiplied by (-1). Hence, if two rows of determinant D are the same, $D = -D$ or $2D = 0$ so that $D = 0$.

Matrices

Let's define a few operations on matrices.

Addition

Let $\mathbf{A} = (a_{ij})$ and $\mathbf{B} = (b_{ij})$ be two $m \times n$ matrices. Then by $\mathbf{A} + \mathbf{B}$, we mean the matrix of elements $(a_{ij} + b_{ij})$.

Multiplication

Let $\mathbf{A} = (a_{ij})$ be $m \times n$ matrix and $\mathbf{B} = (b_{ij})$ be $n \times p$ matrix. Then by $\mathbf{A} \times \mathbf{B}$, we mean a matrix (c_{ik}) where $c_{ik} = \sum_{j=1}^{n} a_{ij}b_{jk}$. That is, (i, k)-th element of the product matrix is obtained by multiplying the elements of the ith row of first matrix with the corresponding element of the kth column of the second matrix and obtaining the sum. The row-column multiplication gives us the product of the matrices. The number of columns of the first matrix must be the same as the number of rows of the second; otherwise, multiplication is not possible. Sometimes $\mathbf{A} \times \mathbf{B}$ is written simply as $\mathbf{AB}$.

Example: Let

$$\mathbf{A} = \begin{bmatrix} 2 & 7 & 1 \\ 3 & 2 & 2 \end{bmatrix}$$

and

$$\mathbf{B} = \begin{bmatrix} 1 & 7 \\ 2 & 1 \\ 1 & 6 \end{bmatrix}$$

Then:

$$\mathbf{AB} = \begin{bmatrix} (2 \times 1) + (7 \times 2) + (1 \times 1) & (2 \times 7) + (7 \times 1) + (1 \times 6) \\ (3 \times 1) + (2 \times 2) + (2 \times 1) & (3 \times 7) + (2 \times 1) + (2 \times 6) \end{bmatrix}$$

$$= \begin{bmatrix} 17 & 27 \\ 9 & 35 \end{bmatrix}$$

The Identity Matrix

A square matrix with ones on the diagonal and zeros as nondiagonal elements is called the *identity* matrix and is denoted by $\mathbf{I}$. For example, an identity matrix of order 5×5:

$$\mathbf{I} = \begin{bmatrix} 1 & 0 & 0 & 0 & 0 \\ 0 & 1 & 0 & 0 & 0 \\ 0 & 0 & 1 & 0 & 0 \\ 0 & 0 & 0 & 1 & 0 \\ 0 & 0 & 0 & 0 & 1 \end{bmatrix}$$

It can be verified by multiplication that $\mathbf{AI} = \mathbf{IA} = \mathbf{A}$. An important matrix operation of "inverse" can be defined with the help of multiplication.

Inverse

A square matrix, $\mathbf{A}$ of order $n \times n$ has an *inverse*, defined as $\mathbf{A}^{-1}$ if:

$$\mathbf{AA}^{-1} = \mathbf{I} = \mathbf{A}^{-1}\mathbf{A}$$

Matrices do not always have an inverse. If the matrix does not have an inverse, it is called *singular*; otherwise, it is *nonsingular*.

Before showing how to obtain an inverse, let's give an application of the operation of the inverse to the solution of simultaneous linear equations. Suppose we have to solve the system:

$$a_1 x_1 + b_1 x_2 + c_1 x_3 = d_1$$
$$a_2 x_1 + b_2 x_2 + c_2 x_3 = d_2$$
$$a_3 x_1 + b_3 x_2 + c_3 x_3 = d_3$$

Let:

$$A = \begin{bmatrix} a_1 & b_1 & c_1 \\ a_2 & b_2 & c_2 \\ a_3 & b_3 & c_3 \end{bmatrix}$$

$$d = \begin{bmatrix} d_1 \\ d_2 \\ d_3 \end{bmatrix}$$

$$x = \begin{bmatrix} x_1 \\ x_2 \\ x_3 \end{bmatrix}$$

Using matrix multiplication, these equations are represented by one equation involving matrices $AX = d$. Hence, if A is nonsingular, multiplying both sides by A^{-1}, we have:

$$A^{-1}Ax = A^{-1}b$$

$$Ix = A^{-1}b$$

$$x = A^{-1}b$$

The theory of determinants can be used to obtain an inverse. Let the determinant of A be denoted by $|A|$. Also let the minors for the elements a_1, b_1, and c_1 be denoted by A_1, B_1, and C_1. That is:

$$A_1 = \begin{vmatrix} b_2 & c_2 \\ b_3 & c_3 \end{vmatrix} \quad B_1 = \begin{vmatrix} a_2 & c_2 \\ a_3 & c_3 \end{vmatrix} \quad C_1 = \begin{vmatrix} a_2 & b_2 \\ a_3 & b_3 \end{vmatrix}$$

and similarly minors A_2, B_2, C_2, and A_3, B_3, and C_3. We then obtain the inverse of the matrix A as:

$$A^{-1} = \begin{bmatrix} \dfrac{A_1}{|A|} & \dfrac{-A_2}{|A|} & \dfrac{A_3}{|A|} \\ \dfrac{-B_1}{|A|} & \dfrac{B_2}{|A|} & \dfrac{-B_3}{|A|} \\ \dfrac{C_1}{|A|} & \dfrac{-C_2}{|A|} & \dfrac{C_3}{|A|} \end{bmatrix}$$

Example: Solve the system of equations:

$$x + y + z = 6$$

$$x + 2y + 3z = 14$$

$$x + 4y + 9z = 36$$

We have:

$$\mathbf{A} = \begin{bmatrix} 1 & 1 & 1 \\ 1 & 2 & 3 \\ 1 & 4 & 9 \end{bmatrix} \qquad b = \begin{bmatrix} 6 \\ 14 \\ 36 \end{bmatrix}$$

Now,

$$\begin{bmatrix} 1 & 1 & 1 \\ 1 & 2 & 3 \\ 1 & 4 & 9 \end{bmatrix}^{-1} = \frac{1}{2} \begin{bmatrix} 6 & -5 & 1 \\ -6 & 8 & -2 \\ 2 & -3 & 1 \end{bmatrix}$$

Hence,

$$\begin{bmatrix} x \\ y \\ z \end{bmatrix} = \frac{1}{2} \begin{bmatrix} 6 & -5 & 1 \\ -6 & 8 & -2 \\ 2 & -3 & 1 \end{bmatrix} \begin{bmatrix} 6 \\ 14 \\ 36 \end{bmatrix} = \begin{bmatrix} 1 \\ 2 \\ 3 \end{bmatrix}$$

Thus $x = 1$, $y = 2$, and $z = 3$ give the solution.

Summary

The representation of multivariate data in more than two dimensions can be done in several ways. Each *vector* of observations is represented by a graph in the form of a *polygonal profile, circular profile, trigonometric series function,* or *Chernoff face.* These graphs allow us to *cluster* data in several groups. The *multinomial distribution* is the generalization of the binomial distribution. Data arising from multinomial distributions can be studied with the help of contingency-table techniques. *Hotelling's T^2* test is used to test hypotheses about the mean of the *multivariate normal distribution. Fisher's linear discriminant functions* allow us to classify a given observation to one of two given multivariate normal populations, providing equal errors of misclassification. When the means and covariance matrices are to be estimated from data for the given multivariate normal populations, we use *Anderson's discriminant function.* Simple operations on *matrices* such as those of *addition, multiplication,* and *inverse* of matrices are given.

References

Anderson, T. W. *Multivariate Statistical Analysis.* New York: John Wiley & Sons, 1958.
Andrews, D. F. Plots of high-dimensional data, *Biometrics,* 1972, *28,* 125–136.
Bliss, C. I. *Statistics in Biology,* vol. II. New York: McGraw-Hill, 1970.
Chambers, John M.; Cleveland, William S.; Kleiner, Beat; and Tukey, Paul A. *Graphical Methods for Data Analysis.* Boston: Duxbury Press, 1983.

Chernoff, H. The use of faces to represent points in n-dimensional space graphically, *J. Amer. Statist. Assoc.*, 1971.

Gnanadesikan, R. *Methods for Statistical Data Analysis of Multivariate Observations*. New York: John Wiley & Sons, 1977.

Johnson, N. L., and Kotz, S. *Continuous Multivariate Distributions*. New York: John Wiley & Sons, 1972.

Morrison, D. F. *Multivariate Statistical Methods*, 2nd ed. New York: McGraw-Hill, 1977.

Rustagi, J. S. Statistical theory in toxicological research, *Proceedings of the Conference on Atmospheric Contamination in Confined Spaces*, Wright-Patterson Air Force Base, 1966, 1015–1020.

Seal, Hilary L. *Multivariate Statistical Analysis for Biologists*. New York: John Wiley & Sons, 1964.

Srivastava, M. S., and Carter, E. M. *An Introduction to Applied Multivariate Statistics*. New York: North-Holland, 1983.

Wainer, Howard. "On multivariate displays," in *Recent Advances in Statistics*, Edited by Rizvi, H.; Rustagi, J. S.; and Siegmund, D. New York: Academic Press, 1983, pp. 469–508.

Answers to Selected Exercises

Chapter 2

1.

Class-interval (minutes)	Frequency
.5–11.5	7
11.5–22.5	13
22.5–33.5	12
33.5–44.5	5
44.5–55.5	2
55.5–66.5	1
Total	40

3. Using 8 classes:

Class-interval (thousands of dollars)	Frequency
.5– 13.5	1
13.5– 26.5	10
26.5– 39.5	12
39.5– 52.5	2
52.5– 65.5	6
65.5– 78.5	4
78.5– 91.5	12
91.5–104.5	3
Total	50

5.

Stem	Leaf
1	7 7 8 7 5 7 5 0 8
2	7 7 8 2 3 7 8 7 7
3	8 2 2 5 7
4	2 8
5	7 6 3 6 3
6	8 5
7	5 3 8
8	3 1 2 5 4 3 3 5 2 1
9	2 1 1 3
10	1 1

7.

Stem	Leaf
5	2
6	2 0 4 4 4 2 4 0 0 8 0 0 6 4 4 8 0 4
7	2 2 2 5 8 2 2 6 6 8 8 2 0 2
8	8 0 0 4 0 0 0 4 0 0 4 0 8 8 0 0 4 0
9	6 4 2 0 6
10	0 8 4 8

Class-interval (beats per minute)	Frequency
51.5– 62.5	9
62.5– 73.5	18
73.5– 84.5	21
84.5– 95.5	6
95.5–106.5	4
106.5–117.5	2
Total	60

13.

Grade	Frequency
A	5
B	8
C	14
D	8
E	5
Total	40

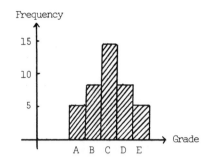

15.   Median = 253.4 mg%
      Modal interval = (275.5 − 296.5) mg%

17.   In thousands of dollars —
      Median = 29.9,   Range = 49.8
      Upper quartile = 29.2,   Lower quartile = 17.8
      Inter-quartile range = 11.4
      Q(.90) = 48.5,   Q(.10) = 11.0

Chapter Exercises

1.   (a) Answer would depend on the number of classes and class-intervals.
     (b) In lbs. − Median = 214, Range = 124
             Upper quartile = 237.75, Lower quartile = 188
     (c)

Stem	Leaf
17	8 3 5 9 1 3 6
18	4 9 9 3 1 9 0 4 5 9 0 3 0
19	4 5 2 0 9 7 1 3
20	1 1 4 9
21	4 4 0 1 6
22	2 5 0 1 7 6 1
23	6 1 0 7 1 6
24	1 4 1 1 6 0
25	8 6 7 6 1 0

     (e) Q(.90) = 256.3 lbs., Q(.10) = 178.7 lbs.

3.   (a) $\bar{x} = 16.04$ oz, s = 1.86 oz, Range = 6.3 oz
     (b) Upper quartile = 17.35 oz, Lower quartile = 14.375 oz.

6.   (a) Median = .49, Mode = .49, Mean = .60
     (b) Upper quartile = .69, Lower quartile = .45.

Chapter 3

1. {Hearts Ace,..., Hearts Deuce,
   Spade Ace,..., Spade Deuce,
   Diamond Ace,..., Diamond Deuce,
   Club Ace,..., Club Deuce}
3. $x: \{97 \le x \le 105\}$
5. (a) $\tilde{A} = \{5, 6, \ldots, 20\}$, $\tilde{B} = \{1, 2, 6, 8, 9, \ldots, 20\}$, $\tilde{C} = \{1, 2, 4, 6, 8, 10, 11, \ldots, 20\}$
   (b) $\{3, 4\}$ (c) $\{1, 2, 3, 4, 5, 7\}$ (d) $\{3\}$ (e) $\{6, 8, 9, \ldots, 20\}$ (f) $\{6, 8, 9, \ldots, 20\}$
7. # no injuries = 55, # with injuries only on face and arms = 10
11. $A \cap B = \phi$
13. $P(A) = 1/2$, $P(B) = 1/4$, $P(C) = 1/4$
15. (a) 1/8, (b) 1/8, (c) 3/8
17. (a) 1/13 (b) 3/13 (c) 2/13 (d) 4/13 (e) 3/13 (f) 0
19. P(Winning = 1/1,000,000, P(Not winning) = 999,999/1,000,000
21. 1/6
23. (a) .003 (b) .999
25. .5
27. (a) .2727 (b) .4242 (c) .0909
29. (a) .819 (b) .064 (c) .52
31. .28
33. C
35. .083
37. (a) 6840; 60; 120
    (b) 35; 28; 270, 725; 635, 013, 550, 000
39. .6

Chapter Exercises

2. .0000568
4. (a) .53 (b) .33
6. P(both not fresh) = .0063, P(one fresh) = .1541, P(both fresh) = .8396
8. .0271
10. P(gun I) = .4, P(gun II) = .35

Chapter 4

1.
x	0	1	2	3
$p_x(x)$	.064	.288	.432	.216

3.
x	0	1	2
$p_x(x)$	.16	.48	.36

5. .033
9. .4096
11. .375
13. (a) 2, 2/13 (b) 5/3, 10/9 (c) 0, 1/2 (d) 3, 2 (e) 3, 3.36

15.

x	$-.50$	9.50	999.50
$p_x(x)$	.97998	.02	.000,02

Mean $= -.28$, Variance $= 21.9516$

17. 1, .98
19. 10, .99
21. .007
23. (a) .491 (b) .560 (c) .265 (d) .509
25. (a) .176 (b) .371 (c) .125 (d) .007
27. .0575

29. $\binom{7}{x}\binom{8}{5-x}\Big/\binom{15}{5}$, x = 0, 1, 2, 3, 4, 5

Mean $= 7/3$, Variance $= 8/9$

31. 200
33. (a) Not independent (b) Independent

Chapter Exercises

2. .2240
4. .3151

Chapter 5

1. (a) Discrete (b) Continuous (c) Continuous (d) Discrete (e) Continuous (f) Discrete
3. (a) 23/24 (b) 2/3 (c) 1/8 (d) 1/3 (e) 1/6 (f) 5/6
5. (a) $f(x) = \begin{cases} 1, & 1 \le x \le 2 \\ 0, & \text{otherwise} \end{cases}$

   (b) 3/16, 3/4
   (c) 1/2, 7/8, 7/8
7. (a) .75 (b) .25 (c) .45
9. (a) .9893 (b) .1968 (c) .8775 (d) .0062 (e) .1954
   (f) .1493 (g) .001 (h) 1.0 (i) .9974 (j) .95
   (k) .0013 (l) .0013 (m) .95 (n) .05 (o) .05
   (p) .95 (q) .025 (r) .975 (s) .9826 (t) .9
11. (a) .5 (b) .0636 (c) .003375

13.

A	B	C	D	F
75–100	68–74	57–67	52–56	0–51

15. (a) .15625 (b) .703 (c) .0173415

17. Expected number $= 240$, Probability $= .2499$

19. (a)

u	2	3	4
$P_U(u)$	.25	.5	.25

(b)

v	1	2	4
$P_V(v)$	.25	.5	.25

(c)

z	.5	1	2
$P_Z(z)$	.25	.5	.25

21. .005, .10

Chapter Execises

2. (a) .1908 (b) .6819 (c) .0098
4. (b) $E(X) = 50$, $Var(X) = 833.33$

Chapter 6

1. $\hat{a} = \bar{x} - \sqrt{3}s$, $\hat{b} = \bar{x} + \sqrt{3}s$
3. 2/3
5. 1.45
7. (a) (10.05, 21.66) (b) (7.7015, 144.5977)
9. (.020, .072)
11. 95% CI is (.226, .390), 99% CI is (.200, .416)
13. 271
15. 35
17.

	90% CI	95% CI
(a)	(6.71, 13.29)	(6.08, 13.92)
(b)	(.62, 1.58)	(.52, 1.68)
(c)	(4.32, 5.68)	(4.18, 5.82)

19. $(-.9472, 1.0272)$
21. $(-1.725, 3.485)$
23. $(-.0367, .1027)$

Chapter Exercises

1. Approved: (2.227, .373), Disapproved: (.627, .773)
3. $(-6.639, 1.979)$
5. For difference: (1.41, 11.09); For variance: (12.05, 43.45)

Chapter 7

1.

	α	β
(a)	.2660	.0304
(b)	.1056	.1056
(c)	.0668	.1587
(d)	.0304	.2660

3.   H   A
      (a) Simple   Composite
      (b) Simple   Composite
      (c) Composite   Composite
      (d) Simple   Composite

5.   Reject H if $\bar{X} < c$ where c is such that $P(\bar{X} < c/\theta = .001) = \alpha$

7.   Reject H if $\bar{X} > 5.33$

μ	5.5	5.75	6.0	6.25
Power	.6950	.8962	.9778	.9971

9.   Reject

11. Reject

13. (a) Reject H if $|\bar{X} - \bar{Y}| > 1.23$ (b) Reject H if $\bar{X} - \bar{Y} < -1.04$

15. Reject

17. Accept

19. Accept

21. Reject

23. Accept

25. Not independent

Chapter Exercises

2.   (a) Reject (b) Accept

Chapter 8

1.   (a) (i) .875 (b) (i) Accept, two-sided level $= .125$

3.   Accept

5.   Reject, level $= .0034$

7.   (b) (i) $T_+ = 7$, $T_- = 21$, Accept, $p = .129$ (ii) accept, $p = .55$
    (iii) Accept, $p = .055$

11. Jumps at observations of size $1/8$ and $D_8 = .411$

13. Expected frequencies: 11, 11, 5.5, 2.5; Reject

15. Expected frequencies: 1, 6, 24, 38, 24, 7; reject

17. $U_x = 20$, $U_y = 22$, Reject

19. $U_x = 20$, $U_y = 5$, Accept

21. $D_{5,5} = 3/5$, Accept

23. Expected frequencies:

11.7	35.2	48.8	17.6	11.7
11.7	35.2	48.8	17.6	11.7
6.6	19.7	27.3	9.8	6.6

    $\chi^2 = 7.32$, Reject

25.   Expected frequencies: 50.4   23.9   24.7
                               42.8   20.3   20.9
                               45.8   21.8   22.4

$\chi^2 = 22.9$, Reject

Chapter Exercises

1.   (a) 7.5 (b) (2,11), $\gamma = .979$ (c) Accept, .021
5.   Collapse last 3 columns, $\chi^2 = 1.43$, Accept

Chapter 9

3. $Y = 30.25 - .12X$
      Residuals: $-.35$ -2.01 1.05 2.95 $-.95$ -1.75 $.85$ $.45$ $-.25$
5. $Y = 20.57 - .4X$
      Residuals: $.2$ 2.0 -2.4 $.4$ $-.6$ 0 $.2$
7. $Y = 23.24 + .87t$
      Residuals: 1.015 -1.48 -1.475 $.28$ 1.285 2.29 $-.33$ -1.575
9.   (a) $Y = 10.5 - .1X$ (b) 11.46
11.   90% CI: $(-.182, -.058)$; $t = 2.365$, Reject
13.   (a) (5.14, 79.33) (b) $t = 2.015$, Reject
15. $Y = 33.6 + .51X$, Reject, 95% CI: (.27, .75)
17. $t = 4.71$, Reject
19. $\hat{y}_{43} = 25.09$, 90% PI: (20.41, 29.77) in thousands of miles
21. $\hat{y}_{84} = 139.7$, 90% PI: (105.1, 174.3)
23.   (a) .98 (b) Reject (c) .966
25. $r = -.35$ $t = .915$, Accept
27. $Y = .167 + 2.708X_1 + 1.292X_2$
      Residuals: 1.542 $.542$ -1.75 $.542$ $-.875$
29. $Y = 32.1124 + .0080X_1 - .5082X_2$
      Residuals: $-.503$ $.628$ $-.121$ $.004$ $.504$ $-.514$
31.   (a) $Y = 112.372 + .717X_1 + 1.763X_2$ (b) $Y = 116.686 + .717X_1$
      $+ .1174X_2^2$
33.   (a) $\hat{\beta}_0 = 28.072$, $\hat{\beta}_1 = -.00266$ (b) $\hat{\beta}_0 = -2.3667$, $\hat{\beta}_1 = 4949.563$
37. $Y = 4.7666 - .0776X_1 - .011X_2$
      Residuals: $.188$ $.363$ $.297$ $-.834$ $.205$ $-.219$

Chapter Exercises

2.   (a) $1.61 < \beta_0 < 4.85$, $-.22 < \beta_1 < .56$ (b) .52 (c) .92
4.   (a) $\hat{\beta}_0 = -6.8$, $\hat{\beta}_1 = 8$, $\hat{b}_2 = -4.4$ (b) .25
6. $Y = -3.7523 - .6508X_1 + 14.4976X_2$

Chapter 10

1. (a) $y_{ij} = \mu + \tau_i + e_{ij}$; $i = 1,2,3,4$, $j = 1,2,3$, with $\&\tau_i = 0$
    (b) $\hat{\mu} = 9.75$, $\hat{t}_1 = -2.75$, $\hat{t}_2 = -1.75$, $\hat{t}_3 = .25$, $\hat{t}_4 = 4.25$
    (c) $F = 5.23$, Reject
3. (b) $\hat{\mu} = 74.67$, $\hat{t}_1 = -6.42$, $\hat{t}_2 = 1.58$, $\hat{t}_3 = 4.83$
    (c) .62, Accept
5. $F = 7.29$, Reject
7. (a) $y_{ij} = \mu + \tau_1 + e_{ij}$; $i = 1, 2, \ldots, n_i$; $n_1 = 3$; $n_2 = 5$, $n_3 = 4$
    with $3\tau_1 + 5\tau_2 + 4\tau_3 = 0$
    (b) $\hat{\mu} = 1.17$, $\hat{t}_1 = 8.83$, $\hat{t}_2 = -5.17$, $\hat{t}_3 = 0.17$
    (c) $F = 21.63$, Reject
9. (b) $F = 1.844$, Accept (c) $f = .094$, Accept
11.

Source	DF	SS	MS	F
Drugs	2	179.170	89.585	.796
Exercises	1	133.337	133.337	1.185
Interaction	2	729.163	364.582	3.241
Error	6	674.997	112.499	
Total	11	1716.667		

None is significant.

13.

Source	DF	SS	MS	F
Laboratories	2	94.500	47.250	1.841
Methods	1	456.333	456.333	17.779
Interaction	2	107.167	53.585	2.088
Error	6	812.000	25.667	
Total	11	812.000		

Methods significantly different at $\alpha = .01$

15. (a) $F = 3.46$, Accept (b) $F = 6.13$, Reject
    (c)

Source	DF	SS	MS	F
Vitamins	2	460.67	230.33	4.56
Error	6	303.33	50.56	
Total	8	764.00		

Vitamins did not affect the weight gain

17. $\hat{\sigma}_a^2 = 932.034$
21. $Q = 9.96$, $p = .009$, Reject
24. $K = 6.99$, $p = .019$, Reject, same as in Exercise 5
25. $Q = 6$, $p = .105$, Accept
27. $Y = 1.285 + .61X$, $F = 1.484$, Accept

Chapter Exercises

1. (a) $F = 55.74$, Reject (b) (38.755, 46.745), (c) $(-42.697, -21.003)$
   (48.505, 56.495),
   (26.827, 33,973)

3. (a) $F = .250$, Accept
5. (a) $F = 1763.13$, Reject (b) $F = 7.08$, Reject

Chapter 11

1. (b) $F = 15.44$, Reject (c) Method D different from A and C
   Method B different from C
3. (b) $F = 2.694$, Accept
5. $F = 4.068$, Accept
7. $F = 21.75$, Reject
9. $F = 77.69$, Reject, Specimens are significantly different
13. $Y_{ijk} = \mu + \alpha_i + \beta_j + \gamma_k + \eta_{ij} + e_{ijk}$; $i = 1, 2$; $j = 1, 2$; $k = 1, 2, 3, 4$
   (Block interactions with factors excluded)
   (a) For temperatures, $F = .165$, Accept; For kilns, $F = .023$, Accept

Chapter Exercises

2. Answer depends on rows and columns permuted.

Source →	Rows	Columns	Treatments	Error	Total
DF →	4	4	4	8	24

4. (a)

Source	DF	SS	MS	F
Treatments	2	250	125	8.33
Rows	2	150	75	5.00
Columns	2	150	75	5.00
Error	2	30	15	
Total	8	580		

   (b) Reject

Chapter 12

1. $\bar{X} = \begin{pmatrix} 281.45 \\ 253.18 \\ 286.00 \end{pmatrix}$ $S = \begin{pmatrix} 7690.07 & 4659.70 & 5723.28 \\ 4659.70 & 6358.76 & 3958.91 \\ 5723.28 & 3958.91 & 5241.00 \end{pmatrix}$

3.

Country	Telephones	Newspapers	Radio	TV Sets
Australia	.522	.308	.412	.345
Canada	.046	.047	.415	.525
Denmark	.709	.460	.288	.626
France	.423	.000	.000	.150
Japan	.499	1.000	.140	.062
Soviet Union	.000	.560	.088	.000
United Kingdom	.486	.601	.224	.426
United States	1.000	.240	1.000	1.000

5. .0005348

7. .0245

11. $T^2 = 4.9876$, $F = 1.36$, Accept

13. π_1 if $-.1017X_1 + .2110X_2 > 1.97$

Appendix Tables

Table I Individual Terms, Binomial Distribution

		p									
n	x	.05	.10	.15	.20	.25	.30	.35	.40	.45	.50
1	0	.9500	.9000	.8500	.8000	.7500	.7000	.6500	.6000	.5500	.5000
	1	.0500	.1000	.1500	.2000	.2500	.3000	.3500	.4000	.4500	.5000
2	0	.9025	.8100	.7225	.6400	.5625	.4900	.4225	.3600	.3025	.2500
	1	.0950	.1800	.2550	.3200	.3750	.4200	.4550	.4800	.4950	.5000
	2	.0025	.0100	.0225	.0400	.0625	.0900	.1225	.1600	.2025	.2500
3	0	.8574	.7290	.6141	.5120	.4219	.3430	.2746	.2160	.1664	.1250
	1	.1354	.2430	.3251	.3840	.4219	.4410	.4436	.4320	.4084	.3750
	2	.0071	.0270	.0574	.0960	.1406	.1890	.2389	.2880	.3341	.3750
	3	.0001	.0010	.0034	.0080	.0156	.0270	.0429	.0640	.0911	.1250
4	0	.8145	.6561	.5220	.4096	.3164	.2401	.1785	.1296	.0915	.0625
	1	.1715	.2916	.3685	.4096	.4219	.4116	.3845	.3456	.2995	.2500
	2	.0135	.0486	.0975	.1536	.2109	.2646	.3105	.3456	.3675	.3750
	3	.0005	.0036	.0115	.0256	.0469	.0756	.1115	.1536	.2005	.2500
	4	.0000	.0001	.0005	.0016	.0039	.0081	.0150	.0256	.0410	.0625
5	0	.7738	.5905	.4437	.3277	.2373	.1681	.1160	.0778	.0503	.0312
	1	.2036	.3280	.3915	.4096	.3955	.3602	.3124	.2592	.2059	.1562
	2	.0214	.0729	.1382	.2048	.2637	.3087	.3364	.3456	.3369	.3125
	3	.0011	.0081	.0244	.0512	.0879	.1323	.1811	.2304	.2757	.3125
	4	.0000	.0004	.0022	.0064	.0146	.0284	.0488	.0768	.1128	.1562
	5	.0000	.0000	.0001	.0003	.0010	.0024	.0053	.0102	.0185	.0312
6	0	.7351	.5314	.3771	.2621	.1780	.1176	.0754	.0467	.0277	.0156
	1	.2321	.3543	.3993	.3932	.3560	.3025	.2437	.1866	.1359	.0938
	2	.0305	.0984	.1762	.2458	.2966	.3241	.3280	.3110	.2780	.2344
	3	.0021	.0146	.0415	.0819	.1318	.1852	.2355	.2765	.3032	.3125
	4	.0001	.0012	.0055	.0154	.0330	.0595	.0951	.1382	.1861	.2344
	5	.0000	.0001	.0004	.0015	.0044	.0102	.0205	.0369	.0609	.0938
	6	.0000	.0000	.0000	.0001	.0002	.0007	.0018	.0041	.0083	.0156

(*Continued*)

Table I (Continued)

n	x						p					
		.05	.10	.15	.20	.25	.30	.35	.40	.45	.50
7	0	.6983	.4783	.3206	.2097	.1335	.0824	.0490	.0280	.0152	.0078
	1	.2573	.3720	.3960	.3670	.3115	.2471	.1848	.1306	.0872	.0547
	2	.0406	.1240	.2097	.2753	.3115	.3177	.2985	.2613	.2140	.1641
	3	.0036	.0230	.0617	.1147	.1730	.2269	.2679	.2903	.2918	.2734
	4	.0002	.0026	.0109	.0287	.0577	.0972	.1442	.1935	.2388	.2734
	5	.0000	.0002	.0012	.0043	.0115	.0250	.0466	.0774	.1172	.1641
	6	.0000	.0000	.0001	.0004	.0013	.0036	.0084	.0172	.0320	.0547
	7	.0000	.0000	.0000	.0000	.0001	.0002	.0006	.0016	.0037	.0078
8	0	.6634	.4305	.2725	.1678	.1001	.0576	.0319	.0168	.0084	.0039
	1	.2793	.3826	.3847	.3355	.2670	.1977	.1373	.0896	.0548	.0312
	2	.0515	.1488	.2376	.2936	.3115	.2965	.2587	.2090	.1569	.1094
	3	.0054	.0331	.0839	.1468	.2076	.2541	.2786	.2787	.2568	.2188
	4	.0004	.0046	.0185	.0459	.0865	.1361	.1875	.2322	.2627	.2734
	5	.0000	.0004	.0026	.0092	.0231	.0467	.0808	.1239	.1719	.2188
	6	.0000	.0000	.0002	.0011	.0038	.0100	.0217	.0413	.0703	.1094
	7	.0000	.0000	.0000	.0001	.0004	.0012	.0033	.0079	.0164	.0312
	8	.0000	.0000	.0000	.0000	.0000	.0001	.0002	.0007	.0017	.0039
9	0	.6302	.3874	.2316	.1342	.0751	.0404	.0207	.0101	.0046	.0020
	1	.2985	.3874	.3679	.3020	.2253	.1556	.1004	.0605	.0339	.0176
	2	.0629	.1722	.2597	.3020	.3003	.2668	.2162	.1612	.1110	.0703
	3	.0077	.0446	.1069	.1762	.2336	.2668	.2716	.2508	.2119	.1641
	4	.0006	.0074	.0283	.0661	.1168	.1715	.2194	.2508	.2600	.2461
	5	.0000	.0008	.0050	.0165	.0389	.0735	.1181	.1672	.2128	.2461
	6	.0000	.0001	.0006	.0028	.0087	.0210	.0424	.0743	.1160	.1641
	7	.0000	.0000	.0000	.0003	.0012	.0039	.0098	.0212	.0407	.0703
	8	.0000	.0000	.0000	.0000	.0001	.0004	.0013	.0035	.0083	.0176
	9	.0000	.0000	.0000	.0000	.0000	.0000	.0001	.0003	.0008	.0020

n	k										
10	0	.5987	.3487	.1969	.1074	.0563	.0282	.0135	.0060	.0025	.0010
	1	.3151	.3874	.3474	.2684	.1877	.1211	.0725	.0403	.0207	.0098
	2	.0746	.1937	.2759	.3020	.2816	.2335	.1757	.1209	.0763	.0439
	3	.0105	.0574	.1298	.2013	.2503	.2668	.2522	.2150	.1665	.1172
	4	.0010	.0112	.0401	.0881	.1460	.2001	.2377	.2508	.2384	.2051
	5	.0001	.0015	.0085	.0264	.0584	.1029	.1536	.2007	.2340	.2461
	6	.0000	.0001	.0012	.0055	.0162	.0368	.0689	.1115	.1596	.2051
	7	.0000	.0000	.0001	.0008	.0031	.0090	.0212	.0425	.0746	.1172
	8	.0000	.0000	.0000	.0001	.0004	.0014	.0043	.0106	.0229	.0439
	9	.0000	.0000	.0000	.0000	.0000	.0001	.0005	.0016	.0042	.0098
	10	.0000	.0000	.0000	.0000	.0000	.0000	.0000	.0001	.0003	.0010
11	0	.5688	.3138	.1673	.0859	.0422	.0198	.0088	.0036	.0014	.0004
	1	.3293	.3835	.3248	.2362	.1549	.0932	.0518	.0266	.0125	.0055
	2	.0867	.2131	.2866	.2953	.2581	.1998	.1395	.0887	.0513	.0269
	3	.0137	.0710	.1517	.2215	.2581	.2568	.2254	.1774	.1259	.0806
	4	.0014	.0158	.0536	.1107	.1721	.2201	.2428	.2365	.2060	.1611
	5	.0001	.0025	.0132	.0388	.0803	.1321	.1830	.2207	.2360	.2256
	6	.0000	.0003	.0023	.0097	.0268	.0566	.0985	.1471	.1931	.2256
	7	.0000	.0000	.0003	.0017	.0064	.0173	.0379	.0701	.1128	.1611
	8	.0000	.0000	.0000	.0002	.0011	.0037	.0102	.0234	.0462	.0806
	9	.0000	.0000	.0000	.0000	.0001	.0005	.0018	.0052	.0126	.0269
	10	.0000	.0000	.0000	.0000	.0000	.0000	.0002	.0007	.0021	.0054
	11	.0000	.0000	.0000	.0000	.0000	.0000	.0000	.0000	.0002	.0005
12	0	.5404	.2824	.1422	.0687	.0317	.0138	.0057	.0022	.0008	.0002
	1	.3413	.3766	.3012	.2062	.1267	.0712	.0368	.0174	.0075	.0029
	2	.0988	.2301	.2924	.2835	.2323	.1678	.1088	.0639	.0339	.0161
	3	.0173	.0852	.1720	.2362	.2581	.2397	.1954	.1419	.0923	.0537
	4	.0021	.0213	.0683	.1329	.1936	.2311	.2367	.2128	.1700	.1208
	5	.0002	.0038	.0193	.0532	.1032	.1585	.2039	.2270	.2225	.1934
	6	.0000	.0005	.0040	.0155	.0401	.0792	.1281	.1766	.2124	.2256
	7	.0000	.0000	.0006	.0033	.0115	.0291	.0591	.1009	.1489	.1934
	8	.0000	.0000	.0001	.0005	.0024	.0078	.0199	.0420	.0762	.1208
	9	.0000	.0000	.0000	.0001	.0004	.0015	.0048	.0125	.0277	.0537
	10	.0000	.0000	.0000	.0000	.0000	.0002	.0008	.0025	.0068	.0161
	11	.0000	.0000	.0000	.0000	.0000	.0000	.0001	.0003	.0010	.0029
	12	.0000	.0000	.0000	.0000	.0000	.0000	.0000	.0000	.0001	.0002

(Continued)

Table I (Continued)

n	x						p					
		.05	.10	.15	.20	.25	.30	.35	.40	.45	.50
13	0	.5133	.2542	.1209	.0550	.0238	.0097	.0037	.0013	.0004	.0001
	1	.3512	.3672	.2774	.1787	.1029	.0540	.0259	.0113	.0045	.0016
	2	.1109	.2448	.2937	.2680	.2059	.1388	.0836	.0453	.0220	.0095
	3	.0214	.0997	.1900	.2457	.2517	.2181	.1651	.1107	.0660	.0349
	4	.0028	.0277	.0838	.1535	.2097	.2337	.2222	.1845	.1350	.0873
	5	.0003	.0055	.0266	.0691	.1258	.1803	.2154	.2214	.1989	.1571
	6	.0000	.0008	.0063	.0230	.0559	.1030	.1546	.1968	.2169	.2095
	7	.0000	.0001	.0011	.0058	.0186	.0442	.0833	.1312	.1775	.2095
	8	.0000	.0000	.0001	.0011	.0047	.0142	.0336	.0656	.1089	.1571
	9	.0000	.0000	.0000	.0001	.0009	.0034	.0101	.0243	.0495	.0873
	10	.0000	.0000	.0000	.0000	.0001	.0006	.0022	.0065	.0162	.0349
	11	.0000	.0000	.0000	.0000	.0000	.0001	.0003	.0012	.0036	.0095
	12	.0000	.0000	.0000	.0000	.0000	.0000	.0000	.0001	.0005	.0016
	13	.0000	.0000	.0000	.0000	.0000	.0000	.0000	.0000	.0000	.0001
14	0	.4877	.2288	.1028	.0440	.0178	.0068	.0024	.0008	.0002	.0001
	1	.3593	.3559	.2539	.1539	.0832	.0407	.0181	.0073	.0027	.0009
	2	.1229	.2570	.2912	.2501	.1802	.1134	.0634	.0317	.0141	.0056
	3	.0259	.1142	.2056	.2501	.2402	.1943	.1366	.0845	.0462	.0222
	4	.0037	.0349	.0998	.1720	.2202	.2290	.2022	.1549	.1040	.0611
	5	.0004	.0078	.0352	.0860	.1468	.1963	.2178	.2066	.1701	.1222
	6	.0000	.0013	.0093	.0322	.0734	.1262	.1759	.2066	.2088	.1833
	7	.0000	.0002	.0019	.0092	.0280	.0618	.1082	.1574	.1952	.2095
	8	.0000	.0000	.0003	.0020	.0082	.0232	.0510	.0918	.1398	.1833
	9	.0000	.0000	.0000	.0003	.0018	.0066	.0183	.0408	.0762	.1222
	10	.0000	.0000	.0000	.0000	.0003	.0014	.0049	.0136	.0312	.0611
	11	.0000	.0000	.0000	.0000	.0000	.0002	.0010	.0033	.0093	.0222
	12	.0000	.0000	.0000	.0000	.0000	.0000	.0001	.0005	.0019	.0056
	13	.0000	.0000	.0000	.0000	.0000	.0000	.0000	.0001	.0002	.0009
	14	.0000	.0000	.0000	.0000	.0000	.0000	.0000	.0000	.0000	.0001

n = 15

x										
0	.4633	.2059	.0874	.0352	.0134	.0047	.0016	.0005	.0001	.0000
1	.3658	.3432	.2312	.1319	.0668	.0305	.0126	.0047	.0016	.0005
2	.1348	.2669	.2856	.2309	.1559	.0916	.0476	.0219	.0090	.0032
3	.0307	.1285	.2184	.2501	.2252	.1700	.1110	.0634	.0318	.0139
4	.0049	.0428	.1156	.1876	.2252	.2186	.1792	.1268	.0780	.0417
5	.0006	.0105	.0449	.1032	.1651	.2061	.2123	.1859	.1404	.0916
6	.0000	.0019	.0132	.0430	.0917	.1472	.1906	.2066	.1914	.1527
7	.0000	.0003	.0030	.0138	.0393	.0811	.1319	.1771	.2013	.1964
8	.0000	.0000	.0005	.0035	.0131	.0348	.0710	.1181	.1647	.1964
9	.0000	.0000	.0001	.0007	.0034	.0116	.0298	.0612	.1048	.1527
10	.0000	.0000	.0000	.0001	.0007	.0030	.0096	.0245	.0515	.0916
11	.0000	.0000	.0000	.0000	.0001	.0006	.0024	.0074	.0191	.0417
12	.0000	.0000	.0000	.0000	.0000	.0001	.0004	.0016	.0052	.0139
13	.0000	.0000	.0000	.0000	.0000	.0000	.0001	.0003	.0010	.0032
14	.0000	.0000	.0000	.0000	.0000	.0000	.0000	.0000	.0001	.0005
15	.0000	.0000	.0000	.0000	.0000	.0000	.0000	.0000	.0000	.0000

n = 20

x										
0	.3585	.1216	.0388	.0115	.0032	.0008	.0002	.0000	.0000	.0000
1	.3774	.2702	.1368	.0576	.0211	.0068	.0020	.0005	.0001	.0000
2	.1887	.2852	.2293	.1369	.0669	.0278	.0100	.0031	.0008	.0002
3	.0596	.1901	.2428	.2054	.1339	.0716	.0323	.0123	.0040	.0011
4	.0133	.0898	.1821	.2182	.1897	.1304	.0738	.0350	.0139	.0046
5	.0022	.0319	.1028	.1746	.2023	.1789	.1272	.0746	.0365	.0148
6	.0003	.0089	.0454	.1091	.1686	.1916	.1712	.1244	.0746	.0370
7	.0000	.0020	.0160	.0545	.1124	.1643	.1844	.1659	.1221	.0739
8	.0000	.0004	.0046	.0222	.0609	.1144	.1614	.1797	.1623	.1201
9	.0000	.0001	.0011	.0074	.0271	.0654	.1158	.1597	.1771	.1602
10	.0000	.0000	.0002	.0020	.0099	.0308	.0686	.1171	.1593	.1762
11	.0000	.0000	.0000	.0005	.0030	.0120	.0336	.0710	.1185	.1602
12	.0000	.0000	.0000	.0001	.0008	.0039	.0136	.0355	.0727	.1201
13	.0000	.0000	.0000	.0000	.0002	.0010	.0045	.0146	.0366	.0739
14	.0000	.0000	.0000	.0000	.0000	.0002	.0012	.0049	.0150	.0370
15	.0000	.0000	.0000	.0000	.0000	.0000	.0003	.0013	.0049	.0148
16	.0000	.0000	.0000	.0000	.0000	.0000	.0000	.0003	.0013	.0046
17	.0000	.0000	.0000	.0000	.0000	.0000	.0000	.0000	.0002	.0011
18	.0000	.0000	.0000	.0000	.0000	.0000	.0000	.0000	.0000	.0002
19	.0000	.0000	.0000	.0000	.0000	.0000	.0000	.0000	.0000	.0000
20	.0000	.0000	.0000	.0000	.0000	.0000	.0000	.0000	.0000	.0000

Table II Individual Terms, Poisson Distribution

m

x	0.1	0.2	0.3	0.4	0.5	0.6	0.7	0.8	0.9	1.0
0	.9048	.8187	.7408	.6703	.6065	.5488	.4966	.4493	.4066	.3679
1	.0905	.1637	.2222	.2681	.3033	.3293	.3476	.3595	.3659	.3679
2	.0045	.0164	.0333	.0536	.0758	.0988	.1217	.1438	.1647	.1839
3	.0002	.0011	.0033	.0072	.0126	.0198	.0284	.0383	.0494	.0613
4	.0000	.0001	.0002	.0007	.0016	.0030	.0050	.0077	.0111	.0153
5	.0000	.0000	.0000	.0001	.0002	.0004	.0007	.0012	.0020	.0031
6	.0000	.0000	.0000	.0000	.0000	.0000	.0001	.0002	.0003	.0005
7	.0000	.0000	.0000	.0000	.0000	.0000	.0000	.0000	.0000	.0001

m

x	1.1	1.2	1.3	1.4	1.5	1.6	1.7	1.8	1.9	2.0
0	.3329	.3012	.2725	.2466	.2231	.2019	.1827	.1653	.1496	.1353
1	.3662	.3614	.3543	.3452	.3347	.3230	.3106	.2975	.2842	.2707
2	.2014	.2169	.2303	.2417	.2510	.2584	.2640	.2678	.2700	.2707
3	.0738	.0867	.0998	.1128	.1255	.1378	.1496	.1607	.1710	.1804
4	.0203	.0260	.0324	.0395	.0471	.0551	.0636	.0723	.0812	.0902
5	.0045	.0062	.0084	.0111	.0141	.0176	.0216	.0260	.0309	.0361
6	.0008	.0012	.0018	.0026	.0035	.0047	.0061	.0078	.0098	.0120
7	.0001	.0002	.0003	.0005	.0008	.0011	.0015	.0020	.0027	.0034
8	.0000	.0000	.0001	.0001	.0001	.0002	.0003	.0005	.0006	.0009
9	.0000	.0000	.0000	.0000	.0000	.0000	.0001	.0001	.0001	.0002

m

x	2.1	2.2	2.3	2.4	2.5	2.6	2.7	2.8	2.9	3.0
0	.1225	.1108	.1003	.0907	.0821	.0743	.0672	.0608	.0550	.0498
1	.2572	.2438	.2306	.2177	.2052	.1931	.1815	.1703	.1596	.1494
2	.2700	.2681	.2652	.2613	.2565	.2510	.2450	.2384	.2314	.2240
3	.1890	.1966	.2033	.2090	.2138	.2176	.2205	.2225	.2237	.2240
4	.0992	.1082	.1169	.1254	.1336	.1414	.1488	.1557	.1622	.1680
5	.0417	.0476	.0538	.0602	.0668	.0735	.0804	.0872	.0940	.1008
6	.0146	.0174	.0206	.0241	.0278	.0319	.0362	.0407	.0455	.0504
7	.0044	.0055	.0068	.0083	.0099	.0118	.0139	.0163	.0188	.0216
8	.0011	.0015	.0019	.0025	.0031	.0038	.0047	.0057	.0068	.0081
9	.0003	.0004	.0005	.0007	.0009	.0011	.0014	.0018	.0022	.0027
10	.0001	.0001	.0001	.0002	.0002	.0003	.0004	.0005	.0006	.0008
11	.0000	.0000	.0000	.0000	.0000	.0001	.0001	.0001	.0002	.0002
12	.0000	.0000	.0000	.0000	.0000	.0000	.0000	.0000	.0000	.0001

m

x	3.1	3.2	3.3	3.4	3.5	3.6	3.7	3.8	3.9	4.0
0	.0450	.0408	.0369	.0334	.0302	.0273	.0247	.0224	.0202	.0183
1	.1397	.1304	.1217	.1135	.1057	.0984	.0915	.0850	.0789	.0733
2	.2165	.2087	.2008	.1929	.1850	.1771	.1692	.1615	.1539	.1465
3	.2237	.2226	.2209	.2186	.2158	.2125	.2087	.2046	.2001	.1954
4	.1734	.1781	.1823	.1858	.1888	.1912	.1931	.1944	.1951	.1954
5	.1075	.1140	.1203	.1264	.1322	.1377	.1429	.1477	.1522	.1563
6	.0555	.0608	.0662	.0716	.0771	.0826	.0881	.0936	.0989	.1042
7	.0246	.0278	.0312	.0348	.0385	.0425	.0466	.0508	.0551	.0595
8	.0095	.0111	.0129	.0148	.0169	.0191	.0215	.0241	.0269	.0298
9	.0033	.0040	.0047	.0056	.0066	.0076	.0089	.0102	.0116	.0132
10	.0010	.0013	.0016	.0019	.0023	.0028	.0033	.0039	.0045	.0053
11	.0003	.0004	.0005	.0006	.0007	.0009	.0011	.0013	.0016	.0019
12	.0001	.0001	.0001	.0002	.0002	.0003	.0003	.0004	.0005	.0006
13	.0000	.0000	.0000	.0000	.0001	.0001	.0001	.0001	.0002	.0002
14	.0000	.0000	.0000	.0000	.0000	.0000	.0000	.0000	.0000	.0001

(Continued)

Table II (Continued)

						m					
x	4.1	4.2	4.3	4.4	4.5	4.6	4.7	4.8	4.9	5.0
0	.0166	.0150	.0136	.0123	.0111	.0101	.0091	.0082	.0074	.0067
1	.0679	.0630	.0583	.0540	.0500	.0462	.0427	.0395	.0365	.0337
2	.1393	.1323	.1254	.1188	.1125	.1063	.1005	.0948	.0894	.0842
3	.1904	.1852	.1798	.1743	.1687	.1631	.1574	.1517	.1460	.1440
4	.1951	.1944	.1933	.1917	.1898	.1875	.1849	.1820	.1789	.1755
5	.1600	.1633	.1662	.1687	.1708	.1725	.1738	.1747	.1753	.1755
6	.1093	.1143	.1191	.1237	.1281	.1323	.1362	.1398	.1432	.1462
7	.0640	.0686	.0732	.0778	.0824	.0869	.0914	.0959	.1002	.1044
8	.0328	.0360	.0393	.0428	.0463	.0500	.0537	.0575	.0614	.0653
9	.0150	.0168	.0188	.0209	.0232	.0255	.0280	.0307	.0334	.0363
10	.0061	.0071	.0081	.0092	.0104	.0118	.0132	.0147	.0164	.0181
11	.0023	.0027	.0032	.0037	.0043	.0049	.0056	.0064	.0073	.0082
12	.0008	.0009	.0011	.0014	.0016	.0019	.0022	.0026	.0030	.0034
13	.0002	.0003	.0004	.0005	.0006	.0007	.0008	.0009	.0011	.0013
14	.0001	.0001	.0001	.0001	.0002	.0002	.0003	.0003	.0004	.0005
15	.0000	.0000	.0000	.0000	.0001	.0001	.0001	.0001	.0001	.0002

(Continued)

Table II (*Continued*)

x	7.0	8.0	9.0	10
0	.0009	.0003	.0001	.0000
1	.0064	.0027	.0011	.0005
2	.0223	.0107	.0050	.0023
3	.0521	.0286	.0150	.0076
4	.0912	.0573	.0337	.0189
5	.1277	.0916	.0607	.0378
6	.1490	.1221	.0911	.0631
7	.1490	.1396	.1171	.0901
8	.1304	.1396	.1318	.1126
9	.1014	.1241	.1318	.1251
10	.0710	.0993	.1186	.1251
11	.0452	.0722	.0970	.1137
12	.0264	.0481	.0728	.0948
13	.0142	.0296	.0504	.0729
14	.0071	.0169	.0324	.0521
15	.0033	.0090	.0194	.0347
16	.0014	.0045	.0109	.0217
17	.0006	.0021	.0058	.0128
18	.0002	.0009	.0029	.0071
19	.0001	.0004	.0014	.0037
20		.0002	.0006	.0019
21		.0001	.0003	.0009
22			.0001	.0004
23				.0002
24				.0001

Table III Area Φ(z) under the normal curve to the left of z

z	.00	.01	.02	.03	.04	.05	.06	.07	.08	.09
0.0	.5000	.5040	.5080	.5120	.5160	.5199	.5239	.5279	.5319	.5359
0.1	.5398	.5438	.5478	.5517	.5557	.5596	.5636	.5675	.5714	.5753
0.2	.5793	.5832	.5871	.5910	.5948	.5987	.6026	.6064	.6103	.6141
0.3	.6179	.6217	.6255	.6293	.6331	.6368	.6406	.6443	.6480	.6517
0.4	.6554	.6591	.6628	.6664	.6700	.6736	.6772	.6808	.6844	.6879
0.5	.6915	.6950	.6985	.7019	.7054	.7088	.7123	.7157	.7190	.7224
0.6	.7257	.7291	.7324	.7357	.7389	.7422	.7454	.7486	.7517	.7549
0.7	.7580	.7611	.7642	.7673	.7704	.7734	.7764	.7794	.7823	.7852
0.8	.7881	.7910	.7939	.7967	.7995	.8023	.8051	.8078	.8106	.8133
0.9	.8159	.8186	.8212	.8238	.8264	.8289	.8315	.8340	.8365	.8389
1.0	.8413	.8438	.8461	.8485	.8508	.8531	.8554	.8577	.8599	.8621
1.1	.8643	.8665	.8686	.8708	.8729	.8749	.8770	.8790	.8810	.8830
1.2	.8849	.8869	.8888	.8907	.8925	.8944	.8962	.8980	.8997	.9015
1.3	.9032	.9049	.9066	.9082	.9099	.9115	.9131	.9147	.9162	.9177
1.4	.9192	.9207	.9222	.9236	.9251	.9265	.9279	.9292	.9306	.9319
1.5	.9332	.9345	.9357	.9370	.9382	.9394	.9406	.9418	.9429	.9441
1.6	.9452	.9463	.9474	.9484	.9495	.9505	.9515	.9525	.9535	.9545
1.7	.9554	.9564	.9573	.9584	.9591	.9599	.9608	.9616	.9625	.9633
1.8	.9641	.9649	.9656	.9664	.9671	.9678	.9686	.9693	.9699	.9706
1.9	.9713	.9719	.9726	.9732	.9738	.9744	.9750	.9756	.9761	.9767
2.0	.9772	.9778	.9783	.9788	.9793	.9798	.9803	.9808	.9812	.9817
2.1	.9821	.9826	.9830	.9834	.9838	.9842	.9846	.9850	.9854	.9857
2.2	.9861	.9864	.9868	.9871	.9875	.9878	.9881	.9884	.9887	.9890
2.3	.9893	.9896	.9898	.9901	.9904	.9906	.9909	.9911	.9913	.9916
2.4	.9918	.9920	.9922	.9925	.9927	.9929	.9931	.9932	.9934	.9936
2.5	.9938	.9940	.9941	.9943	.9945	.9946	.9948	.9949	.9950	.9952
2.6	.9953	.9955	.9956	.9957	.9959	.9960	.9961	.9962	.9963	.9964
2.7	.9965	.9966	.9967	.9968	.9969	.9970	.9971	.9972	.9973	.9974
2.8	.9974	.9975	.9976	.9977	.9977	.9978	.9979	.9979	.9980	.9981
2.9	.9981	.9982	.9982	.9983	.9984	.9984	.9985	.9985	.9986	.9986
3.0	.9987	.9987	.9987	.9988	.9988	.9989	.9989	.9989	.9990	.9990
3.1	.9990	.9991	.9991	.9991	.9992	.9992	.9992	.9992	.9993	.9993
3.2	.9993	.9993	.9994	.9994	.9994	.9994	.9994	.9995	.9995	.9995
3.3	.9995	.9995	.9995	.9996	.9996	.9996	.9996	.9996	.9996	.9997
3.4	.9997	.9997	.9997	.9997	.9997	.9997	.9997	.9997	.9997	.9998

Table IV Chi-Square Distribution

n \ F	.005	.010	.025	.050	.100	.250	.500	.750	.900	.950	.975	.990	.995
1	.0000393	.000157	.000982	.00393	.0158	.102	.455	1.32	2.71	3.84	5.02	6.63	7.88
2	.0100	.0201	.0506	.103	.211	.575	1.39	2.77	4.61	5.99	7.38	9.21	10.6
3	.0717	.115	.216	.352	.584	1.21	2.37	4.11	6.25	7.81	9.35	11.3	12.8
4	.207	.297	.484	.711	1.06	1.92	3.36	5.39	7.78	9.49	11.1	13.3	14.9
5	.412	.554	.831	1.15	1.61	2.67	4.35	6.63	9.24	11.1	12.8	15.1	16.7
6	.676	.872	1.24	1.64	2.20	3.45	5.35	7.84	10.6	12.6	14.4	16.8	18.5
7	.989	1.24	1.69	2.17	2.83	4.25	6.35	9.04	12.0	14.1	16.0	18.5	20.3
8	1.34	1.65	2.18	2.73	3.49	5.07	7.34	10.2	13.4	15.5	17.5	20.1	22.0
9	1.73	2.09	2.70	3.33	4.17	5.90	8.34	11.4	14.7	16.9	19.0	21.7	23.6
10	2.16	2.56	3.25	3.94	4.87	6.74	9.34	12.5	16.0	18.3	20.5	23.2	25.2
11	2.60	3.05	3.82	4.57	5.58	7.58	10.3	13.7	17.3	19.7	21.9	24.7	26.8
12	3.07	3.57	4.40	5.23	6.30	8.44	11.3	14.8	18.5	21.0	23.3	26.2	28.3
13	3.57	4.11	5.01	5.89	7.04	9.30	12.3	16.0	19.8	22.4	24.7	27.7	29.8
14	4.07	4.66	5.63	6.57	7.79	10.2	13.3	17.1	21.1	23.7	26.1	29.1	31.3
15	4.60	5.23	6.26	7.26	8.55	11.0	14.3	18.2	22.3	25.0	27.5	30.6	32.8
16	5.14	5.81	6.91	7.96	9.31	11.9	15.3	19.4	23.5	26.3	28.8	32.0	34.3
17	5.70	6.41	7.56	8.67	10.1	12.8	16.3	20.5	24.8	27.6	30.2	33.4	35.7
18	6.26	7.01	8.23	9.39	10.9	13.7	17.3	21.6	26.0	28.9	31.5	34.8	37.2
19	6.84	7.63	8.91	10.1	11.7	14.6	18.3	22.7	27.2	30.1	32.9	36.2	38.6
20	7.43	8.26	9.59	10.9	12.4	15.5	19.3	23.8	28.4	31.4	34.2	37.6	40.0
21	8.03	8.90	10.3	11.6	13.2	16.3	20.3	24.9	29.6	32.7	35.5	38.9	41.4
22	8.64	9.54	11.0	12.3	14.0	17.2	21.3	26.0	30.8	33.9	36.8	40.3	42.8
23	9.26	10.2	11.7	13.1	14.8	18.1	22.3	27.1	32.0	35.2	38.1	41.6	44.2
24	9.89	10.9	12.4	13.8	15.7	19.0	23.3	28.2	33.2	36.4	39.4	43.0	45.6
25	10.5	11.5	13.1	14.6	16.5	19.9	24.3	29.3	34.4	37.7	40.6	44.3	46.9
26	11.2	12.2	13.8	15.4	17.3	20.8	25.3	30.4	35.6	38.9	41.9	45.6	48.3
27	11.8	12.9	14.6	16.2	18.1	21.7	26.3	31.5	36.7	40.1	43.2	47.0	49.6
28	12.5	13.6	15.3	16.9	18.9	22.7	27.3	32.6	37.9	41.3	44.5	48.3	51.0
29	13.1	14.3	16.0	17.7	19.8	23.6	28.3	33.7	39.1	42.6	45.7	49.6	52.3
30	13.8	15.0	16.8	18.5	20.6	24.5	29.3	34.8	40.3	43.8	47.0	50.9	53.7

Table V Student's *t*-Distribution, Percentage Points

n	.60	.75	.90	.95	.975	.99	.995	.9995
1	.325	1.000	3.078	6.314	12.706	31.821	63.657	636.619
2	.289	.816	1.886	2.920	4.303	6.965	9.925	31.598
3	.277	.765	1.638	2.353	3.182	4.541	5.841	12.941
4	.271	.741	1.533	2.132	2.776	3.747	4.604	8.610
5	.267	.727	1.476	2.015	2.571	3.365	4.032	6.859
6	.265	.718	1.440	1.943	2.447	3.143	3.707	5.959
7	.263	.711	1.415	1.895	2.365	2.998	3.499	5.405
8	.262	.706	1.397	1.860	2.306	2.896	3.355	5.041
9	.261	.703	1.383	1.833	2.262	2.821	3.250	4.781
10	.260	.700	1.372	1.812	2.228	2.764	3.169	4.587
11	.260	.697	1.363	1.796	2.201	2.718	3.106	4.437
12	.259	.695	1.356	1.782	2.179	2.681	3.055	4.318
13	.259	.694	1.350	1.771	2.160	2.650	3.012	4.221
14	.258	.692	1.345	1.761	2.145	2.624	2.977	4.140
15	.258	.691	1.341	1.753	2.131	2.602	2.947	4.073
16	.258	.690	1.337	1.746	2.120	2.583	2.921	4.015
17	.257	.689	1.333	1.740	2.110	2.567	2.898	3.965
18	.257	.688	1.330	1.734	2.101	2.552	2.878	3.922
19	.257	.688	1.328	1.729	2.093	2.539	2.861	3.883
20	.257	.687	1.325	1.725	2.086	2.528	2.845	3.850
21	.257	.686	1.323	1.721	2.080	2.518	2.831	3.819
22	.256	.686	1.321	1.717	2.074	2.508	2.819	3.792
23	.256	.685	1.319	1.714	2.069	2.500	2.807	3.767
24	.256	.685	1.318	1.711	2.064	2.492	2.797	3.745
25	.256	.684	1.316	1.708	2.060	2.485	2.787	3.725
26	.256	.684	1.315	1.706	2.056	2.479	2.779	3.707
27	.256	.684	1.314	1.703	2.052	2.473	2.771	3.690
28	.256	.683	1.313	1.701	2.048	2.467	2.763	3.674
29	.256	.683	1.311	1.699	2.045	2.462	2.756	3.659
30	.256	.683	1.310	1.697	2.042	2.457	2.750	3.646
40	.255	.681	1.303	1.684	2.021	2.423	2.704	3.551
60	.254	.679	1.296	1.671	2.000	2.390	2.660	3.460
120	.254	.677	1.289	1.658	1.980	2.358	2.617	3.373
∞	.253	.674	1.282	1.645	1.960	2.326	2.576	3.291

Table VI Percentage Points, *F*-Distribution

.95

m \ n	1	2	3	4	5	6	7	8	9	10	12	15	20	24	30	40	60	120	∞
1	161.4	199.5	215.7	224.6	230.2	234.0	236.8	238.9	240.5	241.9	243.9	245.9	248.0	249.1	250.1	251.1	252.2	253.3	254.3
2	18.51	19.00	19.16	19.25	19.30	19.33	19.35	19.37	19.38	19.40	19.41	19.43	19.45	19.45	19.46	19.47	19.48	19.49	19.50
3	10.13	9.55	9.28	9.12	9.01	8.94	8.89	8.85	8.81	8.79	8.74	8.70	8.66	8.64	8.62	8.59	8.57	8.55	8.53
4	7.71	6.94	6.59	6.39	6.26	6.16	6.09	6.04	6.00	5.96	5.91	5.86	5.80	5.77	5.75	5.72	5.69	5.66	5.63
5	6.61	5.79	5.41	5.19	5.05	4.95	4.88	4.82	4.77	4.74	4.68	4.62	4.56	4.53	4.50	4.46	4.43	4.40	4.36
6	5.99	5.14	4.76	4.53	4.39	4.28	4.21	4.15	4.10	4.06	4.00	3.94	3.87	3.84	3.81	3.77	3.74	3.70	3.67
7	5.59	4.74	4.35	4.12	3.97	3.87	3.79	3.73	3.68	3.64	3.57	3.51	3.44	3.41	3.38	3.34	3.30	3.27	3.23
8	5.32	4.46	4.07	3.84	3.69	3.58	3.50	3.44	3.39	3.35	3.28	3.22	3.15	3.12	3.08	3.04	3.01	2.97	2.93
9	5.12	4.26	3.86	3.63	3.48	3.37	3.29	3.23	3.18	3.14	3.07	3.01	2.94	2.90	2.86	2.83	2.79	2.75	2.71
10	4.96	4.10	3.71	3.48	3.33	3.22	3.14	3.07	3.02	2.98	2.91	2.85	2.77	2.74	2.70	2.66	2.62	2.58	2.54
11	4.84	3.98	3.59	3.36	3.20	3.09	3.01	2.95	2.90	2.85	2.79	2.72	2.65	2.61	2.57	2.53	2.49	2.45	2.40
12	4.75	3.89	3.49	3.26	3.11	3.00	2.91	2.85	2.80	2.75	2.69	2.62	2.54	2.51	2.47	2.43	2.38	2.34	2.30
13	4.67	3.81	3.41	3.18	3.03	2.92	2.83	2.77	2.71	2.67	2.60	2.53	2.46	2.42	2.38	2.34	2.30	2.25	2.21
14	4.60	3.74	3.34	3.11	2.96	2.85	2.76	2.70	2.65	2.60	2.53	2.46	2.39	2.35	2.31	2.27	2.22	2.18	2.13
15	4.54	3.68	3.29	3.06	2.90	2.79	2.71	2.64	2.59	2.54	2.48	2.40	2.33	2.29	2.25	2.20	2.16	2.11	2.07
16	4.49	3.63	3.24	3.01	2.85	2.74	2.66	2.59	2.54	2.49	2.42	2.35	2.28	2.24	2.19	2.15	2.11	2.06	2.01
17	4.45	3.59	3.20	2.96	2.81	2.70	2.61	2.55	2.49	2.45	2.38	2.31	2.23	2.19	2.15	2.10	2.06	2.01	1.96
18	4.41	3.55	3.16	2.93	2.77	2.66	2.58	2.51	2.46	2.41	2.34	2.27	2.19	2.15	2.11	2.06	2.02	1.97	1.92
19	4.38	3.52	3.13	2.90	2.74	2.63	2.54	2.48	2.42	2.38	2.31	2.23	2.16	2.11	2.07	2.03	1.98	1.93	1.88
20	4.35	3.49	3.10	2.87	2.71	2.60	2.51	2.45	2.39	2.35	2.28	2.20	2.12	2.08	2.04	1.99	1.95	1.90	1.84
21	4.32	3.47	3.07	2.84	2.68	2.57	2.49	2.42	2.37	2.32	2.25	2.18	2.10	2.05	2.01	1.96	1.92	1.87	1.81
22	4.30	3.44	3.05	2.82	2.66	2.55	2.46	2.40	2.34	2.30	2.23	2.15	2.07	2.03	1.98	1.94	1.89	1.84	1.78
23	4.28	3.42	3.03	2.80	2.64	2.53	2.44	2.37	2.32	2.27	2.20	2.13	2.05	2.01	1.96	1.91	1.86	1.81	1.76
24	4.26	3.40	3.01	2.78	2.62	2.51	2.42	2.36	2.30	2.25	2.18	2.11	2.03	1.98	1.94	1.89	1.84	1.79	1.73
25	4.24	3.39	2.99	2.76	2.60	2.49	2.40	2.34	2.28	2.24	2.16	2.09	2.01	1.96	1.92	1.87	1.82	1.77	1.71
26	4.23	3.37	2.98	2.74	2.59	2.47	2.39	2.32	2.27	2.22	2.15	2.07	1.99	1.95	1.90	1.85	1.80	1.75	1.69
27	4.21	3.35	2.96	2.73	2.57	2.46	2.37	2.31	2.25	2.20	2.13	2.06	1.97	1.93	1.88	1.84	1.79	1.73	1.67
28	4.20	3.34	2.95	2.71	2.56	2.45	2.36	2.29	2.24	2.19	2.12	2.04	1.96	1.91	1.87	1.82	1.77	1.71	1.65
29	4.18	3.33	2.93	2.70	2.55	2.43	2.35	2.28	2.22	2.18	2.10	2.03	1.94	1.90	1.85	1.81	1.75	1.70	1.64
30	4.17	3.32	2.92	2.69	2.53	2.42	2.33	2.27	2.21	2.16	2.09	2.01	1.93	1.89	1.84	1.79	1.74	1.68	1.62
40	4.08	3.23	2.84	2.61	2.45	2.34	2.25	2.18	2.12	2.08	2.00	1.92	1.84	1.79	1.74	1.69	1.64	1.58	1.51
60	4.00	3.15	2.76	2.53	2.37	2.25	2.17	2.10	2.04	1.99	1.92	1.84	1.75	1.70	1.65	1.59	1.53	1.47	1.39
120	3.92	3.07	2.68	2.45	2.29	2.17	2.09	2.02	1.96	1.91	1.83	1.75	1.66	1.61	1.55	1.50	1.43	1.35	1.25
∞	3.84	3.00	2.60	2.37	2.21	2.10	2.01	1.94	1.88	1.83	1.75	1.67	1.57	1.52	1.46	1.39	1.32	1.22	1.00

Table VI (Continued)

.975

n \ m	1	2	3	4	5	6	7	8	9	10	12	15	20	24	30	40	60	120	∞
1	647.8	799.5	864.2	899.6	921.8	937.1	948.2	956.7	963.3	968.6	976.7	984.9	993.1	997.2	1001	1006	1010	1014	1018
2	38.51	39.00	39.17	39.25	39.30	39.33	39.36	39.37	39.39	39.40	39.41	39.43	39.45	39.46	39.46	39.47	39.48	39.49	39.50
3	17.44	16.04	15.44	15.10	14.88	14.73	14.62	14.54	14.47	14.42	14.34	14.25	14.17	14.12	14.08	14.04	13.99	13.95	13.90
4	12.22	10.65	9.98	9.60	9.36	9.20	9.07	8.98	8.90	8.84	8.75	8.66	8.56	8.51	8.46	8.41	8.36	8.31	8.26
5	10.01	8.43	7.76	7.39	7.15	6.98	6.85	6.76	6.68	6.62	6.52	6.43	6.33	6.28	6.23	6.18	6.12	6.07	6.02
6	8.81	7.26	6.60	6.23	5.99	5.82	5.70	5.60	5.52	5.46	5.37	5.27	5.17	5.12	5.07	5.01	4.96	4.90	4.85
7	8.07	6.54	5.89	5.52	5.29	5.12	4.99	4.90	4.82	4.76	4.67	4.57	4.47	4.42	4.36	4.31	4.25	4.20	4.14
8	7.57	6.06	5.42	5.05	4.82	4.65	4.53	4.43	4.36	4.30	4.20	4.10	4.00	3.95	3.89	3.84	3.78	3.73	3.67
9	7.21	5.71	5.08	4.72	4.48	4.32	4.20	4.10	4.03	3.96	3.87	3.77	3.67	3.61	3.56	3.51	3.45	3.39	3.33
10	6.94	5.46	4.83	4.47	4.24	4.07	3.95	3.85	3.78	3.72	3.62	3.52	3.42	3.37	3.31	3.26	3.20	3.14	3.08
11	6.72	5.26	4.63	4.28	4.04	3.88	3.76	3.66	3.59	3.53	3.43	3.33	3.23	3.17	3.12	3.06	3.00	2.94	2.88
12	6.55	5.10	4.47	4.12	3.89	3.73	3.61	3.51	3.44	3.37	3.28	3.18	3.07	3.02	2.96	2.91	2.85	2.79	2.72
13	6.41	4.97	4.35	4.00	3.77	3.60	3.48	3.39	3.31	3.25	3.15	3.05	2.95	2.89	2.84	2.78	2.72	2.66	2.60
14	6.30	4.86	4.24	3.89	3.66	3.50	3.38	3.29	3.21	3.15	3.05	2.95	2.84	2.79	2.73	2.67	2.61	2.55	2.49
15	6.20	4.77	4.15	3.80	3.58	3.41	3.29	3.20	3.12	3.06	2.96	2.86	2.76	2.70	2.64	2.59	2.52	2.46	2.40
16	6.12	4.69	4.08	3.73	3.50	3.34	3.22	3.12	3.05	2.99	2.89	2.79	2.68	2.63	2.57	2.51	2.45	2.38	2.32
17	6.04	4.62	4.01	3.66	3.44	3.28	3.16	3.06	2.98	2.92	2.82	2.72	2.62	2.56	2.50	2.44	2.38	2.32	2.25
18	5.98	4.56	3.95	3.61	3.38	3.22	3.10	3.01	2.93	2.87	2.77	2.67	2.56	2.50	2.44	2.38	2.32	2.26	2.19
19	5.92	4.51	3.90	3.56	3.33	3.17	3.05	2.96	2.88	2.82	2.72	2.62	2.51	2.45	2.39	2.33	2.27	2.20	2.13
20	5.87	4.46	3.86	3.51	3.29	3.13	3.01	2.91	2.84	2.77	2.68	2.57	2.46	2.41	2.35	2.29	2.22	2.16	2.09
21	5.83	4.42	3.82	3.48	3.25	3.09	2.97	2.87	2.80	2.73	2.64	2.53	2.42	2.37	2.31	2.25	2.18	2.11	2.04
22	5.79	4.38	3.78	3.44	3.22	3.05	2.93	2.84	2.76	2.70	2.60	2.50	2.39	2.33	2.27	2.21	2.14	2.08	2.00
23	5.75	4.35	3.75	3.41	3.18	3.02	2.90	2.81	2.73	2.67	2.57	2.47	2.36	2.30	2.24	2.18	2.11	2.04	1.97
24	5.72	4.32	3.72	3.38	3.15	2.99	2.87	2.78	2.70	2.64	2.54	2.44	2.33	2.27	2.21	2.15	2.08	2.01	1.94
25	5.69	4.29	3.69	3.35	3.13	2.97	2.85	2.75	2.68	2.61	2.51	2.41	2.30	2.24	2.18	2.12	2.05	1.98	1.91
26	5.66	4.27	3.67	3.33	3.10	2.94	2.82	2.73	2.65	2.59	2.49	2.39	2.28	2.22	2.16	2.09	2.03	1.95	1.88
27	5.63	4.24	3.65	3.31	3.08	2.92	2.80	2.71	2.63	2.57	2.47	2.36	2.25	2.19	2.13	2.07	2.00	1.93	1.85
28	5.61	4.22	3.63	3.29	3.06	2.90	2.78	2.69	2.61	2.55	2.45	2.34	2.23	2.17	2.11	2.05	1.98	1.91	1.83
29	5.59	4.20	3.61	3.27	3.04	2.88	2.76	2.67	2.59	2.53	2.43	2.32	2.21	2.15	2.09	2.03	1.96	1.89	1.81
30	5.57	4.18	3.59	3.25	3.03	2.87	2.75	2.65	2.57	2.51	2.41	2.31	2.20	2.14	2.07	2.01	1.94	1.87	1.79
40	5.42	4.05	3.46	3.13	2.90	2.74	2.62	2.53	2.45	2.39	2.29	2.18	2.07	2.01	1.94	1.88	1.80	1.72	1.64
60	5.29	3.93	3.34	3.01	2.79	2.63	2.51	2.41	2.33	2.27	2.17	2.06	1.94	1.88	1.82	1.74	1.67	1.58	1.48
120	5.15	3.80	3.23	2.89	2.67	2.52	2.39	2.30	2.22	2.16	2.05	1.94	1.82	1.76	1.69	1.61	1.53	1.43	1.31
∞	5.02	3.69	3.12	2.79	2.57	2.41	2.29	2.19	2.11	2.05	1.94	1.83	1.71	1.64	1.57	1.48	1.39	1.27	1.00

Table VI (Continued)

.99

n \ m	1	2	3	4	5	6	7	8	9	10	12	15	20	24	30	40	60	120	∞
1	4052	4999.5	5403	5625	5764	5859	5928	5982	6022	6056	6106	6157	6209	6235	6261	6287	6313	6339	6366
2	98.50	99.00	99.17	99.25	99.30	99.33	99.36	99.37	99.39	99.40	99.42	99.43	99.45	99.46	99.47	99.47	99.48	99.49	99.50
3	34.12	30.82	29.46	28.71	28.24	27.91	27.67	27.49	27.35	27.23	27.05	26.87	26.69	26.60	26.50	26.41	26.32	26.22	26.13
4	21.20	18.00	16.69	15.98	15.52	15.21	14.98	14.80	14.66	14.55	14.37	14.20	14.02	13.93	13.84	13.75	13.65	13.56	13.46
5	16.26	13.27	12.06	11.39	10.97	10.67	10.46	10.29	10.16	10.05	9.89	9.72	9.55	9.47	9.38	9.29	9.20	9.11	9.02
6	13.75	10.92	9.78	9.15	8.75	8.47	8.26	8.10	7.98	7.87	7.72	7.56	7.40	7.31	7.23	7.14	7.06	6.97	6.88
7	12.25	9.55	8.45	7.85	7.46	7.19	6.99	6.84	6.72	6.62	6.47	6.31	6.16	6.07	5.99	5.91	5.82	5.74	5.65
8	11.26	8.65	7.59	7.01	6.63	6.37	6.18	6.03	5.91	5.81	5.67	5.52	5.36	5.28	5.20	5.12	5.03	4.95	4.86
9	10.56	8.02	6.99	6.42	6.06	5.80	5.61	5.47	5.35	5.26	5.11	4.96	4.81	4.73	4.65	4.57	4.48	4.40	4.31
10	10.04	7.56	6.55	5.99	5.64	5.39	5.20	5.06	4.94	4.85	4.71	4.56	4.41	4.33	4.25	4.17	4.08	4.00	3.91
11	9.65	7.21	6.22	5.67	5.32	5.07	4.89	4.74	4.63	4.54	4.40	4.25	4.10	4.02	3.94	3.86	3.78	3.69	3.60
12	9.33	6.93	5.95	5.41	5.06	4.82	4.64	4.50	4.39	4.30	4.16	4.01	3.86	3.78	3.70	3.62	3.54	3.45	3.36
13	9.07	6.70	5.74	5.21	4.86	4.62	4.44	4.30	4.19	4.10	3.96	3.82	3.66	3.59	3.51	3.43	3.34	3.25	3.17
14	8.86	6.51	5.56	5.04	4.69	4.46	4.28	4.14	4.03	3.94	3.80	3.66	3.51	3.43	3.35	3.27	3.18	3.09	3.00
15	8.68	6.36	5.42	4.89	4.56	4.32	4.14	4.00	3.89	3.80	3.67	3.52	3.37	3.29	3.21	3.13	3.05	2.96	2.87
16	8.53	6.23	5.29	4.77	4.44	4.20	4.03	3.89	3.78	3.69	3.55	3.41	3.26	3.18	3.10	3.02	2.93	2.84	2.75
17	8.40	6.11	5.18	4.67	4.34	4.10	3.93	3.79	3.68	3.59	3.46	3.31	3.16	3.08	3.00	2.92	2.83	2.75	2.65
18	8.29	6.01	5.09	4.58	4.25	4.01	3.84	3.71	3.60	3.51	3.37	3.23	3.08	3.00	2.92	2.84	2.75	2.66	2.57
19	8.18	5.93	5.01	4.50	4.17	3.94	3.77	3.63	3.52	3.43	3.30	3.15	3.00	2.92	2.84	2.76	2.67	2.58	2.49
20	8.10	5.85	4.94	4.43	4.10	3.87	3.70	3.56	3.46	3.37	3.23	3.09	2.94	2.86	2.78	2.69	2.61	2.52	2.42
21	8.02	5.78	4.87	4.37	4.04	3.81	3.64	3.51	3.40	3.31	3.17	3.03	2.88	2.80	2.72	2.64	2.55	2.46	2.36
22	7.95	5.72	4.82	4.31	3.99	3.76	3.59	3.45	3.35	3.26	3.12	2.98	2.83	2.75	2.67	2.58	2.50	2.40	2.31
23	7.88	5.66	4.76	4.26	3.94	3.71	3.54	3.41	3.30	3.21	3.07	2.93	2.78	2.70	2.62	2.54	2.45	2.35	2.26
24	7.82	5.61	4.72	4.22	3.90	3.67	3.50	3.36	3.26	3.17	3.03	2.89	2.74	2.66	2.58	2.49	2.40	2.31	2.21
25	7.77	5.57	4.68	4.18	3.85	3.63	3.46	3.32	3.22	3.13	2.99	2.85	2.70	2.62	2.54	2.45	2.36	2.27	2.17
26	7.72	5.53	4.64	4.14	3.82	3.59	3.42	3.29	3.18	3.09	2.96	2.81	2.66	2.58	2.50	2.42	2.33	2.23	2.13
27	7.68	5.49	4.60	4.11	3.78	3.56	3.39	3.26	3.15	3.06	2.93	2.78	2.63	2.55	2.47	2.38	2.29	2.20	2.10
28	7.64	5.45	4.57	4.07	3.75	3.53	3.36	3.23	3.12	3.03	2.90	2.75	2.60	2.52	2.44	2.35	2.26	2.17	2.06
29	7.60	5.42	4.54	4.04	3.73	3.50	3.33	3.20	3.09	3.00	2.87	2.73	2.57	2.49	2.41	2.33	2.23	2.14	2.03
30	7.56	5.39	4.51	4.02	3.70	3.47	3.30	3.17	3.07	2.98	2.84	2.70	2.55	2.47	2.39	2.30	2.21	2.11	2.01
40	7.31	5.18	4.31	3.83	3.51	3.29	3.12	2.99	2.89	2.80	2.66	2.52	2.37	2.29	2.20	2.11	2.02	1.92	1.80
60	7.08	4.98	4.13	3.65	3.34	3.12	2.95	2.82	2.72	2.63	2.50	2.35	2.20	2.12	2.03	1.94	1.84	1.73	1.60
120	6.85	4.79	3.95	3.48	3.17	2.96	2.79	2.66	2.56	2.47	2.34	2.19	2.03	1.95	1.86	1.76	1.66	1.53	1.38
∞	6.63	4.61	3.78	3.32	3.02	2.80	2.64	2.51	2.41	2.32	2.18	2.04	1.88	1.79	1.70	1.59	1.47	1.32	1.00

Table VII   Random Numbers

95673	19543	88787	24433	65904	33551	74156	48065	38021	46480
74699	83908	01636	10964	49671	19779	51751	47927	17071	20272
70964	46447	35542	21274	23133	59964	15201	69740	97523	40853
43881	95137	67017	47898	03066	94299	83206	46872	92561	45597
17298	83805	44977	07091	60002	15728	46555	64622	07046	07096
94358	44992	10243	83217	00432	35232	43151	94761	72008	21788
84733	34751	53178	70870	11553	67251	72441	22549	63872	86555
27231	64931	47501	81075	34915	40922	75190	51229	71198	09826
74058	01486	97119	26790	78232	19669	67726	23840	31358	76326
02073	85738	88152	18375	67759	87106	28086	98353	07890	82849
13425	94256	75965	36623	87373	20687	42116	01786	03205	72838
69336	60414	41509	80611	26156	21857	51306	12693	55495	31218
99829	26683	48063	98001	36302	41358	74959	68716	92629	91443
65332	55608	21062	25375	65793	36247	15535	38840	22463	49372
12611	96265	72622	40180	44743	76773	31743	72756	32350	91186
22350	84735	71650	27600	36726	37979	68523	79521	47438	24225
70550	29182	49634	37272	21755	11056	05818	30061	52264	47081
40264	47942	97587	28802	13850	51816	98508	04436	30403	40009
40820	58402	47756	42273	00206	19714	66010	12570	51211	91832
74458	35792	17575	91216	99362	29454	66163	11517	65206	59509
46445	44109	64962	08813	66088	26043	19323	96816	68625	88192
26825	53912	12377	65681	99184	72798	12001	76178	82032	62195
37787	17836	22366	09090	15502	64203	79524	07225	29282	80327
01850	09763	81675	18728	41239	78901	75171	92338	74263	71451
45026	93368	66406	40339	78710	24857	03402	44700	25633	85993
05365	91765	28390	57363	69137	90608	87436	50299	36386	42989
92200	48616	30259	57543	54630	11377	36038	03942	95074	40907
26920	64882	73326	11730	75555	71631	56031	39323	62212	20701
76616	12670	56849	78288	22511	06086	57123	39359	61089	81535
45672	12722	94260	91843	97630	50406	36517	43575	01428	10835
04799	31526	45819	46640	94612	33026	75223	05604	76273	64496
41793	27126	24953	87463	01581	92186	52904	70915	02432	53242
46757	36369	96190	05797	58470	16021	38645	81241	52793	97916
55272	19065	49227	23226	38656	11896	21525	97572	47983	50040
05466	33569	88498	35202	67652	02301	30448	72718	19724	41239

49251	81598	72663	10875	53196	90907	65415	45497	10395	29588
71016	94188	14369	43123	79152	73727	71719	50386	15205	41576
34010	87907	66345	89062	13092	38945	97165	11689	09036	71932
19562	90990	30889	88923	20913	33536	92467	33345	65474	13767
12974	51394	57811	38840	15122	23827	42564	80293	42518	01842
13568	68468	56238	25776	87214	56388	16245	39322	41467	43511
80928	90362	67884	49942	20573	20679	17218	44961	49882	32796
62212	81543	93254	78969	18973	66963	70019	53685	26709	84100
27143	68153	01084	45781	56655	98385	49705	05481	40592	49747
24397	56650	66567	80216	76267	24651	00794	75032	10105	49700
30157	99348	80868	35088	39480	72015	95042	93319	09869	04823
15161	30403	04924	93707	08514	50133	15744	15517	13607	92836
73005	64826	79875	88689	91319	93111	07487	51463	21486	59773
38834	40095	04605	96708	90133	07000	09747	71216	65398	90824
97243	94836	17184	12353	97593	18647	26251	26321	44325	34845
31692	51607	89056	74472	91284	20263	16039	94491	33767	73915
82997	58320	04852	52595	95514	56543	06636	61291	67504	57205
05043	40582	46051	60261	04996	82256	47375	87507	05112	88489
75781	38768	70475	00601	18378	32077	36523	30843	07057	78326
21033	15175	30741	45814	92222	16704	00197	51267	33224	40276
99092	60991	12571	71753	65214	33885	82939	50723	88987	69761
07204	93373	85112	29610	30375	64836	18459	08235	67650	72390
88859	97254	07771	21393	64657	42013	12753	03028	24224	24918
30497	91407	72900	15699	58653	38063	25072	48698	88083	48040
09726	18075	45852	54968	43743	82050	78412	79456	95032	10984
95330	01985	24128	60514	42539	91907	25694	37097	39566	24043
09760	32388	05601	49923	66126	54146	67213	52234	48381	89442
01534	81967	15337	95831	84643	40792	47562	95494	62087	18064
11234	59350	48368	57195	36287	03046	87136	36057	93913	70080
71056	48762	80221	59683	27504	21121	94711	11807	80882	48359
34208	05374	60304	43178	97247	24875	26259	67622	14657	80354
47132	62839	82198	92445	60650	76219	02772	48651	66449	89213
55685	93302	43019	45861	95493	16106	12783	32748	83533	25440
17803	18184	10510	27159	83008	20544	41665	99439	70606	28974
55045	17219	66737	59080	78489	12626	60661	53733	70062	14289

Table VII (Continued)

01923	33647	98442	59293	83318	33425	76412	87062	01295	11083
07202	76476	71888	54845	17468	41964	68694	59662	55905	26898
68825	68242	95750	11033	58634	78411	08523	19313	29327	47526
68525	06496	17446	41378	32368	82019	66101	56733	43308	82641
80819	33515	97373	43064	16221	99697	37951	07947	12935	49391
64200	96929	26044	49283	56545	67200	21325	85056	51345	06309
30156	29121	75874	42399	41121	90643	19585	06364	47203	19679
50467	14282	89098	66717	14753	73356	47781	34156	82842	00121
53764	83212	26675	64184	64455	29023	03181	13674	08838	83829
81727	35572	95469	36825	81882	95083	68323	14965	34166	32351
30807	55558	96026	97398	21723	86560	52617	07771	61886	48234
75104	23682	78756	72728	85940	57290	75507	78715	01426	02310
06180	62724	36835	80288	25075	32609	33312	21348	87710	55457
22098	34834	66117	36252	82717	50585	43639	79999	07414	84003
13173	64783	20984	11929	18849	26211	77375	49561	96747	67007
75273	36108	55265	15653	82270	99216	27805	60088	06056	97377
89849	65756	44454	04602	14292	74458	57777	35934	05160	26359
91108	43562	18883	16569	49599	73871	67101	12054	56492	15981
51843	01542	17881	12954	94913	39583	94969	61146	35907	72184
02644	23564	85464	62947	92571	89377	85004	84654	20465	86212
38608	83374	74032	62183	08740	05279	30455	31032	71512	16476
43164	28909	88624	14992	85359	10193	32491	14769	63694	92640
80933	52950	45646	36636	05085	28053	27596	54873	68476	65823
67690	96766	69250	19344	47855	43489	77479	62418	54079	40069
68579	17014	25362	15114	30982	27250	29052	71115	83369	46776
46353	39733	44677	50133	26623	15979	10651	04263	34087	67005
30039	09532	52215	09164	20930	88230	43403	63230	83525	93550
89200	92772	42195	91634	39272	46462	76835	27755	03151	75692
58118	57942	14807	68214	76093	47484	24468	91764	52907	16675
97230	33027	70166	43232	98802	70715	30216	35586	18909	79658
75569	40225	17892	88888	01568	90289	82347	74237	06083	29023
26618	17921	46235	08527	81088	71103	28867	48898	11765	20589
21391	95668	15669	06585	07408	94593	74192	16937	30915	69683
27236	78121	37636	37596	36916	19026	83895	37621	38781	34507
64567	03291	55990	07279	31596	50559	31313	18888	66235	43465

50375	34061	52597	59665	23098	51164	05026	79626	86395	56451
88089	03283	86277	86889	03735	67955	43509	24989	12554	06435
59112	26102	34749	75692	11316	90508	12603	06053	68353	87629
12474	66011	25612	35402	82523	21045	38566	94478	44885	62630
90484	58184	14388	54147	13021	61538	96744	38435	42161	30496
57547	52673	14725	25009	59069	42763	81861	77619	24130	33579
39797	27602	13527	14069	47413	82374	56680	53611	56402	01958
42855	74811	14789	73490	90283	75512	75691	32342	94877	94230
16868	45219	73791	60781	34715	92892	55789	02402	08345	82930
45414	56595	48272	85249	24116	34688	99662	26808	36151	26688
98889	26797	77787	18122	66596	81704	76837	25881	80541	88508
14822	73860	76266	46622	96965	24404	41893	04333	07541	73091
33845	48523	11078	43250	07836	15664	69494	09751	82848	28256
55067	40413	16663	28913	11189	02184	91804	98082	35370	38476
86131	39325	62226	40755	72628	67037	44087	85304	26626	68424
28756	87618	42343	61543	46943	48251	61867	37579	00781	98703
76657	98868	55414	37373	71035	62107	97240	39343	65010	94309
57193	01772	93454	07440	11696	64660	86262	63353	41594	58572
06385	29257	96175	17875	32126	05607	87970	87040	41642	11150
89797	95386	22344	24822	35493	40305	20909	24040	68790	09095
84006	34817	52179	59039	94168	48610	04742	83258	23061	25656
07986	72678	13509	99214	45819	18540	83521	92420	16554	26480
84387	22967	93863	37444	86620	70090	98830	84589	53018	68686
87840	74766	99636	61427	83228	19807	42826	20896	35877	67563
88675	65449	55896	39952	29258	17567	57878	83371	09582	69821
12054	98757	48533	80829	29123	75640	29309	23056	99170	67423
87001	44944	55755	35269	76847	64319	06647	30186	91536	83992
61875	67634	51162	29891	11594	56073	67275	47964	49064	86276
44660	01229	11766	19065	79045	84239	90783	47725	74140	84993
48611	36739	00713	96714	75088	41171	19081	58754	17009	47112
59552	78041	56061	66475	88393	29730	27370	48476	06587	88242
62935	52864	64551	56438	02686	99746	86829	00910	38961	89416
21507	77663	52403	41330	71120	47898	67246	80323	64810	83716
80891	52275	26110	60222	94253	07891	67753	19891	51237	11708
26421	60075	91863	75026	36635	67147	95486	56401	59796	41944

Table VII (*Continued*)

03021	43361	52536	10373	07104	86742	72528	64417	32133	66567
42105	81529	16306	67532	94275	45562	85060	44953	50504	78212
19500	57358	62402	42875	94389	66948	33232	60277	31451	95468
01217	39948	55572	50491	04927	99659	00034	25870	72949	97012
03477	74227	05252	37305	81236	22503	96029	55651	27214	40330
09198	25062	23367	32032	63893	91226	74337	44930	39311	83856
53786	97117	65663	66148	02659	67344	23653	92524	38192	86917
72718	84299	32088	74273	83368	90479	32225	93842	73824	26769
24912	20107	14564	42788	05152	65014	08765	03394	49626	03457
72348	93453	71317	91259	36486	76705	80510	72505	72285	00509
61721	03748	38841	69166	01752	51219	78247	70734	10493	25093
98963	01854	20841	52865	31500	09283	07363	71612	77927	61691
23932	43062	33464	15924	24387	75728	31074	53262	70995	42071
15671	70400	88477	85686	91199	11089	56188	53946	30159	59284
73782	89043	22541	79933	57728	87000	97222	97795	16128	55709
83102	95333	72207	71472	52329	37785	52626	28600	13567	45016
46384	82071	00261	22133	43136	74270	98330	82464	44016	03638
58386	47279	85564	56751	02884	47595	53203	37306	69859	22351
31677	34913	92024	95851	73195	55542	52537	17487	95617	64009
90341	27029	35388	10015	54091	25240	28945	51638	55394	00151
53694	56321	51780	85766	24303	91130	08054	93015	83580	78969
14321	34850	57577	81552	88279	30410	36607	00387	97499	10178
05124	43526	80602	08431	39622	90293	58612	37471	28521	98765
04292	67049	37226	77648	62162	92421	33333	76923	38256	78580
09273	84395	97382	42955	25730	55790	91405	34801	64616	44550
15183	69840	80256	57175	06706	71873	79425	03789	29375	04018
88721	47483	15683	85491	46116	03661	47229	63104	29262	61523
43140	44368	16104	46863	53687	53452	78524	02933	56326	83043
91908	49963	85656	41706	18596	42334	71506	69013	64856	48793
96335	20856	03851	47753	99980	24449	39005	58497	03063	68636
87866	14100	55056	01560	58002	26742	43174	30123	39842	37466
78236	79903	57595	30611	49302	87218	77848	38069	70385	36871
36780	08641	20571	13793	09612	27066	20227	50363	44940	38920
18366	06852	71154	94608	70392	04956	21448	09296	19308	61598
45576	91644	74141	23815	82267	28271	81891	31523	20613	46854

89567	21054	47503	25472	75618	13071	81465	31184	23883	03637
23892	65090	87940	33934	31962	59572	77446	83552	31265	38746
31487	44209	43610	01733	05897	66428	34733	44085	58254	37004
46686	41250	54473	98489	59263	22269	45986	60634	51370	17736
71833	75595	31885	08056	46018	99133	90402	28771	43275	34353
72439	98123	65087	40454	51519	56816	30889	15487	81497	75570
77297	37843	69199	76739	18081	58820	85890	87243	84176	35053
33099	33358	55914	34239	47499	14384	13438	39943	86175	28198
95151	61859	09679	83413	91905	36927	18890	57426	28948	02750
74862	50549	49476	62215	87910	43389	28230	64566	36542	10223
10078	82651	08775	29037	36499	33125	98741	71329	21150	07281
56261	21774	62041	84210	04967	23023	88794	40799	96828	01518
06678	84070	69595	48871	86757	92546	81373	54006	67649	56869
34845	41877	95482	84017	32677	11746	69577	07078	12507	58072
80515	53345	11650	03594	64422	97801	26586	82586	54104	26852
32368	38947	68635	82913	60872	11703	40814	67598	28132	64838
80379	42553	16780	91837	61799	70339	45738	71862	56776	15540
72433	90816	97471	61115	17983	26639	45691	24106	42228	00493
87540	87245	61508	31508	36613	00978	03175	34398	14783	88859
92059	84967	49256	33531	79632	66030	15062	01204	23499	32555
12060	41495	77722	23111	04569	70487	87652	27892	93641	51367
39434	85821	57004	00302	24395	67592	50929	27203	56495	44117
49351	81841	20972	12359	52778	14954	28085	90846	75750	72350
87831	38461	72475	26174	54460	45191	88378	78906	57966	61603
25764	17666	45682	74833	92963	48400	13993	70089	69582	82222

Table VIII   Percentage Points, Distribution of the Correlation Coefficient, when $\rho = 0$

$\alpha=$	0.05	0.025	0.01	0.005	0.0025	0.0005	v	$\alpha=$	0.05	0.025	0.01	0.005	0.0025	0.0005
v														
$2\alpha=$	0.1	0.05	0.02	0.01	0.005	0.001		$2\alpha=$	0.1	0.05	0.02	0.01	0.005	0.001
1	0.9877	0.9^2692	0.9^3507	0.9^3877	0.9^4692	0.9^5877	16		0.400	0.468	0.543	0.590	0.631	0.708
2	.9000	.9500	.9800	$.9^2000$	$.9^2500$	$.9^3000$	17		.389	.456	.529	.575	.616	.693
3	.805	.878	.9343	.9587	.9740	$.9^2114$	18		.378	.444	.516	.561	.602	.679
4	.729	.811	.882	.9172	.9417	.9741	19		.369	.433	.503	.549	.589	.665
5	.669	.754	.833	.875	.9056	.9509	20		.360	.423	.492	.537	.576	.652
6	0.621	0.707	0.789	0.834	0.870	0.9249	25		0.323	0.381	0.445	0.487	0.524	0.597
7	.582	.666	.750	.798	.836	.898	30		.296	.349	.409	.449	.484	.554
8	.549	.632	.715	.765	.805	.872	35		.275	.325	.381	.418	.452	.519
9	.521	.602	.685	.735	.776	.847	40		.257	.304	.358	.393	.425	.490
10	.497	.576	.558	.708	.750	.823	45		.243	.288	.338	.372	.403	.465
11	0.476	0.553	0.634	0.684	0.726	0.801	50		0.231	0.273	0.322	0.354	0.354	0.443
12	.457	.532	.612	.661	.703	.780	60		.211	.250	.295	.325	.352	.408
13	.441	.514	.592	.641	.683	.760	70		.195	.232	.274	.302	.327	.380
14	.426	.497	.574	.623	.664	.742	80		.183	.217	.257	.283·	.307	.357
15	.412	.482	.558	.606	.647	.725	90		.173	.205	.242	.267	.290	.338
							100		.164	.195	.230	.254	.276	.321

Note: α is the upper-tail area of the distribution of r appropriate for use in a single-tail test. For a two-tail test, 2α must be used. If r is calculated from n paired observations, enter the table with $v = n - 2$.

Table IX Upper 5 Percent Points of the Studentized Range

The entries are $q_{.05}$, where $P(q < q_{.05}) = .95$

v \ n	2	3	4	5	6	7	8	9	10
1	17.97	26.98	32.82	37.08	40.41	43.12	45.40	47.36	49.07
2	6.08	8.33	9.80	10.88	11.74	12.44	13.03	13.54	13.99
3	4.50	5.91	6.82	7.50	8.04	8.48	8.85	9.18	9.46
4	3.93	5.04	5.76	5.29	6.71	7.05	7.35	7.60	7.83
5	3.64	4.60	5.22	5.67	6.03	6.33	6.58	6.80	6.99
6	3.46	4.34	4.90	5.30	5.63	5.90	6.12	6.32	6.49
7	3.34	4.16	4.68	5.06	5.36	5.61	5.82	6.00	6.16
8	3.26	4.04	4.53	4.89	5.17	5.40	5.60	5.77	5.92
9	3.20	3.95	4.41	4.76	5.02	5.24	5.43	5.59	5.74
10	3.15	3.88	4.33	4.65	4.91	5.12	5.30	5.46	5.60
11	3.11	3.82	4.26	4.57	4.82	5.03	5.20	5.25	5.49
12	3.08	3.77	4.20	4.51	4.75	4.95	5.12	5.27	5.39
13	3.06	3.73	4.15	4.45	4.69	4.88	5.05	5.19	5.32
14	3.03	3.70	4.11	4.41	4.64	4.83	4.99	5.13	5.25
15	3.01	3.67	4.08	4.37	4.59	4.78	4.94	5.08	5.20
16	3.00	3.65	4.05	4.33	4.56	4.74	4.90	5.03	5.15
17	2.98	3.63	4.02	4.30	4.52	4.70	4.86	4.99	5.11
18	2.97	3.61	4.00	4.28	4.49	4.67	4.82	4.96	5.07
19	2.96	3.59	3.98	4.25	4.47	4.65	4.79	4.92	5.04
20	2.95	3.58	3.96	4.23	4.45	4.62	4.77	490	5.01
24	2.92	3.53	3.90	4.17	4.37	4.54	4.68	4.81	4.92
30	2.89	3.49	3.85	4.10	4.30	4.46	4.60	4.72	4.82
40	2.86	3.44	3.79	4.04	4.23	4.39	4.52	4.63	4.73
60	2.86	3.40	3.74	3.98	4.16	4.31	4.44	4.55	4.65
120	2.80	3.36	3.68	3.92	4.10	4.24	4.36	4.47	4.56
∞	2.77	3.31	3.63	3.86	4.03	4.17	4.29	4.39	4.47

v \ n	11	12	13	14	15	16	17	18	19	20
1	50.59	51.96	53.20	54.33	55.36	56.32	57.22	58.04	58.83	59.56
2	14.39	14.75	15.08	15.38	15.65	15.91	16.14	16.37	16.57	16.77
3	9.72	9.95	10.15	10.35	10.53	10.69	10.84	10.98	11.11	11.24
4	8.03	8.21	8.37	8.52	8.66	8.79	8.91	9.03	9.13	9.23
5	7.17	7.32	7.47	7.60	7.72	7.83	7.93	8.03	8.12	8.21
6	6.65	6.79	6.92	7.03	7.14	7.24	7.34	7.43	7.51	7.59
7	6.30	6.43	6.55	6.66	6.76	6.85	6.94	7.02	7.10	7.17
8	6.05	6.18	6.29	6.39	6.48	6.57	6.65	6.73	6.80	6.87
9	5.87	5.98	6.09	6.19	6.28	6.36	6.44	6.51	6.58	6.64
10	5.72	5.83	5.93	6.03	6.11	6.19	6.27	6.34	6.40	6.47
11	5.61	5.71	5.81	5.90	5.98	6.06	6.13	6.20	6.27	6.33
12	5.51	5.61	5.71	5.80	5.88	5.95	6.02	6.09	6.15	6.21
13	5.43	5.53	5.63	5.71	5.79	5.86	5.93	5.99	6.05.	6.11
14	5.36	5.46	5.55	5.64	5.71	5.79	5.85	5.91	5.97	6.03
15	5.31	5.40	5.49	5.57	5.65	5.72	5.78	5.85	5.90	5.96
16	5.26	5.35	5.44	5.52	5.59	5.66	5.73	5.79	5.84	5.90
17	5.21	5.31	5.39	5.47	5.54	5.61	5.67	5.73	5.79	5.84
18	5.17	5.27	5.35	5.43	5.50	5.57	5.63	5.69	5.74	5.79
19	5.14	5.23	5.31	5.39	5.46	5.53	5.59	5.65	5.70	5.75
20	5.11	5.20	5.28	5.36	5.43	5.49	5.55	5.61	5.66	5.71
24	5.01	5.10	5.18	5.25	5.32	5.38	5.44	5.49	5.55	5.59
30	4.92	5.00	5.08	5.15	5.21	5.27	5.33	5.38	5.43	5.47
40	4.82	4.90	4.98	5.04	5.11	5.16	5.22	5.27	5.31	5.36
60	4.73	4.81	4.88	4.94	5.00	5.06	5.11	5.15	5.20	5.24
120	4.64	4.71	4.78	4.84	4.90	4.95	5.00	5.04	5.09	5.13
∞	4.55	4.62	4.68	4.74	4.80	4.85	4.89	4.93	4.97	5.01

Table IX (*Continued*) **Upper 1 Percent Points of the Studentized Range**

The entries are $q_{.01}$, where $P(q < q_{.01}) = .99$

v \ n	2	3	4	5	6	7	8	9	10
1	90.03	135.0	164.3	185.6	202.2	215.2	227.2	237.0	245.6
2	14.04	19.02	22.29	24.72	26.63	28.20	29.53	30.68	31.69
3	8.26	10.62	12.17	13.23	14.24	15.00	154	16.20	16.69
4	6.51	8.12	9.17	9.96	10.58	11.10	11.55	11.93	12.27
5	5.70	6.98	7.80	8.42	8.91	9.32	9.67	9.97	10.24
6	5.24	6.33	7.03	7.56	7.97	8.32	8.61	8.87	9.10
7	4.95	5.92	6.54	7.01	7.37	7.68	7.94	8.17	8.37
8	4.75	5.64	6.20	6.62	6.96	7.24	7.47	7.68	7.86
9	4.60	5.43	5.96	6.35	6.66	6.91	7.13	7.33	7.49
10	4.48	5.27	5.77	6.14	6.43	6.67	6.87	7.05	7.21
11	4.39	5.15	5.62	5.97	6.25	6.48	6.67	6.84	6.99
12	4.32	5.05	5.50	5.84	6.10	6.32	6.51	6.67	6.81
13	4.26	496	5.40	5.73	5.98	6.19	6.37	6.53	6.67
14	4.21	4.89	5.32	5.63	5.88	6.08	6.26	6.41	6.54
15	4.17	4.84	5.25	5.56	5.80	5.99	6.16	6.31	6.44
16	4.13	4.79	5.19	5.49	5.72	5.92	6.08	6.22	6.35
17	4.10	4.74	5.14	5.43	5.66	5.85	6.01	6.15	6.27
18	4.07	4.70	5.09	5.38	5.60	5.79	5.94	6.08	6.20
19	4.05	4.67	5.05	5.33	5.55	5.73	5.89	6.02	6.14
20	4.02	4.64	5.02	5.29	5.51	5.69	5.84	5.97	6.09
24	3.96	4.55	4.91	5.17	5.37	5.54	5.69	5.81	5.92
30	3.89	4.45	4.80	5.05	5.24	5.40	5.54	5.65	5.76
40	3.82	4.37	4.70	4.93	5.11	5.26	5.39	5.50	5.60
60	3.76	4.28	4.59	4.82	4.99	5.13	5.25	5.36	5.45
120	3.70	4.20	4.50	4.71	4.87	5.01	5.12	5.21	5.30
∞	3.64	4.12	4.40	4.60	4.76	4.88	4.99	5.08	5.16

v \ n	11	12	13	14	15	16	17	18	19	20
1	253.2	260.0	266.2	271.8	277.0	281.8	286.3	290.4	294.3	298.0
2	35.59	33.40	34.13	34.81	35.43	36.00	36.53	37.03	37.50	37.95
3	17.13	17.53	17.89	18.22	18.52	18.81	19.07	19.32	19.55	19.77
4	12.57	12.84	13.09	13.32	13.53	13.73	13.91	14.08	14.24	14.40
5	10.48	10.70	10.89	11.08	11.24	11.40	11.55	11.68	11.81	11.93
6	9.30	9.48	9.65	9.81	9.95	10.08	10.21	10.32	10.43	10.54
7	8.55	8.71	8.86	9.00	9.12	9.24	9.35	9.46	9.55	9.65
8	8.03	8.18	8.31	8.44	8.55	8.66	8.76	8.85	8.94	9.03
9	7.65	7.78	7.91	8.03	8.13	8.23	8.33	8.41	8.49	8.57
10	7.36	7.49	7.60	7.71	7.81	7.91	7.99	8.08	8.15	8.23
11	7.13	7.25	7.36	7.46	7.56	7.65	7.73	7.81	7.88	7.95
12	6.94	7.06	7.17	7.26	7.36	7.44	7.52	7.59	7.66	7.73
13	6.79	6.90	7.01	7.10	7.19	7.27	7.35	7.42	7.48	7.55
14	6.66	6.77	6.87	6.96	7.05	7.13	7.20	7.27	7.33	7.39
15	6.55	6.66	6.76	6.84	6.93	7.00	7.07	7.14	7.20	7.26
16	6.46	6.56	6.66	6.74	6.82	6.90	6.97	7.03	7.09	7.15
17	6.38	6.48	6.57	6.66	6.73	6.81	6.87	6.94	7.00	7.05
18	6.31	6.41	6.50	6.58	6.65	6.73	6.79	6.85	6.91	6.97
19	6.25	6.34	6.43	6.51	6.58	6.65	6.72	6.78	6.84	6.89
20	6.19	6.28	6.37	6.45	6.52	6.59	6.65	6.71	6.77	6.82
24	6.02	6.11	6.19	6.26	6.33	6.39	6.45	6.51	6.56	6.61
30	5.85	5.93	6.01	6.08	6.14	6.20	6.26	6.31	6.36	6.41
40	5.69	5.76	5.83	5.90	5.96	6.02	6.07	6.12	6.16	6.21
60	5.53	5.60	5.67	5.73	5.78	5.84	5.89	5.93	5.97	6.01
120	5.37	5.44	5.50	5.56	5.61	5.66	5.71	5.75	5.79	5.83
∞	5.23	5.29	5.35	5.40	5.45	5.49	5.54	5.57	5.61	5.65

Table X   Critical Values of Spearman's Rank Correlation Coefficient

n	$\gamma = 0.10$	$\gamma = 0.05$	$\gamma = 0.02$	$\gamma = 0.01$
5	0.900	—	—	—
6	0.829	0.886	0.943	—
7	0.714	0.786	0.893	—
8	0.643	0.738	0.833	0.881
9	0.600	0.683	0.783	0.833
10	0.564	0.648	0.745	0.794
11	0.523	0.623	0.736	0.818
12	0.497	0.591	0.703	0.780
13	0.475	0.566	0.673	0.745
14	0.457	0.545	0.646	0.716
15	0.441	0.525	0.623	0.689
16	0.425	0.507	0.601	0.666
17	0.412	0.490	0.582	0.645
18	0.399	0.476	0.564	0.625
19	0.388	0.462	0.549	0.608
20	0.377	0.450	0.534	0.591
21	0.368	0.438	0.521	0.576
22	0.359	0.428	0.508	0.562
23	0.351	0.418	0.496	0.549
24	0.343	0.409	0.485	0.537
25	0.336	0.400	0.475	0.526
26	0.329	0.392	0.465	0.515
27	0.323	0.385	0.456	0.505
28	0.317	0.377	0.448	0.496
29	0.311	0.370	0.440	0.487
30	0.305	0.364	0.432	0.478

Table XI   _d_-Factors for Sign Test and Confidence Intervals for the Median

n	d	γ	α"	α'	n	d	γ	α"	α'
3	1	.750	.250	.125	20	4	.997	.003	.001
4	1	.875	.125	.062		5	.988	.012	.006
5	1	.938	.062	.031		6	.959	.041	.021
6	1	.969	.031	.016		7	.885	.115	.058
	2	.781	.219	.109	21	5	.993	.007	.004
7	1	.984	.016	.008		6	.973	.027	.013
	2	.875	.125	.063		7	.922	.078	.039
8	1	.992	.008	.004		8	.811	.189	.095
	2	.930	.070	.035	22	5	.996	.004	.002
	3	.711	.289	.145		6	.983	.017	.008
9	1	.996	.004	.002		7	.948	.052	.026
	2	.961	.039	.020		8	.866	.134	.067
	3	.820	.180	.090	23	5	.997	.003	.001
10	1	.998	.002	.001		6	.989	.011	.005
	2	.979	.021	.011		7	.965	.035	.017
	3	.891	.109	.055		8	.907	.093	.047
11	1	.999	.001	.000		9	.790	.210	.105
	2	.998	.012	.006	24	6	.993	.007	.003
	3	.935	.065	.033		7	.977	.023	.011
	4	.773	.227	.113		8	.936	.064	.032
12	2	.994	.006	.003		9	.848	.152	.076
	3	.961	.039	.019	25	6	.996	.004	.002
	4	.854	.146	.073		7	.985	.015	.007
13	2	.997	.003	.002		8	.957	.043	.022
	3	.978	.022	.011		9	.892	.108	.054
	4	.908	.092	.046	26	7	.991	.009	.005
	5	.733	.267	.133		8	.971	.029	.014
14	2	.998	.002	.001		9	.924	.076	.038
	3	.987	.013	.006		10	.831	.169	.084
	4	.943	.057	.029	27	7	.994	.006	.003
	5	.820	.180	.090		8	.981	.019	.010
15	3	.993	.007	.004		9	.948	.052	.026
	4	.965	.035	.018		10	.878	.122	.061
	5	.882	.118	.059	28	7	.996	.004	.002
16	3	.996	.004	.002		8	.987	.013	.006
	4	.979	.021	.011		9	.964	.036	.018
	5	.923	.077	.038		10	.913	.087	.044
	6	.790	.210	.105		11	.815	.185	.092
17	3	.998	.002	.001	29	8	.992	.008	.004
	4	.987	.013	.006		9	.976	.024	.012
	5	.951	.049	.025		10	.939	.061	.031
	6	.857	.143	.072		11	.864	.136	.068
18	4	.992	.008	.004	30	8	.995	.005	.003
	5	.969	.031	.015		9	.984	.016	.008
	6	.904	.096	.048		10	.957	.043	.021
	7	.762	.238	.119		11	.901	.099	.049
19	4	.996	.004	.002		12	.800	.200	.100
	5	.981	.019	.010					
	6	.936	.064	.032					
	7	.833	.167	.084					

Table XI (*Continued*)

n	d	γ	α″	α′	n	d	γ	α″	α′
31	8	.997	.003	.002	41	12	.996	.004	.002
	9	.989	.011	.005		13	.988	.012	.006
	10	.971	.029	.015		14	.972	.028	.014
	11	.929	.071	.035		15	.940	.060	.030
	12	.850	.150	.075		16	.883	.117	.059
32	9	.993	.007	.004	42	13	.992	.008	.004
	10	.980	.020	.010		14	.980	.020	.010
	11	.950	.050	.025		15	.956	.044	.022
	12	.890	.110	.055		16	.912	.088	.044
33	9	.995	.005	.002		17	.836	.164	.082
	10	.986	.014	.007	43	13	.995	.005	.003
	11	.965	.035	.018		14	.986	.014	.007
	12	.920	.080	.040		15	.968	.032	.016
	13	.837	.163	.081		16	.934	.066	.033
34	10	.991	.009	.005		17	.874	.126	.063
	11	.976	.024	.012	44	14	.990	.010	.005
	12	.942	.058	.029		15	.977	.023	.011
	13	.879	.121	.061		16	.951	.049	.024
35	10	.994	.006	.003		17	.904	.096	.048
	11	.983	.017	.008		18	.826	.174	.087
	12	.959	.041	.020	45	14	.993	.007	.003
	13	.910	.090	.045		15	.984	.016	.008
	14	.825	.175	.088		16	.964	.036	.018
36	10	.996	.004	.002		17	.928	.072	.036
	11	.989	.011	.006		18	.865	.135	.068
	12	.971	.029	.014	46	14	.995	.005	.002
	13	.935	.065	.033		15	.989	.011	.006
	14	.868	.132	.066		16	.974	.026	.013
37	11	.992	.008	.004		17	.946	.054	.027
	12	.980	.020	.010		18	.896	.104	.052
	13	.953	.047	.024	47	15	.992	.008	.004
	14	.901	.099	.049		16	.981	.019	.009
	15	.812	.188	.094		17	.960	.040	.020
38	11	.995	.005	.003		18	.921	.079	.039
	12	.986	.014	.007		19	.956	.144	.072
	13	.966	.034	.017	48	15	.994	.006	.003
	14	.927	.073	.036		16	.987	.013	.007
	15	.857	.143	.072		17	.971	.029	.015
39	12	.991	.009	.005		18	.941	.059	.030
	13	.976	.024	.012		19	.889	.111	.056
	14	.947	.053	.027	49	16	.991	.009	.005
	15	.892	.108	.054		17	.979	.021	.011
40	12	.994	.006	.003		18	.956	.044	.022
	13	.983	.017	.008		19	.915	.085	.043
	14	.962	.038	.019		20	.848	.152	.076
	15	.919	.081	.040	50	16	.993	.007	.003
	16	.846	.154	.077		17	.985	.015	.008
						18	.967	.033	.016
						19	.935	.065	.032
						20	.881	.119	.059

γ = confidence coefficient
$\alpha' = \frac{1}{2}(1 - \gamma)$ = one-sided significance level
$\alpha'' = 2\alpha' = 1 - \gamma$ = two-sided significance level

Table XII *d*-Factors for Wilcoxon Signed Rank Test and Confidence Intervals for the Median

n	d	γ	α″	α′	n	d	γ	α″	α′
3	1	.750	.250	.125	15	16	.992	.008	.004
4	1	.875	.125	.062		17	.990	.010	.005
5	1	.938	.062	.031		26	.952	.048	.024
	2	.875	.125	.063		27	.945	.055	.028
6	1	.969	.031	.016		31	.905	.095	.047
	2	.937	.063	.031		32	.893	.107	.054
	3	.906	.094	.047	16	20	.991	.009	.005
	4	.844	.156	.078		21	.989	.011	.006
7	1	.984	.016	.008		30	.956	.044	.022
	3	.953	.047	.016		31	.949	.051	.025
	4	.922	.078	.039		36	.907	.093	.047
	5	.891	.109	.055		37	.895	.105	.052
8	1	.992	.008	.004	17	24	.991	.009	.005
	2	.984	.016	.008		25	.989	.011	.006
	4	.961	.039	.020		35	.955	.045	.022
	5	.945	.055	.027		36	.949	.051	.025
	6	.922	.078	.039		42	.902	.098	.049
	7	.891	.109	.055		43	.891	.109	.054
9	2	.992	.008	.004	18	28	.991	.009	.005
	3	.988	.012	.006		29	.990	.010	.005
	6	.961	.039	.020		41	.952	.048	.024
	7	.945	.055	.027		42	.946	.054	.027
	9	.902	.098	.049		48	.901	.099	.049
	10	.871	.129	.065		49	.892	.108	.054
10	4	.990	.010	.005	19	33	.991	.009	.005
	5	.986	.014	.007		34	.989	.011	.005
	9	.951	.049	.024		47	.951	.049	.025
	10	.936	.064	.032		48	.945	.055	.027
	11	.916	.084	.042		54	.904	.096	.048
	12	.895	.105	.053		55	.896	.104	.052
11	6	.990	.010	.005	20	38	.991	.009	.005
	7	.986	.014	.007		39	.989	.011	.005
	11	.958	.042	.021		53	.952	.048	.024
	12	.946	.054	.027		54	.947	.053	.027
	14	.917	.083	.042		61	.903	.097	.049
	15	.898	.102	.051		62	.895	.105	.053
12	8	.991	.009	.005	21	43	.991	.009	.005
	9	.988	.012	.006		44	.990	.010	.005
	14	.958	.042	.021		59	.954	.046	.023
	15	.948	.052	.026		60	.950	.050	.025
	18	.908	.092	.046		68	.904	.096	.048
	19	.890	.110	.055		69	.897	.103	.052
13	10	.992	.008	.004	22	49	.991	.009	.005
	11	.990	.010	.005		50	.990	.010	.005
	18	.952	.048	.024		66	.954	.046	.023
	19	.953	.057	.029		67	.950	.050	.025
	22	.906	.094	.047		76	.902	.098	.049
	23	.890	.110	.055		77	.895	.105	.053
14	13	.991	.009	.004	23	55	.991	.009	.005
	14	.989	.011	.005		56	.990	.010	.005
	22	.951	.049	.025		74	.952	.048	.024
	23	.942	.058	.029		75	.948	.052	.026
	26	.909	.091	.045		84	.902	.098	.049
	27	.896	.104	.052		85	.895	.105	.052

Table XII *(Continued)*

n	d	γ	α″	α′	n	d	γ	α″	α′
24	62	.990	.010	.005	25	69	.990	.010	.005
	63	.989	.011	.005		70	.989	.011	.005
	82	.951	.049	.025		90	.952	.048	.024
	83	.947	.053	.026		91	.948	.052	.026
	92	.905	.095	.048		101	.904	.096	.048
	93	.899	.101	.051		102	.899	.101	.051

For $n > 25$ use $d \doteq \frac{1}{2}[\frac{1}{2}n(n+1) + 1 - z\sqrt{n(n+1)(2n+1)/6}]$, where z is read from Table III.
γ = confidence coefficient
$\alpha' = \frac{1}{2}(1 - \gamma)$ = one-sided significance level
$\alpha'' = 2\alpha' = 1 - \gamma$ = two-sided significance level

Table XIII   Critical Values for the Kolmogorov-Smirnov Test of Goodness of fit

Sample Size (n)	Significance Level				
	.20	.15	.10	.05	.01
1	.900	.925	.950	.975	.995
2	.684	.726	.776	.842	.929
3	.565	.597	.642	.708	.829
4	.494	.525	.564	.624	.734
5	.446	.474	.510	.563	.669
6	.410	.436	.470	.521	.618
7	.381	.405	.438	.486	.577
8	.358	.381	.411	.457	.543
9	.339	.360	.388	.432	.514
10	.322	.342	.368	.409	.486
11	.307	.326	.352	.391	.468
12	.295	.313	.338	.375	.450
13	.284	.302	.325	.361	.433
14	.274	.292	.314	.349	.418
15	.266	.283	.304	.338	.404
16	.258	.274	.295	.328	.391
17	.250	.266	.286	.318	.380
18	.244	.259	.278	.309	.370
19	.237	.252	.272	.301	.361
20	.231	.246	.264	.294	.352
25	.21	.22	.24	.264	.32
30	.19	.20	.22	.242	.29
35	.18	.19	.21	.23	.27
Asymptotic Formula:	$\dfrac{1.07}{\sqrt{n}}$	$\dfrac{1.14}{\sqrt{n}}$	$\dfrac{1.22}{\sqrt{n}}$	$\dfrac{1.36}{\sqrt{n}}$	$\dfrac{1.63}{\sqrt{n}}$

Reject the hypothetical distribution $F(x)$ if $D_n = \max|F_n(x) - F(x)|$ exceeds the tabulated value.

Table XIV   **Critical values for the Kolmogorov-Smirnov test of** $H{:}F_1(x) = F_2(x)$

Sample size n_1

Sample size n_2	1	2	3	4	5	6	7	8	9	10	12	15
1	*	*	*	*	*	*	*	*	*	*		
	*	*	*	*	*	*	*	*	*	*		
2			*	*	*	*	*	7/8	16/18	9/10		
			*	*	*	*	*	*	*	*		
3			*	*	12/15	5/6	18/21	18/24	7/9		9/12	
			*	*	*	*	*	*	8/9		11/12	
4				3/4	16/20	9/12	21/28	6/8	27/36	14/20	8/12	
				*	*	10/12	24/28	7/8	32/36	16/20	10/12	
5					4/5	20/30	25/35	27/40	31/45	7/10		10/15
					4/5	25/30	30/35	32/40	36/45	8/10		11/15
6						4/6	29/42	16/24	12/18	19/30	7/12	
						5/6	35/42	18/24	14/18	22/30	9/12	
7							5/7	35/56	40/63	43/70		
							5/7	42/56	47/63	53/70		
8								5/8	45/72	23/40	14/24	
							6/8	54/72	28/40	16/24	
9									5/9	52/90	20/36	
								6/9	62/90	24/36	
10										6/10		15/30
										7/10		19/30
12											6/12	30/60
											7/12	35/60
15												7/15
												8/15

Reject H_0 if

$$D = \max|F_{n_1}(x) - F_{n_2}(x)|$$

exceeds the tabulated value. The upper value gives a level at most .05 and the lower value gives a level at most .01.

Note: Where * appears, do not reject H at the given level.

Table XV *d*-Factors for Wilcoxon-Mann-Whitney Test and Confidence Intervals for the Shift Parameter Δ

γ = confidence coefficient
$\alpha' = \frac{1}{2}(1 - \gamma)$ = one-sided significance level
$\alpha'' = 2\alpha' = 1 - \gamma$ = two-sided significance level

		$m = 3$				$m = 4$		
	d	γ	α''	α'	*d*	γ	α''	α'
$n = 3$	1	.900	.100	.050				
$n = 4$	1	.943	.057	.029	1	.971	.029	.014
	2	.886	.114	.057	2	.943	.057	.029
					3	.886	.114	.057
$n = 5$	1	.964	.036	.018	1	.984	.016	.008
	2	.929	.071	.036	2	.968	.032	.016
	3	.857	.143	.071	3	.937	.063	.032
					4	.889	.111	.056
$n = 6$	1	.976	.024	.012	1	.990	.010	.005
	2	.952	.048	.024	2	.981	.019	.010
	3	.905	.095	.048	3	.962	.038	.019
	4	.833	.167	.083	4	.933	.067	.033
					5	.886	.114	.057
$n = 7$	1	.983	.017	.008	1	.994	.006	.003
	2	.967	.033	.017	2	.988	.012	.006
	3	.933	.067	.033	4	.958	.042	.021
	4	.883	.117	.058	5	.927	.073	.036
					6	.891	.109	.055
$n = 8$	1	.988	.012	.006	2	.992	.008	.004
	3	.952	.048	.024	3	.984	.016	.008
	4	.915	.085	.042	5	.952	.048	.024
	5	.867	.133	.067	6	.927	.073	.036
					7	.891	.109	.055
$n = 9$	1	.991	.009	.005	2	.994	.006	.003
	2	.982	.018	.009	3	.989	.011	.006
	3	.964	.036	.018	5	.966	.034	.017
	4	.936	.064	.032	6	.950	.050	.025
	5	.900	.100	.050	7	.924	.076	.038
					8	.894	.106	.053
$n = 10$	1	.993	.007	.004	3	.992	.008	.004
	2	.986	.014	.007	4	.986	.014	.007
	4	.951	.049	.025	6	.964	.036	.018
	5	.923	.077	.039	7	.946	.054	.027
	6	.888	.112	.056	8	.924	.076	.038
					9	.894	.106	.053
$n = 11$	1	.995	.005	.003	3	.994	.006	.003
	2	.989	.011	.006	4	.990	.010	.005
	4	.962	.038	.019	7	.960	.040	.020
	5	.940	.060	.030	8	.944	.056	.028
	6	.912	.088	.044	9	.922	.078	.039
	7	.874	.126	.063	10	.896	.104	.052
$n = 12$	2	.991	.009	.004	4	.992	.008	.004
	3	.982	.018	.009	5	.987	.013	.007
	5	.952	.048	.024	8	.958	.042	.021
	6	.930	.070	.035	9	.942	.058	.029
	7	.899	.101	.051	10	.922	.078	.039
					11	.897	.103	.052

For sample sizes *m* and *n* beyond the range of this table use $d \doteq \frac{1}{2}[mn + 1 - z\sqrt{mn(m + n + 1)/3}]$, where *z* is read from Table III.

Table XV   (*Continued*)

n		*m* = 5				*m* = 6				*m* = 7				*m* = 8		
	d	γ	α''	α'	*d*	γ	α''	α'	*d*	γ	α''	α'	*d*	γ	α''	α'
n = 5	1	.992	.008	.004												
	2	.984	.016	.008												
	3	.968	.032	.016												
	4	.944	.056	.028												
	5	.905	.095	.048												
	6	.849	.151	.075												
n = 6	2	.991	.009	.004	3	.991	.009	.004								
	3	.983	.017	.009	4	.985	.015	.008								
	4	.970	.030	.015	6	.959	.041	.021								
	5	.948	.052	.026	7	.935	.065	.033								
	6	.918	.082	.041	8	.907	.093	.047								
	7	.874	.126	.063	9	.868	.132	.066								
n = 7	2	.995	.005	.003	4	.992	.008	.004	5	.993	.007	.004				
	3	.990	.010	.005	5	.986	.014	.007	6	.989	.011	.006				
	6	.952	.048	.024	7	.965	.035	.018	9	.952	.038	.019				
	7	.927	.073	.037	8	.949	.051	.026	10	.947	.053	.027				
	8	.984	.106	.053	9	.927	.073	.037	12	.903	.097	.049				
					10	.899	.101	.051	13	.872	.128	.064				
n = 8	3	.994	.006	.003	5	.992	.008	.004	7	.991	.009	.005	8	.993	.007	.004
	4	.989	.011	.005	6	.987	.013	.006	8	.986	.014	.007	9	.990	.010	.005
	7	.955	.045	.023	9	.957	.043	.021	11	.960	.040	.020	14	.950	.050	.025
	8	.935	.065	.033	10	.941	.059	.030	12	.946	.054	.027	15	.935	.065	.033
	9	.907	.093	.047	11	.919	.081	.041	14	.906	.094	.047	16	.917	.083	.042
	10	.873	.127	.064	12	.892	.108	.054	15	.879	.121	.060	17	.895	.105	.052
n = 9	4	.993	.007	.004	6	.992	.008	.004	8	.992	.008	.004	10	.992	.008	.004
	5	.988	.012	.006	7	.988	.012	.006	9	.988	.012	.006	11	.989	.011	.006
	8	.958	.042	.021	11	.950	.050	.025	13	.958	.042	.021	16	.954	.046	.023
	9	.940	.060	.030	12	.934	.066	.033	14	.945	.055	.027	17	.941	.059	.030
	10	.917	.083	.042	13	.912	.088	.044	16	.909	.091	.045	19	.907	.093	.046
	11	.888	.112	.056	14	.887	.113	.057	17	.886	.114	.057	20	.886	.114	.057
n = 10	5	.992	.008	.004	7	.993	.007	.004	10	.990	.010	.005	12	.991	.009	.004
	6	.987	.013	.006	8	.989	.011	.006	11	.986	.014	.007	13	.988	.012	.006
	9	.960	.040	.020	12	.958	.042	.021	15	.957	.043	.022	18	.957	.043	.022
	10	.945	.055	.028	13	.944	.056	.028	16	.945	.055	.027	19	.945	.055	.027
	12	.901	.099	.050	15	.907	.093	.047	18	.912	.088	.044	21	.917	.083	.042
	13	.871	.129	.065	16	.882	.118	.059	19	.891	.109	.054	22	.899	.101	.051
n = 11	6	.991	.009	.004	8	.993	.007	.004	11	.992	.008	.004	14	.991	.009	.005
	7	.987	.013	.007	9	.990	.010	.005	12	.989	.011	.006	15	.988	.012	.006
	10	.962	.038	.019	14	.952	.048	.024	17	.956	.044	.022	20	.959	.041	.020
	11	.948	.052	.026	15	.938	.062	.031	18	.944	.056	.028	21	.949	.051	.025
	13	.910	.090	.045	17	.902	.098	.049	20	.915	.085	.043	24	.909	.091	.045
	14	.885	.115	.058	18	.878	.122	.061	21	.896	.104	.052	25	.891	.109	.054
n = 12	7	.991	.009	.005	10	.990	.010	.005	13	.990	.010	.005	16	.990	.010	.005
	8	.986	.014	.007	11	.987	.013	.007	14	.987	.013	.007	17	.988	.012	.006
	12	.952	.048	.024	15	.959	.041	.021	19	.955	.045	.023	23	.953	.047	.024
	13	.936	.064	.032	16	.947	.053	.026	20	.944	.056	.028	24	.943	.057	.029
	14	.918	.082	.041	18	.917	.083	.042	22	.917	.083	.042	27	.902	.098	.049
	15	.896	.104	.052	19	.898	.102	.051	23	.900	.100	.050	28	.885	.115	.058

Table XV (Continued)

		m = 9				m = 10				m = 11				m = 12		
	d	γ	α″	α′	d	γ	α″	α′	d	γ	α″	α′	d	γ	α″	α′
n = 9	12	.992	.008	.004												
	13	.989	.011	.005												
	18	.960	.040	.020												
	19	.950	.050	.025												
	22	.906	.094	.047												
	23	.887	.113	.057												
n = 10	14	.992	.008	.004	17	.991	.009	.005								
	15	.990	.010	.005	18	.989	.011	.006								
	21	.957	.043	.022	24	.957	.043	.022								
	22	.947	.053	.027	25	.948	.052	.026								
	25	.905	.095	.047	28	.911	.089	.045								
	26	.887	.113	.056	29	.895	.105	.053								
n = 11	17	.990	.010	.005	19	.992	.008	.004	22	.992	.008	.004				
	18	.988	.012	.006	20	.990	.010	.005	23	.989	.011	.005				
	24	.954	.046	.023	27	.957	.043	.022	31	.953	.047	.024				
	25	.944	.056	.028	28	.949	.051	.026	32	.944	.056	.028				
	28	.905	.095	.048	32	.901	.099	.049	35	.912	.088	.044				
	29	.888	.112	.056	33	.886	.114	.057	36	.899	.101	.051				
n = 12	19	.991	.009	.005	22	.991	.009	.005	25	.991	.009	.004	28	.992	.008	.004
	20	.988	.012	.006	23	.989	.011	.006	26	.989	.011	.005	29	.990	.010	.005
	27	.951	.049	.025	30	.957	.043	.021	34	.956	.044	.022	38	.955	.045	.023
	28	.942	.058	.029	31	.950	.050	.025	35	.949	.051	.026	39	.948	.052	.026
	31	.905	.095	.048	35	.907	.093	.047	39	.909	.091	.045	43	.911	.089	.044
	32	.889	.111	.056	36	.893	.107	.054	40	.896	.104	.052	44	.899	.101	.050

Table XVI   Probabilities Associated with Values as Large as Observed Values of K in the Kruskal-Wallis One-Way Analysis of Variance by Ranks

n_1	n_2	n_3	K	p	n_1	n_2	n_3	K	p
2	1	1	2.7000	.500	4	3	2	6.4444	.008
								6.3000	.011
2	2	1	3.6000	.200				5.4444	.046
								5.4000	.051
2	2	2	4.5714	.067				4.5111	.098
			3.7143	.200				4.4444	.102
3	1	1	3.2000	.300	4	3	3	6.7455	.010
								6.7091	.013
3	2	1	4.2857	.100				5.7909	.046
			3.8571	.133				5.7273	.050
								4.7091	.092
3	2	2	5.3572	.029				4.7000	.101
			4.7143	.048					
			4.5000	.067	4	4	1	6.6667	.010
			4.4643	.105				6.1667	.022
								4.9667	.048
3	3	1	5.1429	.043				4.8667	.054
			4.5714	.100				4.1667	.082
			4.0000	.129				4.0667	.102
3	3	2	6.2500	.011	4	4	2	7.0364	.006
			5.3611	.032				6.8727	.011
			5.1389	.061				5.4545	.046
			4.5556	.100				5.2364	.052
			4.2500	.121				4.5545	.098
								4.4455	.103
3	3	3	7.2000	.004					
			6.4889	.011	4	4	3	7.1439	.010
			5.6889	.029				7.1364	.011
			5.6000	.050				5.5985	.049
			5.0667	.086				5.5758	.051
			4.6222	.100				4.5455	.099
								4.773	.102
4	1	1	3.5714	.200					
					4	4	4	7.6538	.008
4	2	1	4.8214	.057				7.5385	.011
			4.5000	.076				5.6923	.049
			4.0179	.114				5.6538	.054
								4.6539	.097
4	2	2	6.0000	.014				4.5001	.104
			5.3333	.033					
			5.1250	.052	5	1	1	3.8571	.143
			4.4583	.100					
			4.1667	.105	5	2	1	5.2500	.036
								5.0000	.048
4	3	1	5.8333	.021				4.4500	.071
			5.2083	.050				4.2000	.095
			5.0000	.057				4.0500	.119
			4.0556	.093					
			3.8889	.129					

Table XVI (*Continued*)

Samples sizes					Samples sizes				
n_1	n_2	n_3	K	p	n_1	n_2	n_3	K	p
5	2	2	6.5333	.008				5.6308	.050
			6.1333	.013				4.5487	.099
			5.1600	.034				4.5231	.103
			5.0400	.056					
			4.3733	.090	5	4	4	7.7604	.009
			4.2933	.122				7.7440	.011
								5.6571	.049
5	3	1	6.4000	.012				5.6176	.050
			4.9600	.048				4.6187	.100
			4.8711	.052				4.5527	.102
			4.0178	.095					
			3.8400	.123	5	5	1	7.3091	.009
								6.8364	.011
5	3	2	6.9091	.009				5.1273	.046
			6.8218	.010				4.9091	.053
			5.2509	.049				4.1091	.086
			5.1055	.052				4.0364	.105
			4.6509	.091					
			4.4945	.101	5	5	2	7.3385	.010
								7.2692	.010
5	3	3	7.0788	.009				5.3385	.047
			6.9818	.011				5.2462	.051
			5.6485	.049				4.6231	.097
			5.5152	.051				4.5077	.100
			4.5333	.097					
			4.4121	.109	5	5	3	7.5780	.010
								7.5429	.010
5	4	1	6.9545	.008				5.7055	.046
			6.8400	.011				5.6264	.051
			4.9855	.044				4.5451	.100
			4.8600	.056				4.5363	.102
			3.9873	.098					
			3.9600	.102	5	5	4	7.8229	.010
								7.7914	.010
5	4	2	7.2045	.009				5.6657	.049
			7.1182	.010				5.6429	.050
			5.2727	.049				4.5229	.099
			5.2682	.050				4.5200	.101
			4.5409	.098					
			4.5182	.101	5	5	5	8.0000	.009
								7.9800	.010
5	4	3	7.4449	.010				5.7800	.049
			7.3949	.011				5.6600	.051
			5.6564	.049				4.5600	.100
								4.5000	.102

Table XVII   Critical Values of Friedman Statistic

t	b	.90	.95	.975	.99	.995
3	3	6.000	6.000	—	—	—
3	4	6.000	6.500	8.000	8.000	8.000
3	5	5.200	6.400	7.600	8.400	10.000
3	6	5.333	7.000	8.333	9.000	10.333
3	7	5.429	7.143	7.714	8.857	10.286
3	8	5.250	6.250	7.750	9.000	9.750
3	9	5.556	6.222	8.000	8.667	10.667
4	2	6.000	6.000	—	—	—
4	3	6.600	7.400	8.200	9.000	9.000
4	4	6.300	7.800	8.400	9.600	10.200

Acknowledgements for Table

Table IV. Reprinted from *Biometrika Tables for Statisticians*, Vol. I, by E. S. Pearson and H. O. Hartley, 1970. By permission of *Biometrika* Trustees.
Table V. Reprinted from *Biometrika Tables for Statisticians*, Vol. I, by E. S. Pearson and H. O. Hartley, 1970. By permission of *Biometrika* Trustees.
Table VI. Reprinted from *Biometrika Tables for Statisticians*, Vol. I, by E. S. Pearson and H. O. Hartley, 1970. By permission of *Biometrika* Trustees.
Table VII. *A Million Random Digits with 100,000 Normal Deviates*, by The Rand Corporation. New York: The Free Press, 1955. Copyright 1955 and 1983 by the Rand Corporation. Reprinted from pp. 115, 258, 262, and 360. Used by permission.
Table VIII. Reprinted from *Biometrika Tables for Statisticians*, Vol. I, by E. S. Pearson and H. O. Hartley, 1970. By permission of *Biometrika* Trustees.
Table IX. Adapted from H. Leon Harter, Tables of range and studentized range, *Annals of Mathematical Statistics*, 1960, 31, 1122–47. By permission of author and publisher.
Table X. Adapted from E. G. Olds, Distribution of sums of sq of rank differences for small number of individuals, *Annals of Mathematical Statistics*, 1938, 9, 133–148. By permission of author and publisher.
Table XI. Reprinted from Gottfried Noether, *Introduction to Statistics: A Fresh Approach.* Boston: Houghton Mifflin Company, 1971. Used by permission.
Table XII. Reprinted from Gottfried Noether, *Introduction to Statistics: A Fresh Approach*, Boston: Houghton Mifflin Company, 1971. Used by permission.
Table XIII. Adapted from Frank Massey, Jr., The Kolmogorov-Smirnov test for goodness of fit, *Journal of American Statistical Association*, 1951, 46, 68–78. Used by permission.
Table XIV. Adapted from F. J. Massey, Jr., Distribution table for the deviation between two sample cumulatives, *Annals of Mathematical Statistics*, 1952, 23, 435–441. By permission of author and publisher.
Table XV. Reprinted from Gottfried Noether, *Introduction to Statistics: A Fresh Approach.* Boston: Houghton Mifflin Company, 1971. Used by permission.
Table XVI. Adapted from W. H. Kruskal and W. A. Wallis, Use of ranks in one-criterion variance analysis, *Journal of American Statistical Association*, 1952, 47, 581–621. Used by permission.
Table XVII. Adapted from M. Friedman, The use of ranks to avoid the assumption of normality implicit in the analysis of variance, *Journal of American Statistical Association*, 1937, 52, 675–701. Used by permission.

Index

DATE DUE
REMINDER

MAY 16 2006

**Please do not remove
this date due slip.**